Heat and Mass Transfer in Energy Systems

Heat and Mass Transfer in Energy Systems

Special Issue Editors

Alessandro Mauro
Nicola Massarotti

MDPI • Basel • Beijing • Wuhan • Barcelona • Belgrade

MDPI

Special Issue Editors
Alessandro Mauro
Università degli Studi di Napoli
"Parthenope"
Italy

Nicola Massarotti
Università degli Studi di Napoli
"Parthenope"
Italy

Editorial Office
MDPI
St. Alban-Anlage 66
4052 Basel, Switzerland

This is a reprint of articles from the Special Issue published online in the open access journal *Energies* (ISSN 1996-1073) from 2018 to 2019 (available at: https://www.mdpi.com/journal/energies/special_issues/heat_mass)

For citation purposes, cite each article independently as indicated on the article page online and as indicated below:

LastName, A.A.; LastName, B.B.; LastName, C.C. Article Title. *Journal Name* **Year**, *Article Number*, Page Range.

ISBN 978-3-03921-982-7 (Pbk)
ISBN 978-3-03921-983-4 (PDF)

Contents

About the Special Issue Editors

Alessandro Mauro is a tenured Associate Professor of Applied Thermodynamics and Thermal Science and technical manager of LaTEC lab at the Department of Engineering of the University of Naples "Parthenope". His Ph.D. thesis was awarded with the Emerald Engineering Outstanding Doctoral Research Award in the category Numerical Heat Transfer & CFD. He is the author of more than 80 papers in the fields of thermo-fluid dynamics and energy systems.

Nicola Massarotti is a Full Professor of Fluid and Thermal Sciences. He got his Ph.D. at the University of Wales Swansea. He is the author of more than 150 papers in the fields of thermo-fluid dynamics and energy systems and he is the head of the Laboratory of Thermo-fluid dynamics, Energy, and HVAC systems (LaTEC) at the Department of Engineering of the University of Naples "Parthenope".

Preface to "Heat and Mass Transfer in Energy Systems"

In recent years, the interest of the scientific community towards efficient energy systems has significantly increased. One of the reasons is certainly related to the change in the temperature of the planet, which has increased by 0.76 °C with respect to preindustrial levels, according to the Intergovernmental Panel on Climate Change (IPCC), and is still increasing. The European Union considers it vital to prevent global warming from exceeding 2 °C with respect to pre-industrial levels, as it has been proven that this will result in irreversible and potentially catastrophic changes. These changes in climate are mainly caused by the greenhouse gas emissions related to human activities, and can be drastically reduced by employing energy systems for the heating and cooling of buildings, as well as for power production, characterized by high efficiency levels and/or based on renewable energy sources.

This Special Issue, published in the *Energies* journal, includes 13 contributions from across the world, including a wide range of applications such as hybrid residential renewable energy systems, desiccant-based air handling units, heat exchanges for engine WHR, solar chimney systems, and other interesting topics.

Finally, we wish to express our deep gratitude to all the authors and reviewers who have significantly contributed to this special issue. Our sincere thanks also go to the editorial team of MDPI and Energies for giving us the opportunity to publish this book and for helping in all possible ways, especially Ms. Julyn Li for her precious support and availability.

Alessandro Mauro, Nicola Massarotti
Special Issue Editors

energies

MDPI

Article

Ancillary Services Provided by Hybrid Residential Renewable Energy Systems through Thermal and Electrochemical Storage Systems

Lorenzo Bartolucci, Stefano Cordiner, Vincenzo Mulone * and Marina Santarelli

Department of Industrial Engineering, University of Rome Tor Vergata, via del Politecnico 1, 00133 Rome, Italy; lorenzo.bartolucci@uniroma2.it (L.B.); cordiner@uniroma2.it (S.C.); marina.santarelli@uniroma2.it (M.S.)
* Correspondence: mulone@uniroma2.it

Received: 21 March 2019; Accepted: 11 June 2019; Published: 24 June 2019

Abstract: Energy Management System (EMS) optimal strategies have shown great potential to match the fluctuating energy production from renewables with an electric demand profile, which opens the way to a deeper penetration of renewable energy sources (RES) into the electric system. At a single building level, however, handling of different energy sources to fulfill both thermal and electric requirements is still a challenging task. The present work describes the potential of an EMS based on Model Predictive Control (MPC) strategies to both maximize the RES exploitation and serve as an ancillary service for the grid when a Heat Pump (HP) coupled with a Thermal Energy Storage (TES) is used in a residential Hybrid Renewable Energy System (HRES). Cost savings up to 30% as well as a reduction of the purchased energy unbalance with the grid (about 15%–20% depending on the season) have been achieved. Moreover, the thermal energy storage leads to a more efficient and reliable use of the Heat Pump by generally decreasing the load factor smoothing the power output. The proposed control strategy allows to have a more stable room temperature, with evident benefits also in terms of thermal comfort.

Keywords: renewables; ancillary services; hybrid systems; thermal storage; energy storage; microgrids; heat pump; model predictive control; optimization

---

1. Introduction

In 2016, the residential sector accounted for 25.4% of final European energy consumptions [1,2] and about 65% was employed for heating and cooling purposes. According to Eurostat 2017, the thermal demand is mainly met with fossil fuel contributing to greenhouse gas emissions. The European Union (EU) has already launched a number of initiatives in order to reduce the environmental impact of thermal loads and increase the energy efficiency of buildings [3]. These objectives can be achieved by improving heat losses performances of envelopes, with a lever effect on efficiency in the final use.

A possible solution to reduce the impact of thermal energy production on fossil fuel consumptions while decreasing emissions is to exploit the thermal power generation potential from renewables. Nevertheless, the massive deployment of Renewable Energy Systems (RES) introduces few critical issues to maintain system reliability, and, thus, programmable energy sources and Energy Storage Systems (ESS) represent effective means to optimize the systems.

In this context, Hybrid Renewable Energy Systems (HRES) with Heat Pumps (HP) could be a viable solution to decarbonize the heating system and exploit the fluctuating generation profiles from renewable sources. However, in order to effectively match the instantaneous user thermal and electric demand with the stochastic production and minimize at the same time the operative costs, a proper design of the Energy Management System (EMS) control strategy taking into account the smart use of renewable heating is key [4–9].

Studies in literature have extensively proven Model Predictive Control strategies to be suitable control methods for EMS, allowing taking control actions by considering the evolution of the state variables of the systems over a time horizon rather than instantaneously [10–17]. The decision variables are in fact optimized by considering not only the current state, but also future predictable events, which can affect the system behavior. According to the literature, the most influential parameters to take into account are the weather conditions and the users' energy consumption profiles/characteristics [18]. The former affect the system reliability whenever RES are used due to the fluctuating production patterns. The latter introduce variability and forecast errors in the load profile, making coupling between load and production challenging. To overcome those issues, several research studies on MPC strategy for thermal management purposes have been carried out in the last decade, focusing on different aspects [19–29].

Energy demand control by setting a variable room temperature set-point has been often suggested due to the relatively easy implementation [19,20]. Different approaches have been presented, dealing mainly with parameters and selection of control variables, according to the objective function defined. Bruni et al. in Reference [19] proposed a MPC strategy aimed at minimizing the total energy consumption and, as a consequence, the total operating costs of the system, while limiting the comfort violations. Killian et al. in Reference [20] developed a flexible control scheme, which allows us to tune the weight of the cost function in a multi-objective optimization framework. The control variables have been optimized according to the users' decisions for either minimizing the total operation cost, the environmental impact, or the comfort violations. As the authors in References [19,20] showed, the minimization of costs and of comfort violations are traded-off one each other. Therefore, the performance of the EMS is strictly dependent on the optimal control problem formulation and the target chosen.

Few works [21–24] have investigated the flexibility offered by optimally controlled hybrid generators including a heat pump and a gas boiler for residential applications. The second generator can, in fact, prevent the heat pump from operating at inefficient operating conditions, with a clear advantage in terms of efficiency. However, D'Ettorre et al. in Reference [25] showed that only the presence of a Thermal Energy Storage (TES) allows for a consistent cost reduction (up to 8% gain) and in the consumption of primary energy. In fact, it enables the decoupling between heat demand and production, moving the heat generation to off-peak hours, or to periods with RES overproduction, or to periods with high efficiency HP operation. Authors in Reference [25] performed a sensitivity analysis of cost savings with respect to both the horizon used in the MPC strategy and the storage capacity. They showed that, for a given forecast window, there is an optimum tank capacity to achieve the maximum cost savings when taking into account losses of the storage system to the environment.

Some studies [26–29] have focused on the flexibility potential of coupling the thermal and the electric load, when both are optimally controlled by an EMS. Authors in Reference [27] and Reference [28] illustrated how building thermal dynamics and an ESS can be managed in order to reduce the renewable energy curtailment, mitigate the energy unbalance penalty, and increase the expected profit of a given microgrid. However, they did not consider TES as an additional source of flexibility to avoid the curtailment of the energy produced by renewables for maintaining the target comfort conditions. Comodi et al. in Reference [29] demonstrated that more consistent yearly energy-savings and higher level of self-consumption could be achieved by increasing the integration of electrical and thermal storage. Moreover, they implemented a design tool for a residential microgrid with different storage strategies and system configurations, not taking into account the potential of the system to provide ancillary services to the grid while maintaining comfort conditions.

The purpose of this study is then to fill the gap between the studies presented in the literature, identifying an optimal economic solution while maintaining the comfort requirements and attending a predefined energy exchange profile with the grid. To this aim, special focus has been given to the representation of both the thermal and electric load profiles to lever on the flexibility of time-deferrable appliances for demand response features, extending the validity of the previous work presented [30].

Similarly to most of the cited papers, the optimization algorithm used in the Model Predictive controller is a Mixed-Integer-Linear-Programming (MILP), since, as demonstrated by Gelleschus et al. in Reference [31], this technique is characterized by high accuracy and computational efficiency. The MILP-based optimization framework, despite the mandatory simplification to linearize the system, allows to speed up, which is key to perform relatively long simulations. In this work, two representative months for the winter and the summer season are studied, to identify two opposite operating conditions for the HP and then the overall HRES.

The following structure is adopted in this work. The residential HRES and the different system configurations are described in Section 2. Section 3 describes the mathematical modelling of the building thermal system and the input data used to run the simulations. Sections 4 and 5 are dedicated to the discussion of the results and to comment on the main findings of the work, respectively.

2. HRES Description

The layout of the system considered in this paper is shown in Figure 1. It refers to a residential building connected to the electric grid and equipped with smart metering sensors, renewable power supply systems (PV panels), and an Electrochemical Energy Storage system (EES). A hydrogen fueled Proton Exchange Membrane (PEM) Fuel Cell is used as controllable power supply. Fuel cells are, in fact, ideal for residential applications since they are characterized by low noise and near zero emissions. An air-source heat pump and a Thermal Energy Storage system meet the heating and cooling loads.

A Micro Grid Central Controller (MGCC) is in charge of the optimal management of the energy fluxes of the HRES based on the current state of the system (actual energy demand and production, State of Charge of the EES, temperature of water in the TES, and power output of the Fuel Cell and power absorption of the Heat Pump), on the information collected from the weather station (ambient temperature and solar irradiance) and on the energy exchanged with the grid.

The electric load model of the system is presented in Reference [30] and only a detailed description of the thermal circuit is outlined in the following sections.

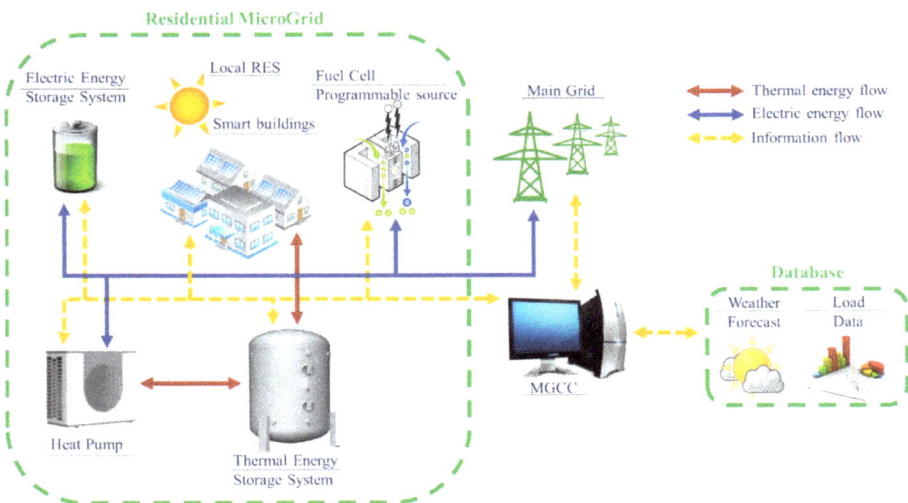

Figure 1. Scheme of the residential hybrid renewable energy system.

Thermal Model of Building and System

Special attention has been given to represent the thermal load as a function of the ambient temperature by taking into account the thermal inertia of the building. To this aim, the heat flux between each apartment and the surrounding environment has been defined by considering the thermophysical properties (density, heat capacities, and heat exchange coefficients) of the building envelope materials as well as the properties of the radiant heating/cooling floor (Figure 2). The internal room temperature is determined by evaluating the parallel heat fluxes from the external ambient via the roof, the walls, and the windows, and from the ground pipe via the floor. The thermal inertia of the building has been modelled by lumped mass blocks, operating as thermal storage systems.

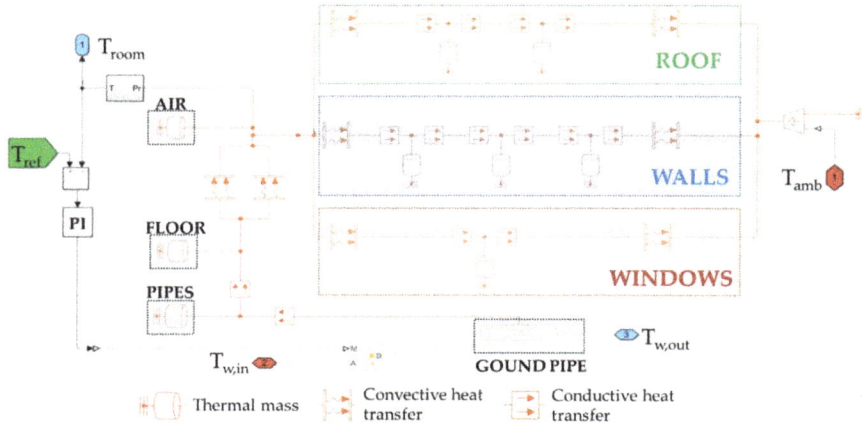

Figure 2. Thermal physical model.

The occupancy load has been neglected in this study since the focus is on the control of the thermal load and its effects on electric energy production and electric load matching. Predicted and real thermal load profiles have been calculated with the simulation of the same model with forecasted and real weather conditions, respectively. The differences between forecasted and real thermal load are due to the differences between the expected and the real weather profiles. The thermal demand profile generated for a typical week in the summer and winter is reported in Figure 3.

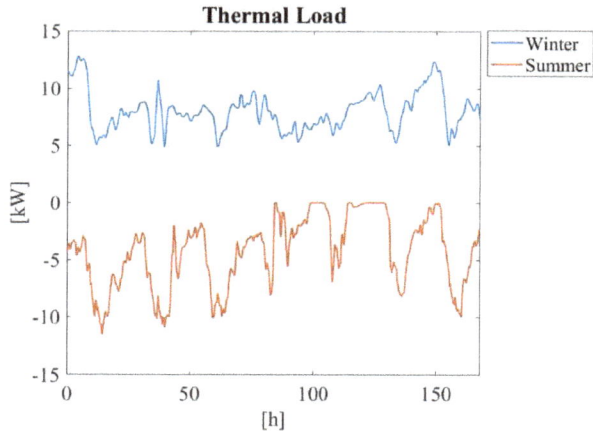

Figure 3. Weekly thermal load profiles for the winter and summer season.

Two different configurations have been considered (Figure 4) to evaluate the effect of the TES on the HRES operation flexibility and its capability to provide ancillary services to the grid. In the first

4

one, the heat pump feeds directly the thermal load without the TES (a), while, in the second one, HP and TES are connected in series to the load (b).

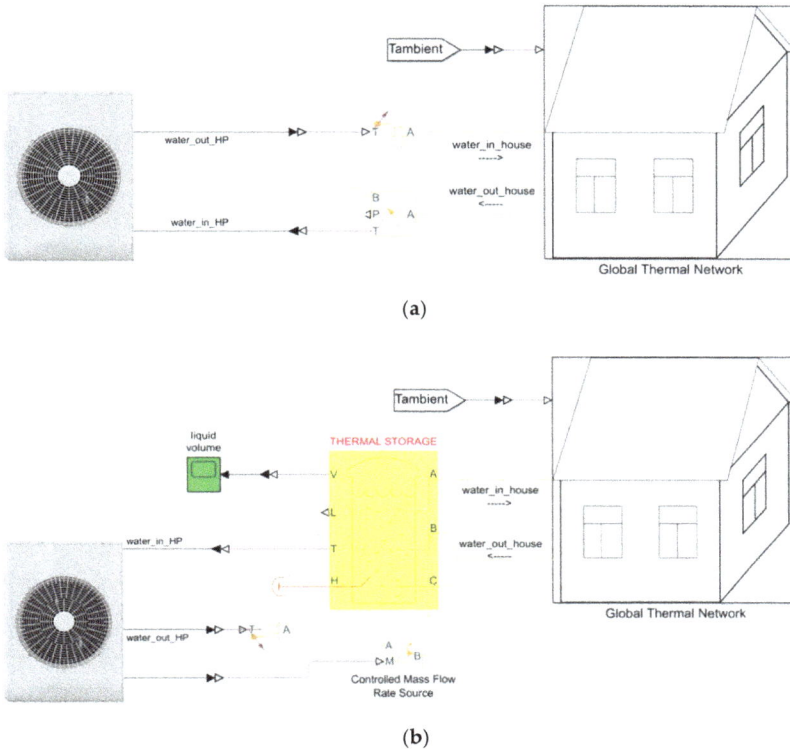

(a)

(b)

Figure 4. Scheme of the thermal subsystem of the HRES. (**a**) System configuration without the TES. (**b**) System configuration with TES.

Without thermal storage, the heat pump works with constant outlet water temperature setpoint and variable mass flow rate, as defined by a PI controller. Constraints to the maximum and minimum power output of the HP have been considered as well.

In the second configuration, the heat pump feeds the thermal storage at variable temperature and fixed maximum flow rate. The HP thermal energy delivered to the TES is controlled by the MGCC at each time step of the optimization in agreement with the operative constraints, while the thermal fluxes between the TES and the house are controlled with the same PI considered and used for the case (a).

3. HRES Control Strategy

Electric and thermal energy fluxes of the HRES are managed by the MGCC featuring an MPC strategy. In a general MPC approach, control actions are defined with the aim of minimizing—or maximizing—objective functions, whose evaluation is predicted using a mathematical model of the system behavior. In particular, the controller solves the set of optimal control variables for a period of time from t (current state) and t + CP (where CP is the control period), based on the initial state and on an estimation of the external disturbance parameters (in this paper weather and load forecasts). Control actions determine the evolution of the system state at the time t + 1, used as input for the forthcoming optimization process. The effectiveness of the optimization is highly dependent on the definition of the constraints and of the objective function. The first ones are needed in order to ensure the physical and technical feasibility of the optimal solution found. For this study, a minimization of the operative costs and penalties for unbalanced energy is proposed as a target of the controller system. In particular, the unbalance energy is evaluated as the difference between a reference profile—communicated to the grid operator the day ahead—and the actual energy exchanged. Details of the implementation of this control strategy have been discussed in Reference [32], but, for the sake of completeness, the description of the mathematical formulation of the optimization algorithm is presented in Appendix A; an overview of the models of the heat pump and the Thermal Energy Storage is reported in the following sections.

3.1. Heat Pump Model

The characteristic performance parameter of a Heat Pump is the Coefficient Of Performance (COP) (named as Energy Efficiency Ratio (EER) when the HP is in the cooling operating mode), which is defined as the ratio between the thermal power output and the electrical power input. The COP depends on the operative conditions of the HP, in terms of external air temperature, water setpoint temperature, inlet water temperature and load factor. The relationship between these variables is not linear and a simplification has to be performed to reduce the complexity of the model without affecting its accuracy. In this study, the COP has been considered dependent only on external temperature, since the HP performance is mostly sensitive to this parameter. However, the HP has been operated at a maximum flow rate in order to guarantee the minimum output water temperature and, as a consequence, the maximum value of the COP for a given thermal load and ambient condition. In the MILP algorithm constraints related to HP (Equations (1) and (2)) are formulated according to the manufacturer's datasheet [20].

$$P_{HP,t} = \frac{Q_{HP,t}}{COP\left(t_{amb}\right)} \tag{1}$$

$$\delta_{on} \cdot P_{HP,min}(t_{amb}) \leq P_{HP,t} \leq P_{HP,max}(t_{amb}) \cdot \delta_{on} \tag{2}$$

The electric power required by the HP is a control variable of the MILP algorithm defined in the range between a minimum and a maximum load factor (40% and 100% respectively). These limits are input variables of the control strategy, since they are temperature dependent, and they are updated at each time step of the optimization based on actual ambient conditions.

The operating status of the HP at each time step is described by a binary variable δ_{on}. Two further binary variables have been considered ($\delta_{startup}$, $\delta_{shutdown}$) with a penalization cost to limit the number of start-ups and shutdowns.

3.2. Thermal Energy Storage System Model

The thermal energy storage system is modelled as a perfectly-mixed storage tank and all thermal losses are neglected. According to these assumptions, the thermal energy balance in the TES, expressed as a linear finite difference equation, can be calculated using the equation below.

$$t_{TES,t} = t_{TES,t-1} + \frac{Q_{HP,t}}{m_{TES} \cdot c_p} \cdot \Delta t - \frac{Q_{load,t}}{m_{TES} \cdot c_p} \cdot \Delta t \tag{3}$$

The charging process is controlled by the MILP algorithm, which defines the thermal energy to be delivered to the tank ($Q_{HP,t}$). The discharging process, instead, is controlled with a PI controller, which is in charge to calculate the water mass flow rate in the underfloor heat exchanger in order to maintain the target indoor temperature. $Q_{load,t}$ represents the thermal power request of the building at the time step t.

Temperature limits (Equation (4)) have been set in the range of the operative working conditions for radiant floor heating and cooling system.

$$t_{TES,min} \leq t_{TES,t} \leq t_{TES,max} \tag{4}$$

3.3. Simulation Specifications

The residential HRES considered is an eight-apartment building (90 m² each) located in Rome. A PV system of 34.2 kWp powers the whole building, while a Fuel Cell system with rated power 1.2 kW and an EES with maximum power 3.5 kW and a storage capacity of 280Ah@48V is installed in each apartment. The electrically-driven thermal energy source is an air-water heat pump with 19.4 kW and 19.5 kW nominal heating and cooling thermal power, respectively. The TES is a tank of 2000 L of water and a thermal capacity (evaluated according to Equation (5)) of 27 kWh in the winter and 20 kWh in the summer.

$$Q_{cap} = \frac{V_{TES} \cdot \rho_w \cdot c_p \cdot (t_{TES,max} - t_{TES,min})}{3600} \tag{5}$$

Three test cases have been studied to analyze the performances of the two different configurations presented in the HRES description section.

Effects of the TES may be studied comparing a first case, with no TES installed, and a second case, with the TES installed, and the COP value used in the optimization fixed and equal to 3.

A third case has been considered to evaluate the impact of a variable COP on the optimal control actions taken by the MILP algorithm. In particular, the dependence of the COP on ambient temperature has been taken into account, which is in agreement with the values reported in the manufacturer datasheet (Table 1 [33]). Figure 5 shows the values the EER for a typical ambient temperature pattern in a summer day.

A setpoint indoor temperature of 20 °C in the winter and 24 °C in the summer has been chosen, and the total thermal energy demand for all the simulated cases is equal to 2023 kWh in the summer and 6941 kWh in the winter. The maximum and minimum storage temperatures have also been kept the same among the different tests: in the range of 7 °C to 16 °C in cooling mode and 28 °C to 40 °C in the heating mode. These values have been chosen inside the operative range of the heat pump, as reported in the manufacturer datasheet.

The simulations have been run for a time-period of a month for the winter and the summer season, using data collected at the weather station located at the University of Rome "Tor Vergata" as real-time inputs, whereas data provided by the meteorological Service of the Italian Air Force have been used as weather forecast [34]. Specific information about the numerical weather prediction model used has been given in Appendix B.

It is worth highlighting the role of the influence of temperature prediction accuracy on system behavior, as it directly affects the thermal demand of the apartments as well as the operating limits of the heat pump. As a direct consequence, inaccurate temperature forecast would give different thermal loads and heat pump requests for the day-ahead and real time simulations, penalizing the match of power demand and attending the predefined energy profile to be exchanged with the grid.

Table 1. Manufacturer datasheet [33] of the COP and EER as a function of the ambient air temperature.

Heating		Cooling	
Tamb (°C)	COP	Tamb (°C)	EER
−10/−10.5	1.98	20	5.23
−7/−8	2.13	25	4.40
0/−0.6	2.47	30	3.73
2/1.1	2.59	35	3.13
7/6	3.23	40	2.64
10/8.2	3.40	45	2.30
15/13	3.84		
18/14	3.81		

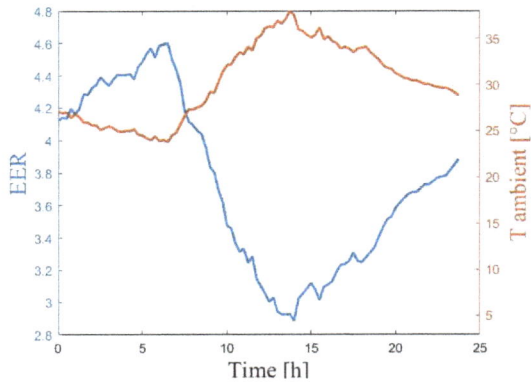

Figure 5. Daily value of the EER as a function of a typical temperature pattern in the summer season.

The minimization of the operative costs has been performed by the MILP algorithm according to Equation (6), considering a two-rate time-of-day tariff in Italy for the purchasing cost (c_{buy}) of electricity from the grid (0.17 €/kWh from 8 p.m. to 8 a.m. and 0.21 €/kWh for the remaining hours), and assuming a selling (c_{sell}) and unbalancing cost (c_{unb}) equal to half of the purchased energy price. As suggested in Reference [35], the operating cost associated with the battery discharge (c_{dch}) is 0.035 €/kWh, whereas hydrogen has been considered at 6.5 €/kg when assuming an average efficiency of 46% of the Fuel Cell. In order to avoid frequent startup and shutdown of the Heat Pump, a cost of 0.25 €/kWh has been associated to the switch on the binary variable ($c_{HP,startup}$).

$$Objective\ function\quad = c_{buy} \cdot P_{buy} - c_{sell} \cdot P_{sell} + c_{unb} \cdot P_{unb} + c_{dch} \cdot P_{dch} + c_{FC} \cdot P_{FC} \\ + c_{HP,startup} \cdot \delta_{HP,startup} \tag{6}$$

4. Analysis of Results

Since the objective function of the optimization algorithm is the minimization of the total operational costs, in the first analysis, the three system configurations have been compared in terms of economic performances. Figure 6 shows the cash flow for the winter and summer season. The contribution of the grid, the battery, and the Fuel Cell to the total amounts are reported in Table 2.

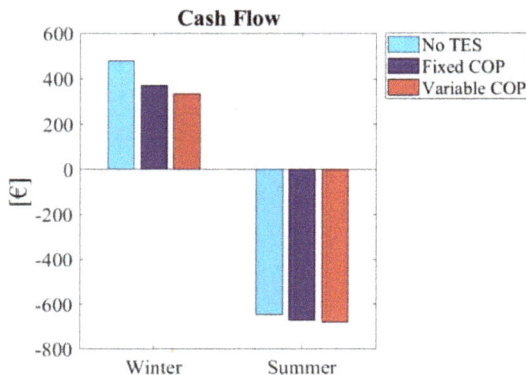

Figure 6. Cash flow for the three test cases: without TES, with TES and fixed COP, and with TES and variable COP.

Table 2. Cost composition for the three test cases in the winter and the summer season.

Season	Test Case	Grid Sold (€)	Grid Bought (€)	Battery (€)	Fuel Cell (€)
Winter					
	No TES	−5.6	206.8	49.4	227.6
	Fixed COP	−6.8	238.8	42.3	96.9
	Variable COP	−19.1	220.7	42	91.1
Summer					
	No TES	−722.4	2.1	58	15.7
	Fixed COP	−749.5	15.4	52.2	9.8
	Variable COP	−750.8	7	52.8	10.3

The monthly winter cost has been decreased by 22.4% and 30% with respect to the first test case for fixed and variable COP, respectively. This result is due to the reduced use of the Fuel Cell when the heat pump operation is controlled by the MILP algorithm, which is thus able to take full advantage of the flexibility offered by the thermal storage, shifting and modulating the HP power. This difference in cost savings is less evident during the summer season as the day-ahead scheduling of the heat pump is more accurate than in the winter case, and the system does not need to use the programmable energy source to avoid the energy unbalance with the grid. However, in the summer, there is a 4% and 5% net income increase for the fixed and variable COP case with respect to the no TES case. The greater revenue is due to the more effective use of the thermal and electric energy storage systems. As shown in Table 2, in the second and in the third case the operational costs of battery and Fuel Cell are lower while the energy sold to the grid are greater than in the first case. The economic analysis highlights another important outcome considering the dependence of the COP on the external temperature to improve the HRES performance. In fact, lower total operational costs and higher profits for both the winter and summer tests have been obtained for the variable COP case. This result, which is also confirmed in Reference [36], is due to the capability of the MPC to schedule the HP operation to exploit the best operative conditions.

The increased adaptability of the system to weather uncertainty with a TES has also been evaluated in terms of energy unbalance with the grid. Figure 7 shows such a parameter in terms of the percentage of (a) the total amount of energy exchange with the grid and (b) of the energy sold and purchased in the summer (absolute values reported in Table 3).

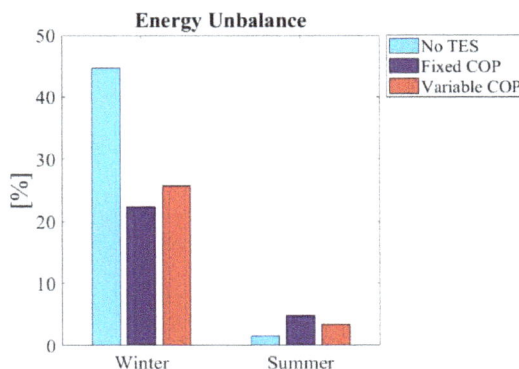

(a)

Figure 7. *Cont.*

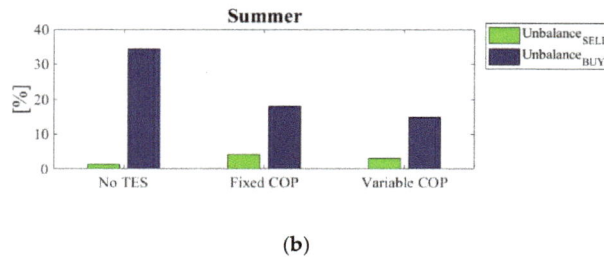

(b)

Figure 7. (a) Total energy unbalance for the three cases for the winter and the summer season. (b) Energy unbalance sold and bought for the three test cases for the summer season.

Table 3. Energy sold, purchased, and exchanged with the grid for all the test cases.

Season	Test Case	Energy Sold (kWh)	Energy Purchased (kWh)
Winter			
	No TES	24.5	2405.5
	Fixed COP	32.2	2772
	Variable COP	90.8	2563.6
Summer			
	No TES	3441.4	23.9
	Fixed COP	3571.1	180.5
	Variable COP	3578.1	81.9

In the winter, the integration of the thermal storage gives a noteworthy reduction of the energy unbalance, in the order of 20%. In the summer, the overall energy unbalance exchanged with the grid is lower for the "No TES" case, but this is mainly due to the greater amount of energy sold for the "TES" cases. Figure 7b shows the percentage of energy unbalance with respect to the overall energy sold and purchased. It can be observed that the greater unbalance for the "TES" cases is due to the energy sold while confirming the increased capability of absorbing load fluctuations, as indicated by the lower percentage of energy corresponding to the energy purchased.

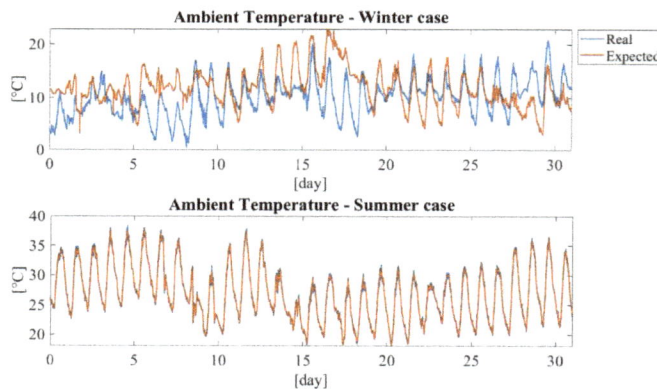

Figure 8. Difference between real and expected (forecast) ambient temperature in the winter and the summer.

The load forecast accuracy is also important toward the obtainment of high HRES performance, as already demonstrated in a previous work [30]. In this study, this aspect is related to the fact that thermal load depends on the ambient temperature. As shown in Figure 8, in the winter season, the real ambient temperature profile is rather different than the expected one, due to high weather condition variability in this period of the year in Rome. Thus, the thermal load forecast can easily get compromised. This difference in the load prediction is the main reason why a high energy unbalance occurs.

The number of HP startups and the statistical distributions of the load factor have been analyzed in order to evaluate the MGCC performance in terms of HP operation strategy (Figure 9).

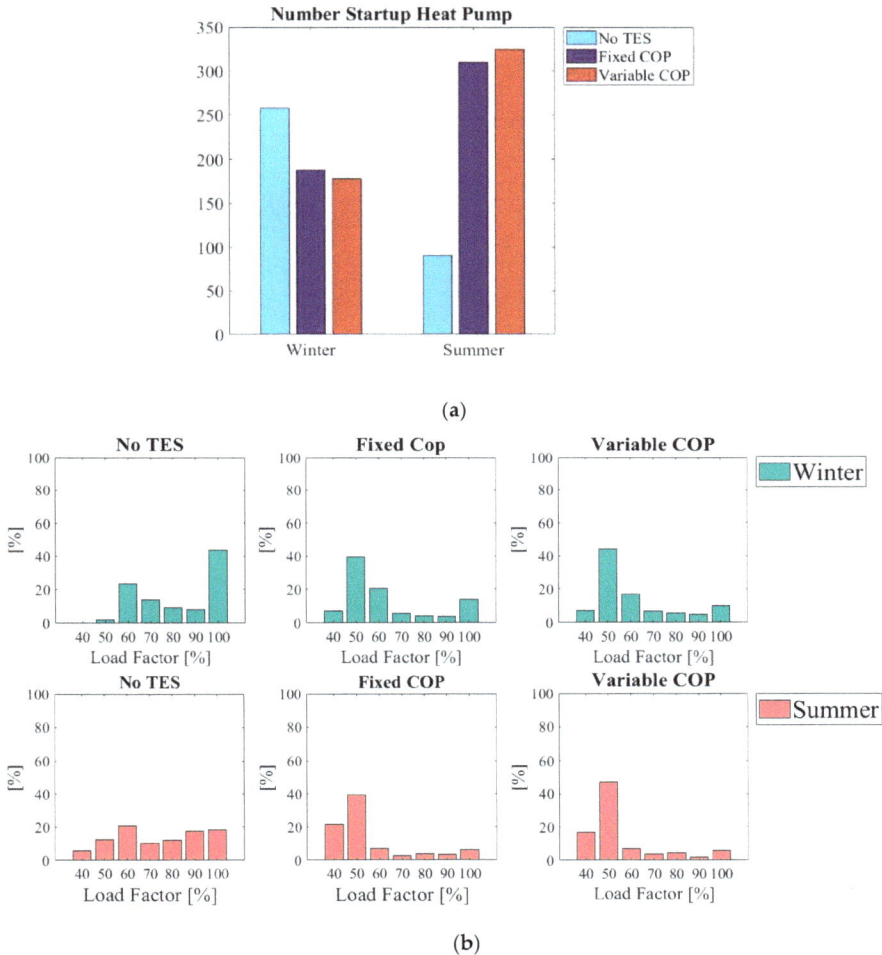

(a)

(b)

Figure 9. (**a**) Number of startups of the Heat Pump for the three cases in the winter and the summer. (**b**) Statistical Distribution of the Load Factor of the Heat Pump for the three cases in the winter and the summer.

In the winter, a TES allows for reducing the number of HP startups (Figure 9a) and decreasing the average HP load factor, as shown in Figure 9b. The statistical trend is similar in the summer, while the number of HP startups is increased for the TES installed cases with respect to a No TES case.

These differences can be motivated by the capability of the thermal storage to decouple the thermal load from the HP operation. Without TES, the HP is forced by the PI controller to follow the thermal power request, which results in a greater variability of the HP operating conditions, often falling in high power setpoints. As a result, the HP works at a higher load factor if compared with the TES cases.

This HP behavior affects the indoor ambient temperature fluctuating around the setpoint, as illustrated in Figure 10a. With TES, the thermal heating power is continuously drawn from the water tank, according to the PI signals. Thus, the indoor temperature pattern results smoother. The performance of the system to comply with the comfort conditions has been evaluated in terms of a standard deviation of the real indoor temperature if compared with the reference profile. Figure 10b shows that, in the second and in the third case, the comfort conditions are attended more accurately than in the first case for both the winter and the summer cases.

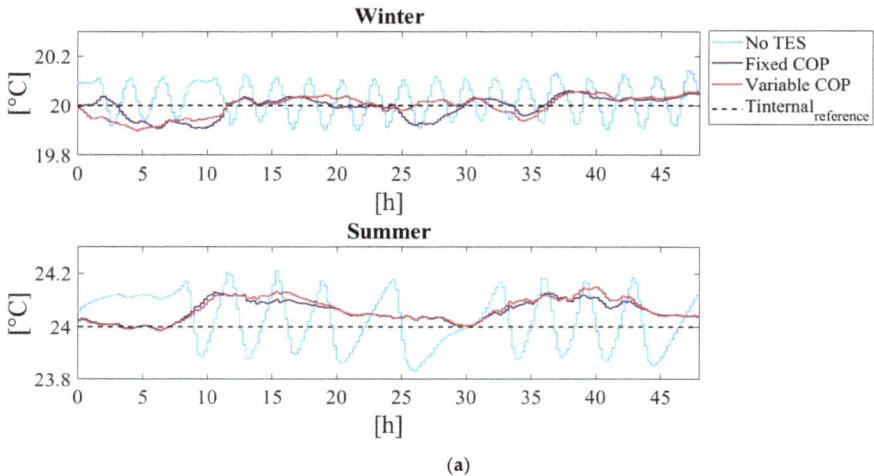

(a)

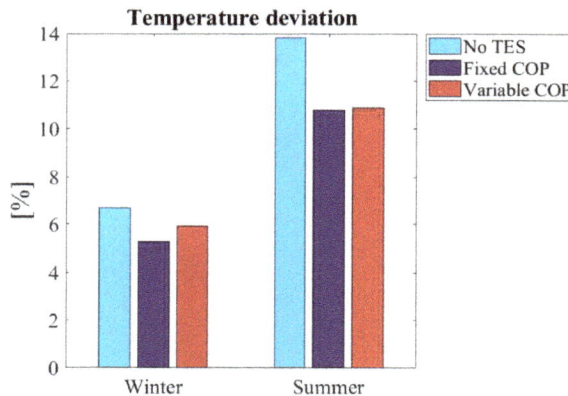

(b)

Figure 10. (a) Real indoor ambient temperature for the three test cases in the winter and the summer. (b) Ambient temperature deviation from the reference temperature for the three test cases in the winter and the summer.

5. Conclusions

A study on the potential of optimally controlled residential Hybrid Renewable Energy System to serve as ancillary services to the grid while guaranteeing comfort conditions has been presented.

Particular focus has been given to the representation of both thermal and electric load profiles, to exploit the flexibility of time-deferrable appliances and systems for demand response features, and to the effects of the inclusion of a Thermal Energy Storage (TES). The main findings of the work can be listed as follows.

- The installation of a TES in the microgrid allows for cost savings up to 30% in the winter and give increased profits in the order of 5% in the summer.
- In the winter, a strong reduction of energy unbalance (in the order of 20%) has been achieved including a TES. In the summer, despite the overall unbalanced energy exchanged with the grid is lower for the "No TES" case, a greater value in terms of unexpected energy purchased is observed.
- In the winter, the TES leads to reduce the number of HP startups while decreasing the average load factor. The load factor trend is similar in the summer, while the number of HP startups is increased for the TES cases with respect to the cases without TES.
- Benefits in the stabilization of comfort conditions have also been achieved thanks to the TES. Room temperature has more stable profiles as standard deviation gets lower by a margin of 2% to 5%.

Author Contributions: Coding and simulations, L.B. and M.S. Writing—original draft preparation, M.S. Writing—review and editing, L.B., S.C., and V.M. Supervision, S.C. and V.M.

Funding: This research received no external funding

Acknowledgments: The authors would like to thank the CNMCA (Italian Air Force Meteorological Service) for having provided the weather forecasts. MSc student Verdiana Orsi is acknowledged for her contribution to the results shown in this paper.

Conflicts of Interest: The authors declare no conflict of interest.

Appendix A

The objective function is solved using the following canonical MILP formulation.

$$\min f^T X \tag{A1}$$

subjected to:

$$\begin{cases} AX \le b \\ A_{eq}X = b_{eq} \\ L_b \le X \le U_b \end{cases} \tag{A2}$$

where the vector X contains the control variables associated with the appliances, the grid, the battery, the fuel cell, and the heat pump. The vector f^T is composed of the cost related to each control variable. L_b and U_b define the upper and lower limits for the values of the elements of X.

The linear equality constraint (3) can be represented by $A_{1,eq}X = b_{1,eq}$ where:

$$A_{1,eq} = \left[A_{appl1,1} \; A_{appl2,1} \cdots \; A_{appl1,n} \; A_{appl2,n} \; -A_{fromgrid} A_{togrid} A_{ch} \; - \; A_{dch} \; - \; A_{fc} A_{hp} \ldots 0 \right]_{Nx((a+b)n+23N)} \tag{A3}$$

$$b_{1,eq} = \left[\; E_{disp} \; \right]_{Nx1} \tag{A4}$$

where N is the number of time-steps a day.

$$A_{buy}, A_{sell}, A_{ch}, A_{dch}, A_{fc}, A_{hp} \tag{A5}$$

are $N \times N$ identity matrices.

$$A_{appl1}, A_{appl2} \tag{A6}$$

are the shifting matrices of the n appliances (in this study, two per apartment) with dimensions $N \times a$ and $N \times b$, respectively. They are defined according to their power profile.

E_{disp} is defined as the difference between the energy production and the load.

The single activation a day of the appliances can be expressed by $A_{2,eq}X = b_{2,eq}$

$$A_{2,eq} = \begin{bmatrix} A_{appl1,1} & 0 & 0 & 0 & \cdots & 0 & 0 & 0 & \cdots & 0 \\ 0 & A_{appl2,1} & 0 & 0 & \cdots & 0 & 0 & 0 & \cdots & 0 \\ \vdots & \vdots & \vdots & \ddots & & \vdots & \vdots & \vdots & \ddots & \vdots \\ 0 & 0 & 0 & 0 & A_{appl1,n} & 0 & 0 & 0 & \cdots & 0 \\ \vdots & \vdots & \vdots & \vdots & & 0 & A_{appl2,n} \backslash & 0 & 0 & \cdots & 0 \end{bmatrix}_{nx((a+b)n+23N)} \tag{A7}$$

$$b_{2,eq} = \begin{bmatrix} \delta_{pr,1} \\ \delta_{pr,2} \\ \vdots \\ \delta_{pr,n} \end{bmatrix}_{nx1}$$

where δ_{pr} is the binary variable representing the activation request of the user.

A_{appl1}, A_{appl2}, are ones' vectors of dimension a, b, respectively.

The energy balance of the battery can be expressed in the form $A_{3,eq}X = b_{3,eq}$.

$$A_{3,eq} = \begin{bmatrix} 0 & \cdots & 0 & \frac{A_{ch}}{E_c} & -\frac{A_{dch}}{E_c} & SOC & 0 & \cdots & 0 \end{bmatrix}_{Nx((a+b)n+23N)} \tag{A8}$$

$$SOC = \begin{bmatrix} 1 & 0 & 0 & \cdots & 0 & 0 \\ -1 & 1 & 0 & \cdots & 0 & 0 \\ 0 & -1 & 1 & \cdots & 0 & 0 \\ \vdots & \vdots & \vdots & \ddots & \ddots & \vdots \\ 0 & 0 & 0 & 0 & -1 & 1 \end{bmatrix}_{NxN} \tag{A9}$$

$$b_{3,eq} = \begin{bmatrix} -SOC_0 \\ 0 \\ \vdots \\ 0 \end{bmatrix}_{Nx1} \tag{A10}$$

The thermal balance in the thermal storage can be expressed in the form $A_{4,eq}X = b_{4,eq}$:

$$A_{4,eq} = \begin{bmatrix} 0 & \cdots & 0 & \frac{Q_{HP,t}}{m_{TES} \cdot c_p} \Delta t & -\frac{Q_{load,t}}{m_{TES} \cdot c_p} \Delta t & t_{TES} & 0 & \cdots & 0 \end{bmatrix}_{Nx((a+b)n+23N)} \tag{A11}$$

$$t_{TES} = \begin{bmatrix} 1 & 0 & 0 & \cdots & 0 & 0 \\ -1 & 1 & 0 & \cdots & 0 & 0 \\ 0 & -1 & 1 & \cdots & 0 & 0 \\ \vdots & \vdots & \vdots & \ddots & \ddots & \vdots \\ 0 & 0 & 0 & 0 & -1 & 1 \end{bmatrix}_{NxN} \tag{A12}$$

$$b_{4,eq} = \begin{bmatrix} -t_{TES,0} \\ 0 \\ \vdots \\ 0 \end{bmatrix}_{Nx1} \tag{A13}$$

The definition of the power unbalance is expressed by $A_{5,eq}X = b_{5,eq}$:

$$A_{5,eq} = \begin{bmatrix} 0 & \cdots & 0 & -A_{fromgrid} & A_{togrid} & 0 & \cdots & 0 & A_{unb+} & -A_{unb-} \end{bmatrix}_{Nx((a+b+)n+23N)} \tag{A14}$$

$$b_{5,eq} = \begin{bmatrix} E_{grid\_ref} \end{bmatrix}_{Nx1} \tag{A15}$$

where A_{unb+}, A_{unb-} are $N \times N$ identity matrices.

Then, the canonical linear equality constraint $A_{eq}X = b_{eq}$ becomes

$$\begin{bmatrix} A_{1,eq} \\ A_{2,eq} \\ A_{3,eq} \\ A_{4,eq} \\ A_{5,eq} \end{bmatrix}_{(n+3N)x((a+b)n+18N)} X = \begin{bmatrix} b_{1,eq} \\ b_{2,eq} \\ b_{3,eq} \\ b_{4,eq} \\ b_{5,eq} \end{bmatrix}_{(n+3N)x1} \tag{A16}$$

In a similar way, all the linear inequality constraints such as in Equation (2) can be represented in the form $A_1 X \leq b_1$.

$$A_1 = \begin{bmatrix} 0 & \cdots & 0 & P_{HP,t} & -P_{HP,max} & 0 & \cdots & 0 \\ 0 & \cdots & 0 & -P_{HP,t} & -P_{HP,min} & 0 & \cdots & 0 \end{bmatrix}_{2Nx((a+b)n+23N)} \tag{A17}$$

$$b_1 = \begin{bmatrix} 0 & \cdots & 0 \end{bmatrix}_{2Nx1} \tag{A18}$$

where $P_{HP,max}$ $P_{HP,min}$ are $N \times N$ identity matrices.

Mutual exclusive conditions can be defined as $A_2 X \leq b_2$.

$$A_2 = \begin{bmatrix} 0 & \cdots & 0 & A_{ch} & A_{dch} & 0 & \cdots & 0 \end{bmatrix}_{Nx((a+b)n+23N)} \tag{A19}$$

$$b_2 = \begin{bmatrix} 1 & \cdots & 1 \end{bmatrix}_{Nx1} \tag{A20}$$

Then, the canonical linear inequality constraint $AX \leq b$ assumes the form below.

$$\begin{bmatrix} A_1 \\ A_2 \end{bmatrix}_{yN((a+b)n+18N)} X = \begin{bmatrix} b_1 \\ b_2 \end{bmatrix}_{yNx1} \tag{A21}$$

Appendix B

The numerical weather prediction (NWP) model used by the meteorological Service of the Italian Air Force is based on a four-step process: data collection, data assimilation, prediction of future states with the NWP model, data post-processing, and validation of the previsions.

Data Collection

Weather data collection is done within the collaboration framework of the World Weather Watch (WWW) program enacted by the World Meteorological Organization (WMO). Observations have to be performed in each country, according to the Global Observing System (GOS) requirements, and results are shared with other countries through the Global Telecommunication System (GTS).

Data Assimilation

Real-time measurements in different locations within a predefined time horizon are analyzed through a three-dimensional variational assimilation algorithm (3D-VAR) to look for the best estimation of the true state of the atmosphere. Recently, an advanced data assimilation system has been setup based on a stochastic Kalman filter called Local Ensemble Transform Kalman Filter (LETKF). The results of the analysis performed on real data with the LETKF are then used as input for the deterministic Ensemble Prediction System (EPS) and the probabilistic EPS. Similar systems are used by the Canadian Meteorological Center and UK Met Office, while German, Japanese, and French National Weather Services employ a 4D-VAR algorithm.

Weather Forecast and Data Post-Processing

Once the initial state of the atmosphere and any boundary conditions are known, the numerical forecast is obtained using modeling equations describing the behavior of the atmospheric circulation. Post-processing techniques are used to have additional information and to further interpret results.

The Meteorological Operative Center of the Italian Air Force (COMET) employs a non-hydrostatic model developed within the Consortium for Small-Scale Modelling, including Germany, Swiss, Italy, Greece, Poland, Romania, Russia, and Israel. Such a model is used in two deterministic configurations.

- COSMO-ME: equations integrated up to 72 h on a grid with a 5-km step and 45 vertical levels. It covers part of the central-southern Europe and the Mediterranean basin with four runs a day (00, 06, 12, and 18 UTC). The initial state is the result of the probabilistic data assimilation analysis performed by the COMET and the boundary conditions are defined by the European Center for Medium-Range Weather Forecast's models.
- COSMO-IT: equations integrated up to 30/48 h on a grid connected to COSMO-ME with a 2.2 km-step and 65 vertical levels. It covers Italy with four runs per day (00, 06, 12, and 18 UTC)) and uses as the initial state the fields of analysis produced by the very high resolution COMET assimilation system.

COMET is equipped with a probabilistic prevision system that is able to determine the uncertainty related to the deterministic prevision. The model is also used in two other probabilistic configurations.

- COSMO-ME EPS, consisting of 40 + 1 members integrated on a grid with a 7-km step and 45 vertical levels, covering central-southern Europe and the Mediterranean basin, with two runs per day (00 and 12 UTC), for forecasts up to 72 h.
- COSMO-IT EPS (pre-operational), consisting of 20 + 1 integrated members on a grid with a 2.2 km and 65 vertical steps, covering Italy, with two runs per day (00 and 12 UTC), for forecasts up to 48 h.

Validation

The forecast obtained with the numerical model for a certain time-lapse is compared with measurements to evaluate some statistical values such as average error or standard deviation. The Italian Air Force has developed the Versus system, which is a tool for the validation of the numerical forecast. It allows us to perform statistical validations for all the operating models at the COMET and on the global IFS reference model of the ECMWF.

Forecast Accuracy

The meteorological Service of the Italian Air Force performs a statistical evaluation on the weather forecast accuracy every three months. Results are then analyzed in terms of some performance parameters (standard deviation, mean error, and root-mean-square error). Temperature can be forecasted with a mean absolute error of 4 °C within 72 h and the confidence interval of 5% has an amplitude of 8 °C. Detailed description of the validation method versus trimestral reports are available in Reference [37].

References

1. Eurostat, Consumption of Energy, Statistics Explained Website, Data extracted in June 2017. Available online: http://ec.europa.eu/eurostat/statistics-explained/index.php/Consumption_of_energy (accessed on 1 March 2019).
2. Energy Consumption in Households. Available online: https://ec.europa.eu/eurostat/statistics-explained/index.php?title=Energy_consumption_in_households (accessed on 1 March 2019).
3. European Parliament and Council. *Directive 2010/31/EU on the Energy Performance of Buildings*; European Parliament and Council: Brussels, Belgium, 2010.

4. Kim, N.K.; Shim, M.H.; Won, D. Building Energy Management Strategy Using an HVAC System and Energy Storage System. *Energies* **2018**, *11*, 2690. [CrossRef]
5. Hafeez, G.; Javaid, N.; Iqbal, S.; Khan, F.A. Optimal Residential Load Scheduling Under Utility and Rooftop Photovoltaic Units. *Energies* **2018**, *11*, 611. [CrossRef]
6. Javaid, N.; Ahmed, F.; Ullah, I.; Abid, S.; Abdul, W.; Alamri, A.; Almogren, A.S. Towards Cost and Comfort Based Hybrid Optimization for Residential Load Scheduling in a Smart Grid. *Energies* **2017**, *10*, 1546. [CrossRef]
7. Aslam, S.; Iqbal, Z.; Javaid, N.; Khan, Z.A.; Aurangzeb, K.; Haider, S.I. Towards Efficient Energy Management of Smart Buildings Exploiting Heuristic Optimization with Real Time and Critical Peak Pricing Schemes. *Energies* **2017**, *10*, 2065. [CrossRef]
8. Iqbal, Z.; Javaid, N.; Mohsin, S.M.; Akber, S.M.A.; Afzal, M.K.; Ishmanov, F. Performance Analysis of Hybridization of Heuristic Techniques for Residential Load Scheduling. *Energies* **2018**, *11*, 286. [CrossRef]
9. Park, L.; Jang, Y.; Bae, H.; Lee, J.; Park, C.Y.; Cho, S. Automated Energy Scheduling Algorithms for Residential Demand Response Systems. *Energies* **2017**, *10*, 1326. [CrossRef]
10. He, M.F.; Zhang, F.X.; Huang, Y.; Chen, J.; Wang, J.; Wang, R. A distributed demand side management algorithm for smart grid. *Energies* **2019**, *12*, 426. [CrossRef]
11. Lemus, F.D.S.; Minor Popocatl, O.; Aguilar Mejia, R. Tapia Olvera, Optimal Economic Dispatch in Microgrids with Renewable Energy Sources. *Energies* **2019**, *12*, 181. [CrossRef]
12. Ruiz, G.R.; Segarra, E.L.; Bandera, C.F. Model Predictive Control Optimization via Genetic Algorithm Using a Detailed Building Energy Model. *Energies* **2019**, *12*, 34. [CrossRef]
13. Kontes, G.D.; Giannakis, G.I.; Sanchez, V.; de Augustin Chamacho, P.; Romero Amortrortu, A.; Panagiotidou, N.; Rovas, D.V.; Steiger, S.; Mutschler, C.; Gruen, G. Simulation-Based Evaluation and Optimization of Control Strategies in Buildings. *Energies* **2018**, *11*, 3376. [CrossRef]
14. Barata, F.; Igreja, J. Energy Management in Buildings with Intermittent and Limited Renewable Resources. *Energies* **2018**, *11*, 2748. [CrossRef]
15. Izawa, A.; Fripp, M. Multi-Objective Control of Air Conditioning Improves Cost, Comfort and System Energy Balance. *Energies* **2018**, *11*, 2373. [CrossRef]
16. Jin, D.; Christopher, W.; James, N.; Teja, K. Occupancy-Based HVAC Control with Short-Term Occupancy Prediction Algorithms for Energy-Efficient Buildings. *Energies* **2018**, *11*, 2427.
17. Serale, G.; Fiorentini, M.; Capozzoli, A.; Bernardini, D.; Bemporad, A. Model Predictive Control (MPC) for Enhancing Building and HVAC System Energy Efficiency: Problem Formulation, Applications and Opportunities. *Energies* **2018**, *11*, 631. [CrossRef]
18. Robillart, M.; Schalbart, P.; Chaplais, F.; Peuportier, B. Model reduction and model predictive control of energy-efficient buildings for electrical heating load shifting. *J. Process. Control.* **2018**, *74*, 23–34. [CrossRef]
19. Bruni, G.; Cordiner, S.; Mulone, V.; Rocco, V.; Spagnolo, F. A study on the energy management in domestic micro-grids based on Model Predictive Control strategies. *Energy Convers. Manag.* **2015**, *102*, 50–58. [CrossRef]
20. Killian, M.; Zauner, M.; Kozek, M. Comprehensive smart home energy management system using mixed-integer quadratic-programming. *Appl. Energy* **2018**, *222*, 662–672. [CrossRef]
21. D'Ettorre, F.; de Rosa, M.; Conti, P.; Schito, E.; Testi, D.; Finn, D.P. Economic assessment of flexibility offered by an optimally controlled hybrid heat pump generator: A case study for residential building. *Energy Procedia* **2018**, *148*, 1222–1229. [CrossRef]
22. di Perna, C.; Magri, G.; Giuliani, G.; Serenelli, G. Experimental assessment and dynamic analysis of a hybrid generator composed of an air source heat pump coupled with a condensing gas boiler in a residential building. *Appl. Therm. Eng.* **2015**, *76*, 86–97. [CrossRef]
23. Klein, K.; Huchtemann, K.; Müller, D. Numerical study on hybrid heat pump systems in existing buildings. *Energy Build.* **2014**, *69*, 193–201. [CrossRef]
24. Bagarella, G.; Lazzarin, R.; Noro, M. Annual simulation, energy and economic analysis of hybrid heat pump systems for residential buildings. *Appl. Therm. Eng.* **2016**, *99*, 485–494. [CrossRef]
25. D'Ettorre, F.; Conti, P.; Schito, E.; Testi, D. Model predictive control of a hybrid heat pump system and impact of the prediction horizon on cost-saving potential and optimal storage capacity. *Appl. Therm. Eng.* **2019**, *148*, 524–535. [CrossRef]

26. Mohammadi, S.; Mohammadi, A. Stochastic scenario-based model and investigating size of battery energy storage and thermal energy storage for micro-grid. *Int. J. Electr. Power Energy Syst.* **2014**, *61*, 531–546. [CrossRef]

27. Nguyen, D.T.; Le, L.B. Optimal Bidding Strategy for Microgrids Considering Renewable Energy and Building Thermal Dynamics. *IEEE Trans. Smart Grid* **2014**, *5*, 1608–1620. [CrossRef]

28. Korkas, C.D.; Baldi, S.; Michailidis, I.; Kosmatopoulos, E.B. Occupancy-based demand response and thermal comfort optimization in microgrids with renewable energy sources and energy storage. *Appl. Energy* **2016**, *163*, 93–104. [CrossRef]

29. Comodi, G.; Giantomassi, A.; Severini, M.; Squartini, S.; Ferracuti, F.; Fonti, A.; Cesarini, D.N.; Morodo, M.; Polonara, F. Multi-apartment residential microgrid with electrical and thermal storage devices: Experimental analysis and simulation of energy management strategies. *Appl. Energy* **2015**, *137*, 854–866. [CrossRef]

30. Bartolucci, L.; Cordiner, S.; Mulone, V.; Santarelli, M. Short-therm forecasting method to improve the performance of a model predictive control strategy for a residential hybrid renewable energy system. *Energy* **2019**, *172*, 997–1004. [CrossRef]

31. Gelleschus, R.; Böttiger, M.; Stange, P.; Bocklisch, T. Comparison of optimization solvers in the model predictive control of a PV-battery-heat pump system. *Energy Procedia* **2018**, *155*, 524–535. [CrossRef]

32. Bartolucci, L.; Cordiner, S.; Mulone, V.; Rocco, V.; Rossi, J.L. Renewable source penetration and microgrids: Effects of MILP–Based control strategies. *Energy* **2018**, *152*, 416–426. [CrossRef]

33. ELFOEnergy Extended Inverter, SERIE WSAN-XIN 81-171. Available online: http://portal.clivet.it/products/app.jsp# (accessed on 1 March 2019).

34. Available online: http://www.meteoam.it/ta/previsione/482/ROMA (accessed on 1 March 2019).

35. Sousa, T.; Morais, H.; Castro, R.; Vale, Z. Evaluation of different initial solution algorithms to be used in the heuristics optimization to solve the energy resources scheduling in smart grid. *Appl. Soft Comput.* **2016**, *48*, 491–506. [CrossRef]

36. Verhelst, C.; Logist, F.; Impe, J.V.; Helsen, L. Study of the optimal control problem formulation for modulating air-to-water heat pumps connected to a residential floor heating system. *Energy Build.* **2012**, *45*, 43–53. [CrossRef]

37. Available online: http://www.meteoam.it/page/verifiche-modelli (accessed on 1 March 2019).

energies

MDPI

Article

Design and Development of Innovative Protracted-Finned Counter Flow Heat Exchanger (PFCHE) for an Engine WHR and Its Impact on Exhaust Emissions

Rajesh Ravi * and Senthilkumar Pachamuthu

Department of Automobile Engineering, Madras Institute of Technology, Anna University, Tamil Nadu 600044, India; mit_senthil@yahoo.com
* Correspondence: rajesh4mech@gmail.com; Tel.: +91-9942338792

Received: 20 September 2018; Accepted: 4 October 2018; Published: 11 October 2018

Abstract: This article describes and evaluates an Organic Rankine Cycle (ORC) for waste heat recovery system both theoretically as well as experimentally. Based on the thermodynamic analysis of the exhaust gas temperature identified at different locations of the exhaust manifold of an engine, the double-pipe, internally–externally protruded, finned counter flow heat exchanger was innovatively designed and installed in diesel engine for exhaust waste heat recovery (WHR). The tests were conducted to find the performance of heat recovery system by varying the fin geometries of the heat exchanger. The effect of heat exchanger on emission parameters is investigated and presented in this work. The experimental results demonstrated that the amount of heat transfer rate, the effectiveness of heat exchange rand the brake thermal efficiency improved with an increase in length and number of the fins. A significant reduction was observed in all major emissions after the implementation of catalytic-coated, protracted finned counter flow heat exchanger. It also demonstrated the possibility of electric power production using steam turbo-electric-generator setup driven by the recovered exhaust heat energy.

Keywords: waste heat recovery; exhaust steam; heat exchanger; protracted fin; turbo-electric generator; exhaust emissions

---

1. Introduction

Global energy demand is increasing every day due to excess population, transportation of people and products across the nations, and for industrial purposes. In order to overcome the present deficiency situation, there is a need for effective techniques to leverage the maximum amount of available energy. Improvement in the efficiency of internal combustion engines plays a great hardship for the researchers. A diesel engine utilizes a maximum of 30% from the fuel energy whereas the rest is lost due to cooling and exhaust gases. The engine crankshaft receives less than 30% of the generated energy. About 40% of the fuel energy gets wasted in exhaust gas whereas the remaining amount of heat energy goes unused in cooling system as well as during friction losses [1]. In the present research work, innovative steps have been taken to recover the heat wasted from the engine exhaust gas. Thermo electric generator, turbo-compounding, rankine cycle, Organic Rankine Cycle, gas turbine cycle, exhaust gas recirculation, automotive air conditioning, six stroke engine concept are the different techniques available to utilize engine WHR [2].

Based on the literature review, in order to utilize the waste heat energy available in the exhaust gas, different types of heat exchangers and Organic Rankine Cycles are used. Previous research works mainly focused in harvesting the exhaust heat energy. However, studies related to heat exchanger with innovative heat recovery and simultaneous reduction in emissions were not carried out. This necessitates the invention of novel conceptual design on exhaust heat recovery heat exchanger for the betterment of energy recovery and diesel engine exhaust emission reductions. In the present work, an internally–externally protruded and finned counter flow heat exchanger was designed, fabricated, and experimented in order to utilize the exhaust heat. The waste heat recovered was utilized to generate power using turbo-electric-generator set up and the emission characteristics were also investigated. The current paper discusses about the experimental set up and the methodology to conduct experiments. The PFCHE design and its parametric properties were estimated. The detailed results, discussion regarding the PFCHE-based waste heat recovery and its impact on engine exhaust emissions are discussed in the sections below.

2. Literature Review

In every phase of research and development in WHR, there is a well-defined need exist for exploration and interpretation of the technical literature. The first step in research work is to conduct an extensive review of the related works conducted earlier. The review process comprises of specific information, detailed survey, and preliminary review. This paper reviewed a number of peer-reviewed journal articles to illustrate various fields of hypotheses subjects. A summary of closely-related literature is presented in Table 1.

Table 1. Literature summary of closely related work.

Ref. No.	Authors	Year	Objective and Outcomes
[3]	W Gu et al.	2009	Recovering low and medium-temperature heat (that ranges from 60 °C to 200 °C) from sources which include industrial waste heat, geothermal energy, solar heat, biomass, and so on, is an important sustainable method to solve the energy crisis.
[4]	Borsukiewicz-Gozdur et al.	2007	Organic Rankine Cycle (ORC) systems are feasible for power generation from these low and medium-temperature heat sources. This cycle consists of elements such as boiler, condenser, expander, pump, and working fluid. Organic fluids are more comfortable to work at a low temperature source compared to other fluids.
[5]	Damiana Chinese et al.	2004	The Organic Rankine Cycle boiler receives heat energy from engine exhaust gases and it converts working fluid into steam energy. Steam expands at turbine and produces the mechanical rotation of the output shaft which generates power. The exhaust of the turbine is supplied to a condenser for phase change. Finally, the working fluid gets circulated to boiler for the same kind of repeated operations through the pump.
[6]	Uilli Drescher et al.	2007	A procedure was created to calculate ORC efficiency with sufficient accuracy, based on the design institute for physical properties and also to find appropriate fluids for ORC in biomass power plants.
[7,8]	Rieder de Oliveira Netoetal et al.	2016	A study was conducted upon waste heat energy recovery from internal combustion engines using Organic Rankine Cycle. A technical and economic study was conducted in this work in order to increase the efficiency of electricity production, and thus reduce the fuel consumption as well as emission of polluting gases from internal combustion engines. In order to achieve it, two Organic Rankine Cycle sets were suggested. The first one was facing deployment in water shortage areas (Organic Rankine Cycle using a cooling tower for the condensing system) and another one with water supply condenser made by urban water net. Both simulated systems were able to increase electricity production by almost 20% when toluene was used as working fluid.

Table 1. *Cont.*

Ref. No.	Authors	Year	Objective and Outcomes
[9]	Muhammad Fairuz Remeli et al.	2014	A theoretical model was developed to extract the exhaust waste heat and simultaneous power generation by utilizing a thermo-electric generator and heat pipe. The theoretical system was developed to measure the performance of heat pipes and thermo-electric generator based on heat pipe in numbers, heat input, and air flow rate. The heat transfer rate has reached its maximum level when more rows of heat pipe and thermo-electric generator module were installed. It was concluded that the design was able to recover 10.39 kW of the electrical power and 2 kW of the heat input used to extract 1.35 kW of thermal energy from the engine.
[10]	Mastrullo R et al.	2015	Modelled and optimized a shell and lowered mini-tube heat exchanger for internal combustion engine which was installed with Organic Rankine Cycle. In order to achieve better energy conversion process, heat exchangers were designed with less weight and refrigerant charge. Two engines were installed in order to design and optimize the heat exchanger for waste heat recovery process.
[11]	M. Hatami et al.	2014	Aimed at obtaining a numerical study of the finned type of heat exchangers for waste heat recovery from internal combustion engines. The proposed technique of Hatami et al. used water as the working fluid for compression ignition engines whereas for spark ignition engine, it used water with ethylene glycol as working fluid. In this work, numerical designs were carried out successfully for extracting exhaust heat from internal combustion engines. The discussions were very clear in this study about the impacts on heat recovery due to fin numbers, length, and thickness. The study concluded with improved heat transfer rate and positive energy recovery results.
[12]	Bock Choon Pak et al.	2003	An experimental study was conducted to investigate the effects on air side fouling and cleaning of various condenser coils. The results stated that the amount of dust deposits mostly depends on fin geometry of the heat exchangers.
[13,14]	Chen Bei and H. G. Zhang et al.	2015 2013	A numerical model was established for heat recovery from the exhaust gas of an engine. Engine exhaust gas mass flow rate and exhaust gas temperature values were taken for the analysis from heavy duty diesel truck engine and light duty passenger car engine. It was found that under any working conditions, the efficiency of engine is increased with the combination of Organic Rankine Cycle when compared to original engine performance.
[15,16]	Heng Chen and Yu jin et al.	2015 2013	In their experimental results, it was revealed that the pressure drop increases with the increase of fin height and fin width.
[17]	Songsong Song et al.	2015	Waste heat recovery has great potential in terms of increasing the efficiency and optimizing the fuel consumption. The conservative steam power cycle is applied in general industrial power plants widely; however, the performance of the Rankine cycle is not suitable to tap the energy from low-temperature waste heat source. In order to increase the efficiency and sound economic performance of energy sources, ORC (Organic Rankine Cycle) is used widely to tap low temperature heat sources such as solar energy, geothermal energy and industrial waste heat, and convert it into useful power.

Table 1. *Cont.*

Ref. No.	Authors	Year	Objective and Outcomes
[18]	Marco Altosole et al.	2017	The exhaust heat recovery works on Organic Rankine Cycle concept which is targeted at improving the overall efficiency of the diesel engine.
[19]	Pablo Fernandez-Yanez et al.	2018	Deployed thermoelectric generators to convert thermal energy recovered from exhaust gases to electrical energy.
[20,21]	Pablo Fernandez-Yanez et al.	2018	The experimental investigation was carried out in gasoline and light-duty diesel engines. The possibilities of energy recovery were determined at higher loads with speed and concentric tube heat exchanger. The electric-turbo generators harvested high amount of exhaust energy at high load operating modes. The electric-turbo generator recovered power seven times higher than the thermoelectric generator.

3. Experimental Setup and Methodology

A single cylinder with four stroke, water-cooled, and naturally-aspirated diesel engine was designed to generate 3.7 kW power at 1500 rpm which was utilized for waste heat recovery and emission analysis experiments. The technical specifications of the engine are tabulated in Table 2. Air flow was determined accurately by measuring the pressure drop across a sharp edge orifice of the air surge chamber using U-tube manometer. The diesel flow was measured using a burette arrangement by noting the time of fixed volume of diesel consumed by the engine. A water-cooled piezoelectric pressure transducer was fixed onto the cylinder head to record the pressure variations in cathode-ray oscilloscope screen along with crank angle encoder. The exhaust gas temperature was measured by a chromel–alumel K-type thermocouple. The exhaust gas constituents such as HC, CO, CO_2, and O_2 were measured using an AVL 444N model gas analyzer (Gurgaon, New Delhi, India). NO_x emission was measured using heated vacuum NO_x analyzer (Camberley, Surrey, England) whereas the smoke emission was measured by an AVL smoke meter (Gurgaon, New Delhi, India). Based on the thermodynamic analyses of exhaust gases' temperature at different locations in the exhaust manifold of an engine, the protracted internally–externally finned counter flow heat exchanger was designed and implemented in diesel engine for exhaust heat recovery with water as the working fluid. A schematic representation of the experimental arrangement is shown in the Figure 1. The engine was started using diesel fuel and allowed to warm up. The amount of the injected diesel fuel got automatically varied due to the governor attached to it and this maintained the engine speed at 1500 rpm throughout the experiment. The exhaust heat recovery and emission analysis were performed for different engine loading conditions.

Table 2. Technical specifications of engine.

Description	Type
Engine	Vertical Single Cylinder, Water cooled 4-stroke "Kirloskar Diesel engine"
Rated Power	3.7 kW at 1500 RPM
Bore × Stroke	80 mm × 110 mm
Displacement Volume	553 cc
Compression Ratio	16.5:1
Dynamometer	Rope-Brake Dynamometer
Fuel injection release pressure	200 bar
Specific fuel consumption	40 g/kW·h
Fuel injection timing	27° BTDC
Nozzle	M1CO; DLL110S 1630
Injector type	Mechanical injector
Type of Lubrication	Splash type
Lubricating Oil	SAE30/SAE40
Overall Dimensions	W2000 × D2500 × H1500 mm

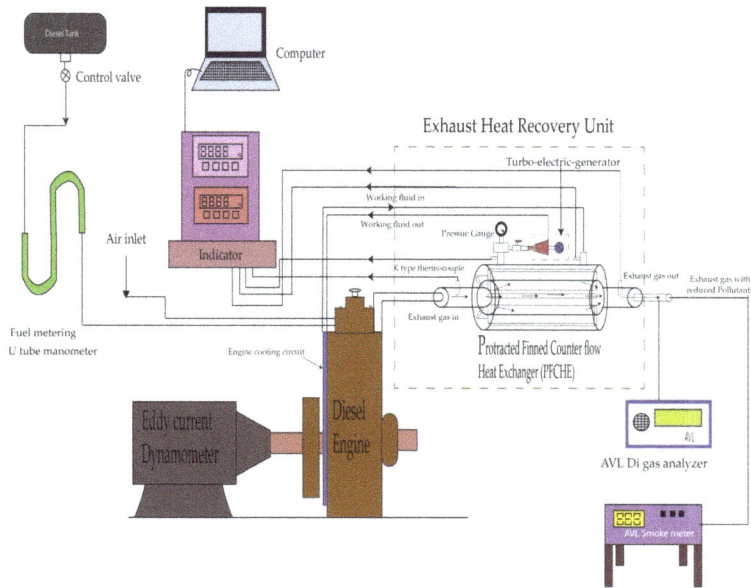

Figure 1. Schematic diagram of experimental setup.

Furthermore, the turbine generator setup was facilitated to examine the efficiency of the heat recovery system. The exhaust gas temperature was measured at different locations in order to locate the heat recovered by the heat exchanger to achieve the maximum effectiveness of heat exchanger which is installed in the diesel engine. As shown in the Figure 2, the exhaust gas temperature decreases as the distance from engine mouth increases. It is mainly due to the loss of heat energy to surroundings with increase in travel distance. At full load condition, the engine acquires more amount of air fuel mixture and thus involves in high rate of combustion which results in higher exhaust gas temperature compared to other engine conditions.

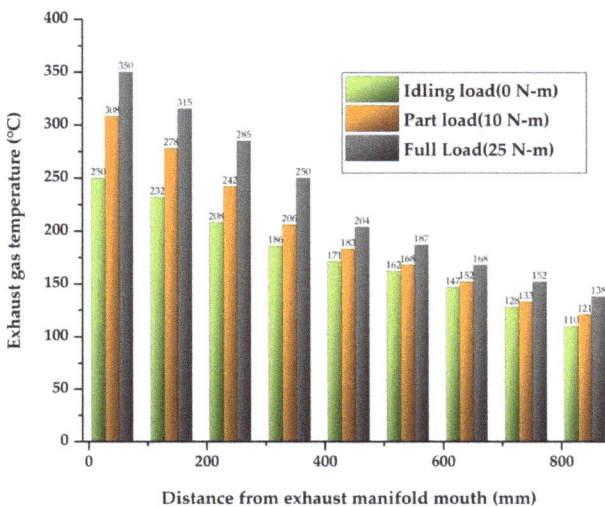

Figure 2. Exhaust gases temperature at different locations.

3.1. Protracted Finned Counter Flow Heat Exchanger Design Parameters

The protracted-finned heat exchanger design is primarily aimed at recovering internal combustion exhaust heat energy. While enumerating the heat exchanger design, the exhaust gas mass flow rate was calculated for a selected diesel engine by adding the mass flow rate of air and fuel for single cylinder diesel engine. The working fluid used in heat exchanger plays a vital role in extracting the heat. The properties of working fluids (i.e., water as cold fluid and engine exhaust gas as hot fluid) are listed in Table 3. Aluminium is selected as the heat exchanger material by considering its favorable thermal conductivity value of 204.4 W/(m·K).

Table 3. Properties of hot and cold fluids.

Input Parameters	Symbols	Hot Fluid (Exhaust Gas)	Symbols	Cold Fluid (Water)	Units
Inlet Temperature	T_{hi}	235	T_{ci}	32	
Outlet Temperature	T_{ho}	124	T_{co}	106	
Thermal Conductivity	K_h	0.0404	K_c	0.6	W/m·K
Specific Heat Capacity	C_{ph}	1030	C_{pc}	4182	J/(kg·K)
Viscosity (Absolute)	μ_h	0.000027	μ_c	0.0006	(N·s)/m^2
Density	ρ_h	0.696	ρ_c	998	kg/m^3
Mass Flow Rate	m_h	0.009336	m_c	0.0054	kg/s

3.1.1. Convection Heat Transfer Coefficient of Heat Exchanger

The heat transfer coefficient formulates the required length and size of the heat exchanger including developing and fully-developed regions to recover exhaust waste heat energy. Numerous parameters were considered and the data was calculated to find the heat transfer coefficient for inner pipe and the outer pipe of a typical heat exchanger [22]. Table 4 shows the determined values of various parameters which are essential to find out the heat transfer coefficient.

Table 4. Heat transfer calculation parameters for heat exchanger.

Parameters	Symbols	Inner Pipe (i) Hot Fluid (Exhaust Gas)	Symbols	Outer Pipe (a) Cold Fluid (Water)	Units
Velocity of fluid	V_{hi}	4.069	V_{ca}	0.0015	m/s
Fluid Flow Area	A_i	0.00322	A_a	0.0103	m^2
Hydraulic Diameter	D_{hi}	0.064	D_{ha}	0.064	m
Reynolds Number	Re_i	6713.34	Re_a	156.01	-
Prandtl Number	Pr_i	0.68	Pr_a	4.182	-
Friction Factor	f_i	0.0353	f_a	0.41	-
Nusselt Number	Nu_i	19.63	Nu_a	5.49	-
Heat Transfer coefficient	h_i	12.40	h_a	51.47	W/(m^2·K)

3.1.2. Heat Transfer Surface Area of Protracted Finned Heat Exchanger

The convective heat transfer area has to be calculated in order to determine the heat transfer rate as well as other important thermal parameters. The equations used for different conditions are as follows. If fins are added into exhaust gas (internally–externally) side to promote the boundary layer separation, it enhances the increased heat transfer for this experimental work and the total heat transfer surface area is calculated using the expression as follows

$$A_s = A_{total} = A_{unfinned} + N_f A_{fin} \tag{1}$$

The determined values of the heat transfer area are tabulated in Table 5. The heat exchanger fins' performance parameters like efficiency and effectiveness are getting increased due to its construction material that possess higher thermal conductivity characteristics.

Table 5. Heat transfer surface area of finned heat exchanger.

Description	Symbols	Hot Fluid (Exhaust Gas)	Symbols	Cold Fluid (Water)	Units
Finned Area	A_{fi}	0.54162	A_{fa}	0.54162	m^2
Un finned (Inner pipe) Area	A_{bi}	0.171069	A_{ba}	0.189911	m^2
Total surface Area	A_{ti}	0.71268	A_{ta}	0.731531	m^2

3.1.3. Overallheat Transfer Co-Efficient

The overall heat transfer coefficient is primarily influenced by thickness and thermal conductivity of the media through which heat is getting transferred effectively. The larger the coefficient, the easier heat gets transferred from its source to the working fluid being heated. In a heat exchanger, the overall heat transfer co-efficient (U) for protracted finned double-pipe heat exchanger was calculated using the following equation for the particular design considered [23]. The determined overall co-efficient value was 10.257 W/(m$^2 \cdot$K).

$$\frac{1}{UA_s} = R_{total} = \frac{1}{h_i A_i} + \frac{R_{fi}}{A_i} + \frac{\ln\left(\frac{d_o}{d_i}\right)}{2\pi L k} + \frac{R_{fo}}{A_o} + \frac{t_i}{kA_{fi}} + \frac{t_o}{kA_{fo}} + \frac{1}{h_o A_o} \tag{2}$$

3.1.4. Effectiveness of the Designed Finned Heat Exchanger

The effectiveness of the heat exchanger remains the 'performance measuring parameter' of the component which was designed and utilized for the heat recovery. It is the ratio of the actual heat transfer rate for a heat exchanger to the maximum possible heat transfer rate [24].

$$\text{Effectiveness } (\varepsilon) = \frac{Q_{actual}}{Q_{max}} = \frac{C_h(T_{hi} - T_{ho})}{C_{min}(T_{hi} - T_{ci})} = \frac{C_c(T_{co} - T_{ci})}{C_{min}(T_{hi} - T_{ci})} = \frac{1 - e^{-NTU(1-C_r)}}{1 - C_r e^{-NTU(1-c_r)}} \tag{3}$$

In general, it is possible to express effectiveness as a function of Number of Transfer Units, NTU; the heat capacity rate ratio, C_r; and the flow arrangement in the heat exchanger,

$$\text{Number Transfer Unit (NTU)} = \frac{UA_s}{C_{min}} \tag{4}$$

The first dimensionless parameter is nothing but Heat Capacity Ratio, the percentage of the minimum-to-the-maximum value of Heat Capacity Rate for the exhaust gas and water. The heat capacity ratio of a fluid is determination of its capability to liberate or take up heat. This is calculated for both fluids as the product of the mass flow rate times the specific heat capacity of the fluid,

$$\text{Heat capacity ratio } (C_r) = \frac{C_{min}}{C_{max}} \tag{5}$$

By substituting all the determined values in the effectiveness Equation (3), the protracted internally–externally finned counter flow double-pipe heat exchanger effectiveness (ε) in % was calculated as 75.67. The effectiveness of the designed double pipe protracted finned type of heat exchanger shows a considerable improvement when compared with the design carried outfor double pipe heat exchanger without fins [25].

3.1.5. Geometry Design Parameters of Finned Heat Exchanger

Based on the above designed procedure, Table 6 lists out the parameters obtained for the protruded finned counter flow heat exchanger. Without affecting the heat transfer co-efficient, the optimized number of fins were coated with diesel oxidation catalysts so as to reduce the engine emissions. The Figure 3 depict various designed parameters of the heat exchanger. According to the design, the heat exchanger was fabricated and fitted in the diesel engine exhaust manifold.

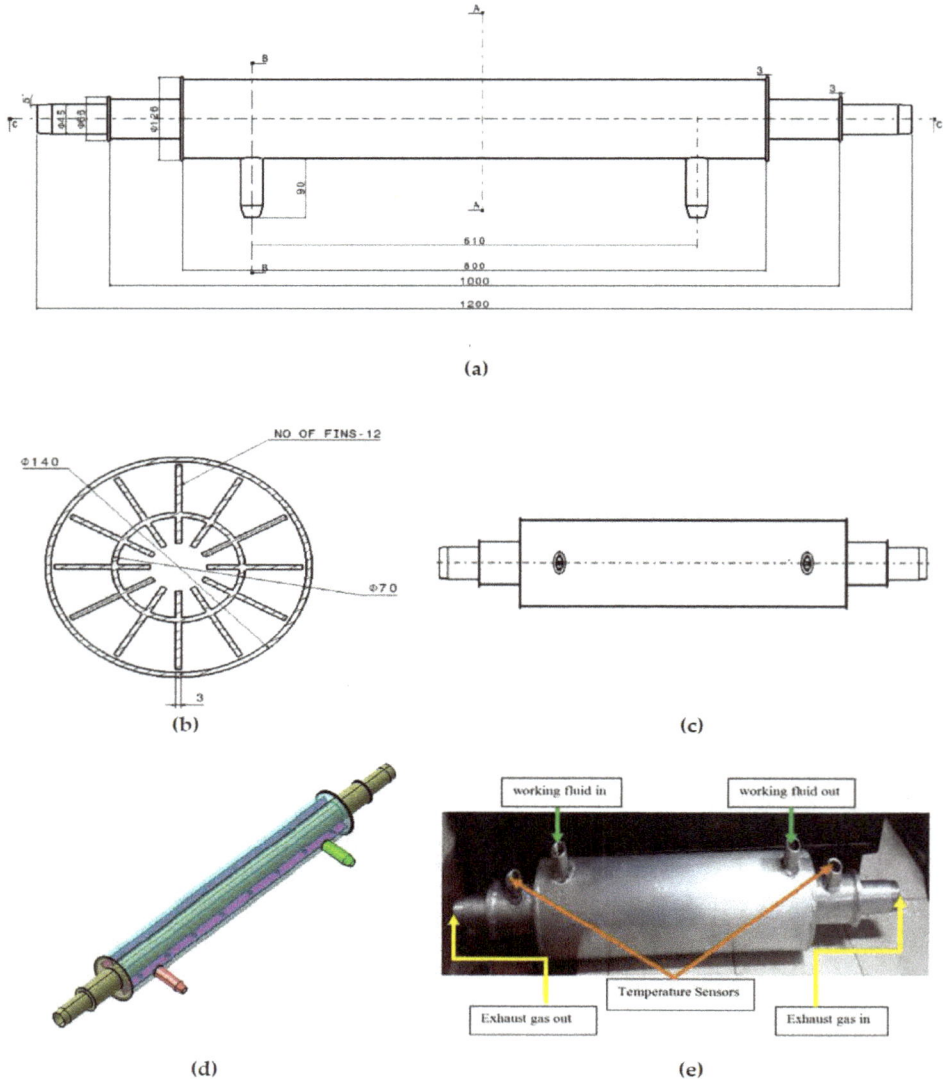

(a)

(b)

(c)

(d)

(e)

Figure 3. Schematic sketch of 2-Dimensional, 3-Dimensional and fabricated heat exchanger views of PFCHE. (**a**) Front view, (**b**) Side view, (**c**) Top view, (**d**) Isometric view, (**e**) Fabricated view.

Table 6. Dimensions of heat exchanger.

Parameters	Symbols	Inside	Outside	Units
Height	H	0.03	0.03	m
Thickness	T	0.003	0.003	m
Cross section Area	A	0.03	0.03	m^2
Perimeter	P	2.006	2.006	m
Length of fin	L_f	1.0	1.0	m
Number fins	N_f	12	12	-
Finned heat transfer Surface Area Convection	A_f	0.5416	0.5416	m^2
Tube area available for heat transfer in finned tube heat exchanger	A_b	0.17106	0.1899	m^2
Total area of finned tube heat exchanger	A_t	0.7127	0.7315	m^2
Inside diameters of pipes	d_i, D_i	0.064	0.114	m
Outside diameters of pipes	d_o, D_o	0.07	0.12	m

3.2. Error Analysis

Errors and uncertainties are usually unavoidable in any research and it is associated with various primary experimental measurements and the calculations of performance parameters. It can occur during selection of instruments, conditioning, environmental factors, calibration, observation, test planning, and while taking readings. Uncertainty analysis is usually carried out to prove the precision and accuracy of the experiments. Percentage uncertainties of different parameters like performance, emission, and combustion parameters were calculated using the percentage uncertainties of various instruments tabulated in Table 7. The total percentage of uncertainty in this experiment is determined by using an Equation (6)

Total percentage of uncertainty

$$
\begin{aligned}
&= \{(TFC_{UC})^2 + (BP_{UC})^2 + (SFEC_{UC})^2 + (BTE_{UC})^2 + (CO_{2UC})^2 \\
&+ (HC_{UC})^2 + (NOx_{UC})^2 + (Smoke_{UC})^2 + (EGT_{UC})^2 + (PP_{UC})^2 + (\epsilon)^2 \\
&+ (HT_{UC})^2 + (PO_{UC})^2\} = \pm 2.61.
\end{aligned} \quad (6)
$$

Using the appropriate calculation procedure, the total uncertainty for the entire experiment was obtained to be $\pm 2.61\%$. An uncertainty analysis was carried out using the method put forth by Holman [26,27].

Table 7. List of devices, its range, accuracy, and percentage.

S. Number	Devices	Range	Accuracy	(%) Uncertainty
1	Exhaust gas temperature indicator	0–900 °C	+0.1°C to −0.1°C	+0.15 to −0.15
2	Gas analyzer	CO (0–10%) CO$_2$ (0–20%) HC (0–10,000 ppm) NO$_x$ (0–5000 ppm)	+0.02% to −0.02% +0.03% to −0.02% +20 ppm to −20 ppm +10 ppm to −10 ppm	+0.2 to −0.2 +0.15 to −0.1 +0.2 to −0.2
3	Smoke level measuring instrument	437C,IP52(0 to 100%)	+0.1 to −0.1	+1 to −1
4	Speed measuring unit	0–10,000 ppm	+10 rpm to −10 rpm	+0.1 to −0.1
5	Burette for fuel measurement	-	+0.1 cm$_3$ to −0.1 cm$_3$ +0.6 s to −0.6 s	+1 to −1 +0.2 to −0.2
6	Pressure pickup	-	+1 o to −1 o	+0.2 to −0.2
7	Manometer	0–110 bar	+0.1 kg to −0.1 kg	+0.1 to −0.1

4. Results and Discussion

An innovative PFCHE was designed and fabricated to extract and utilize in the production of useful power output and simultaneously to reduce the engine emissions. The following topics discuss about the performance of PFCHE and emission characteristics of the engine.

4.1. Fin Geometry Effect on Exhaust Gas Outlet Temperature

The effect of fin geometry on exhaust gas temperature is shown in the Figure 4. As shown in the plot, the final temperature of the exhaust gas decreases as the number and length of the fin increases due to high availability of heat transfer surface area. About 57% of the heat was recovered when the number of fins and the fin length of heat exchanger were modified from 6 to 12 and 0.6 m to 1.0 m respectively.

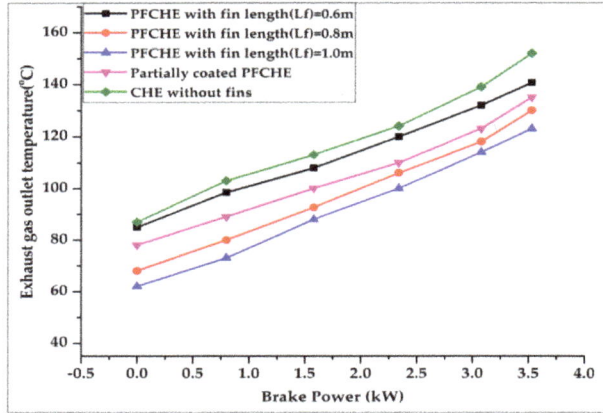

Figure 4. Variation of exhaust gas outlet temperature with fin geometry.

4.2. Energy Recovered by the Working Fluid

The amount of heat absorbed by the working fluid shows an increasing trend with respect to increase in the geometric parameters such as number and length of the fin. As shown in Figure 5, the maximum water outlet temperature was attained when the fin length was 1.0 m with a total of 12 fins. About 37% additional heat absorption was observed, when the fin number and the fin length varied from 6 to 12 and 0.6 m to 1.0 m respectively, due to improved heat transfer area.

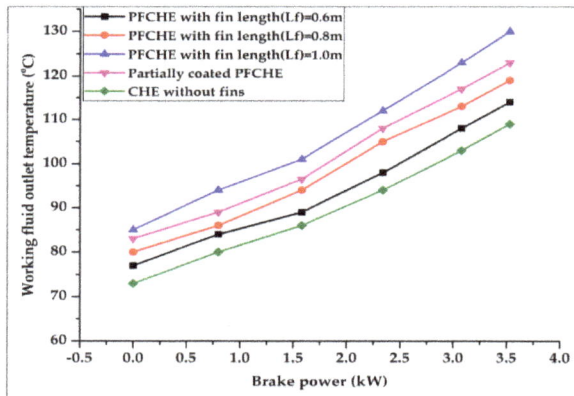

Figure 5. Variation of working fluid outlet temperature with fin geometry.

4.3. Heat Transfer Rate of Working Fluid

Figure 6 shows the variation in heat transfer rate with fin geometry. The rate of heat transfer in the working fluid was determined according to Log-Mean Temperature Difference method (LMTD).

The maximum heat transfer of 550 W was obtained for 1.0 m fin length with 12 fins compared to all other fin geometries. About 39% high amount of heat transfer rate was observed in PFCHE with fin length of 1.0 m when compared with CHE without fins. High heat transfer rate was achieved for 1.0 m fin length when compared with partially-coated heat exchanger. The recovered exhaust energy was used to run the turbine and in-turn the turbo-electric generator. The results showed that this set up can produce 0.06 kW and 0.55 kW of power when the turbine speed is 1500 rpm and 3500 rpm respectively.

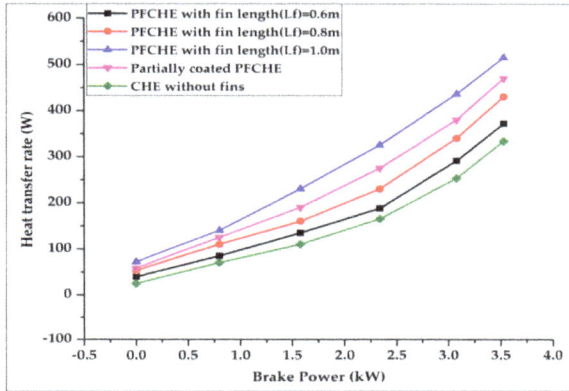

Figure 6. Variation of heat transfer rate with fin geometry.

4.4. Effectiveness of the Heat Exchanger

As the number of fins and length increases, the transfer area also increases due to which the actual heat transfer rate increases which results in the increased effectiveness of heat exchanger. It is evident from the Figure 7, that the effectiveness of heat exchanger increases with increased length and number of fins. The maximum amount of effectiveness, that is, 76%, was found for 1.0 m fin length and 12 fins compared to other fin geometries. The overall performance of the heat exchanger got improved by fin and geometrical configurations of the heat exchanger as advised by Mohd Zeeshan [28].

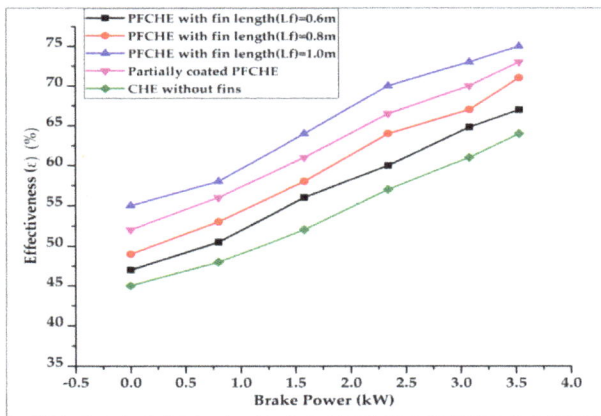

Figure 7. Effectiveness of heat exchanger with fin geometry.

4.5. Brake Thermal Efficiency

The variation of brake thermal efficiency with brake power for different heat exchanger parameters is shown in Figure 8. The brake thermal efficiency was calculated from the ratio of brake power and additional power produced by WHR to the total energy input contributed by the diesel fuel. The brake thermal efficiency in waste heat recovery technique at full load is found to be 37% at PFCHE without coating, 34.5% at heat exchanger without fins, 36% at partially coated PCFHE. It was observed that the overall brake thermal efficiency increased about 5% by utilizing the waste heat energy recovered from the exhaust gas.

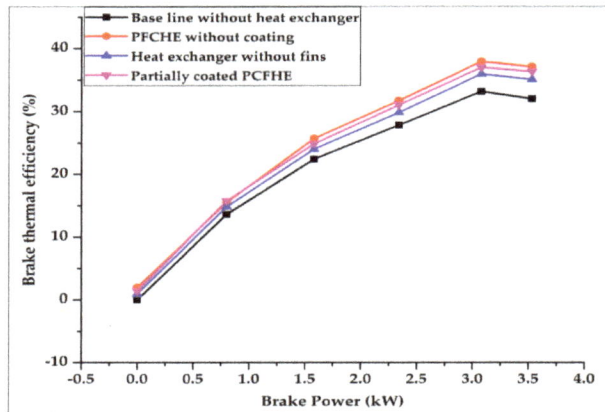

Figure 8. Effect of waste heat recovery (WHR) on brake thermal efficiency.

4.6. Engine Emission Parameters

4.6.1. Hydro Carbons

Hydrocarbon emission is the emission of a combination of unburned fuels due to low temperature that occurs near the cylinder wall. Hydrocarbons consists of thousands of species such as alkanes, alkenes, and aromatics. Diesel engines normally release low levels of hydrocarbons compared to petrol engines. The impact of heat exchanger on hydrocarbon emission is shown in Figure 9. As shown in the plot, there is a significant increase in hydrocarbon when brake power was increased. The hydrocarbon emission was comparatively higher when the emission test was conducted without finned heat exchanger set up, whereas it was lower when the emission test was conducted with finned heat exchanger. The partially-coated finned heat exchanger resulted in reduced hydro carbon level due to oxidation effect [29].

4.6.2. Carbon Monoxide

Carbon monoxide emission mainly occurs due to incomplete combustion of air and fuel. Particularly, it can be caused at beginning as well as at instantaneous acceleration of engine during which the rich mixtures are required. In rich air–fuel mixtures, due to air shortage and reactant concentration, all the carbon cannot be converted into CO_2 due to which CO is produced. Figure 10 depicts the comparison of CO emission from an engine exhaust 'with finned' and 'without finned' the use of heat exchanger. When the test was conducted without installing finned heat exchanger, CO was highly emitted than when using heat exchanger with fins. With the protracted finned heat exchanger, carbon monoxide emissions started to reduce at high load conditions due to lattice oxygen donation from the diesel oxidation catalyst and its reaction with exhaust gases [30].

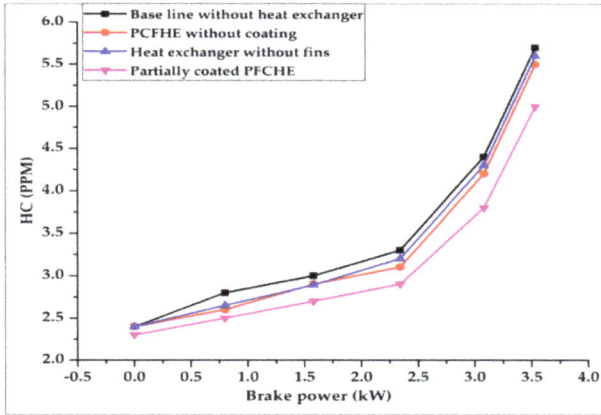

Figure 9. Effect of heat exchanger on HC emission.

4.6.3. Nitrogen Oxides

Nitrogen oxides are known as nitrogen monoxide (NO) and nitrogen dioxide (NO_2). NO constitutes about 85–95% of NO_x. Figure 11 shows the comparison of NO_x emission variations in the experimentation conducted between 'with finned heat exchanger' and 'without finned heat exchanger'. Heat exchanger-based heat recovery technique can reduce considerable amount of NO_x emission when compared with experimental results retrieved from 'without using fins' in the heat exchanger. This reduction might be due to the recovery of thermal energy from exhaust gases and the reaction of exhaust gases with oxidation catalyst used in heat exchanger fins [31].

4.6.4. Carbon Dioxide

The variation of carbon dioxide emission in the exhaust manifold using heat exchanger is plotted in the Figure 12. It shows that there is a slight increase in CO_2 emission when using PFCHE in the diesel engine exhaust compared to heat exchanger without using fins. Waste heat recovery system and its design allow to reduce half of the CO_2 emissions in the diesel engines, when compared to other techniques [32]. It may be due to the oxidation effect of the catalysts used in the PFCHE.

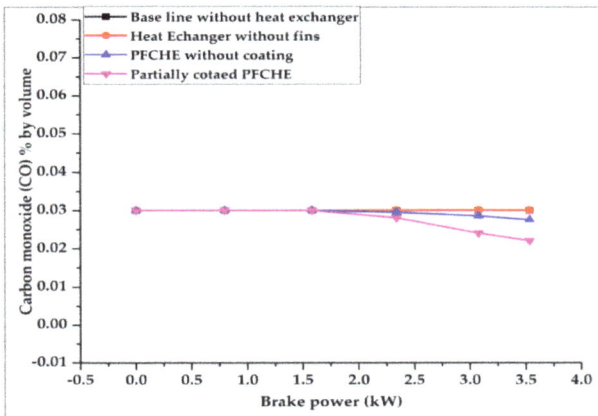

Figure 10. Effect of heat exchanger on CO emission.

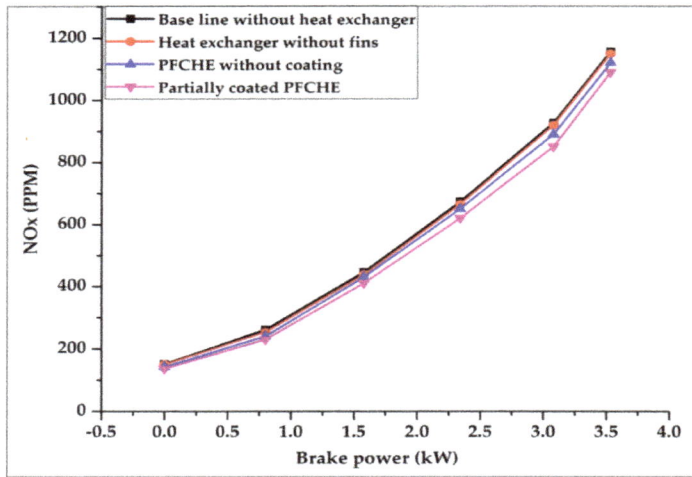

Figure 11. Effect of heat exchanger on NO$_x$ emission.

4.6.5. Smoke Intensity

The comparative results of smoke intensity is indicated as Hatridge Smoke Units (HSU)and is shown in Figure 13. Variation of smoke emissions is relatively less at low power operations due to a smaller amount of fuel particle being present in the burning process. The impact of the implementation of 'protracted both internally–externally finned counter flow heat exchanger with partial coating' in waste heat recovery resulted in low smoke emissions when compared with diesel fuel. This is due to the diesel oxidation catalyst used in part of the heat exchanger fins and reactions with exhaust gases. The smoke was formed due to pyrolysis of PAH (Polycyclic Aromatic Hydrocarbons) and it predominated at full load operating conditions [33].

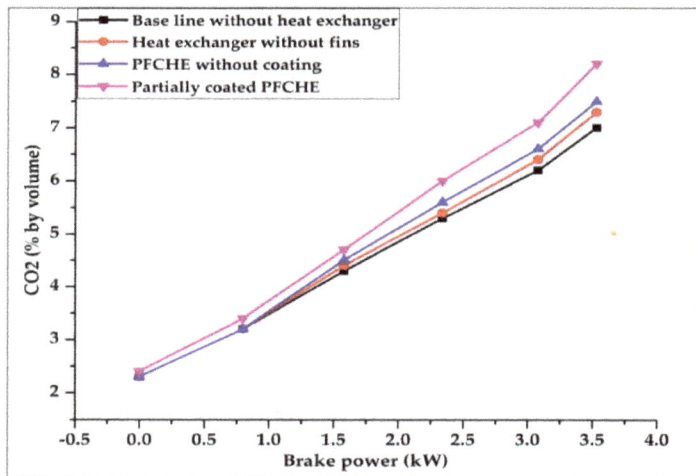

Figure 12. Effect of heat exchanger on CO$_2$ emission.

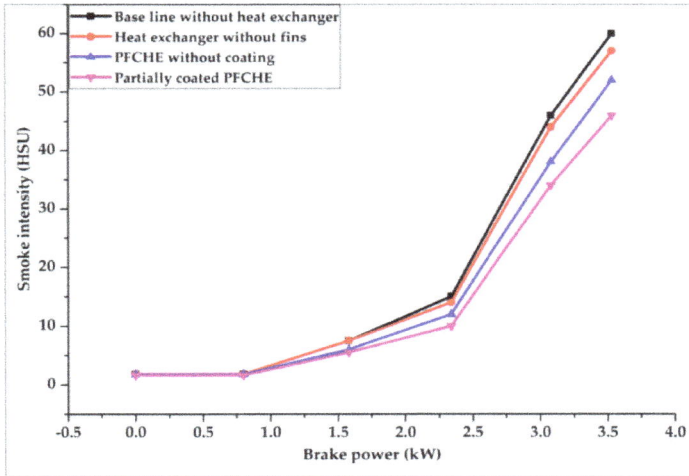

Figure 13. Effect of heat exchanger on smoke intensity.

5. Conclusions

In internal combustion engines (ICEs), About 70% of the heat energy produced by pistons is lost due to the exhaust and cooling process. In this study, an innovative double-pipe, both internally–externally PFCHE was designed and adopted to recover the heat from engine exhaust gas and in parallel, it was also aimed at reducing the harmful exhaust emissions. The following conclusions can be drawn from the experimental results.

➤ The analytical results indicated a positive notion about the overall efficiency of Organic Rankine Cycle heat recovery system. The experimental results demonstrated that from the heat exchanger without fins, the amount of heat extracted was 335 W. With innovative finned heat exchanger, the amount of heat recovered was 550 W. High amount of heat transfer was achieved, namely, 39%, when the fin length varied from 0.6 m to 1.0 m and when the number of fins were increased from 6 to 12.

➤ By comparison, the effectiveness of the PFCHE was able to reach its full load operation from 71% to 75%. The effectiveness of the heat exchanger without fins was 10–13% lesser than that of the effectiveness with 1.0 m length finned heat exchanger. It was also revealed that, as the fin numbers and its length increases, the heat transfer rate increases which further resulted in improved performance of the heat recovery system and increased brake thermal efficiency from 32% to 37%.The developed heat recovery system can produce 0.06 kW and 0.55 kW of power when the turbines execute at 1500 rpm and 3500 rpm respectively.

➤ The application of partially-coated, protracted, and finned heat exchanger showed a considerable reduction in the engine emissions due to diesel oxidation catalysts coating. A reduction in HC, CO and NO_x emissions was observed with partially coated PFCHE. The reduction in HC and NO_x emissions at full load were 16% and 7% respectively when compared to base line engine operation without heat exchanger. Smoke got decreased from 60 HSU to 45 HSU at full load operation of the engine. The reduction could be attributed due to the properties of material used for heat exchanger and exhaust gas reactions.

On the whole, it can be concluded that DOC-coated PFCHE resulted in increase in the heat transfer rate, effectiveness and brake thermal efficiency when compared to the heat exchanger without fins. HC, CO, NO_x and smoke emissions were less than the base line engine operation without heat exchanger. As per the current study results, it is found that the partially-coated PFCHE could be a suitable heat exchanger to recover exhaust heat energy and can reduce engine exhaust emissions. In future, the area of research has to be explored in depth towards the integration of engine coolant heat and exhaust using various power plant cycles. It should also focus on performance improvement under different working fluids that can make better use of a low temperature exhaust heat recovery system and minimize the installation cost along with the size of WHR system.

Author Contributions: R.R. contributed in designing and experimental fabrication of the research work, conducting experiments, analyses of the results, manuscript development, organization and presentation of the work. Further, R.R. contributed in editing and improving the paper both in language as well as technical aspects and responded to revisions. S.P. contributed in conceptualization, methodology for conducting experiments, parametric analysis, results comparison, implementation of the research and peer-reviewing of the manuscript. Further, S.P. supervised the project with knowledge and experience in waste heat recovery.

Funding: This research received no external funding.

Acknowledgments: The authors are grateful to Madras Institute of Technology(MIT)and the Department of Automobile Engineering for their precious co-operation and support.

Conflicts of Interest: The authors declare no conflict of interest.

Nomenclatures

ρ	fluid density (kg/m^3)
A_s	heat transfer surface area (m^2)
U	overall heat transfer coefficient (W/m^2·K)
Q	rate of heat transfer (W)
Q_{actual}	actual heat transfer (W)
Q_{max}	maximum possible heat transfer (W)
D	cylinder bore (m)
L	stroke length (m)
N	engine speed (rpm)
m_a	mass flow rate of air (kg/s)
m_f	mass flow rate of fuel (kg/s)
C_p	specific heat capacity (J/kg·K)
μ	viscosity (N·s/m^2)
d_i	inner pipe inside diameter (m)
d_o	inner pipe outside diameter (m)
V_h & V_c	velocity of hot and cold fluids (m/s)
K_h & K_c	thermal conductivity of hot and cold fluids (m/s)
T_{hi} & T_{ci}	inlet temperature of hot and cold fluids (K)
T_{ho} & T_{co}	outlet temperature hot, cold fluids (K)
Re	Reynolds number
Pr_i	Prandtl number
Nu	Nusselt number
f	friction factor
h	convection heat transfer co-efficient (W/m^2·K)
D_h	hydraulic diameter (m)
Nf	number of fins
H	height of fin (m)
T	thickness of fin (m)
A	fin cross section area (m^2)
P	perimeter of fin (m)
L_f	length of fin (m)
A_f	finned surface area for heat transfer (m^2)
A_b	un-finned area of pipe (m^2)
A_t	total area of finned tube heat exchanger (m^2)

R_{total}	total thermal resistance $(m^2 \cdot K/W)$
$R_{fi} \& R_{fo}$	fouling factors of inner pipe and outer pipe $(m^2 \cdot K/W)$
R_{wall}	resistance of wall $(m^2 \cdot K/W)$
C_c	heat capacity of cold fluid $(J/kg \cdot K)$
C_h	heat capacity of hot fluid $(J/kg \cdot K)$
C_r	heat capacity ratio
ε	effectiveness of heat exchanger

Acronyms

ICEs	internal combustion engines
WHR	waste heat recovery
TEG	turbo electric generator
ORC	organic rankine cycle
PFCHE	protracted finned counter flow heat exchanger
CHE	counter flow heat exchanger
LMTD	log mean temperature difference
NTU	number of transfer units
EGT	exhaust gas temperature
HT	heat transfer
BP	brake power
PP	pressure pickup
PO	power output
BTE	brake thermal efficiency
HC	hydro carbons
CO	carbon monoxide
NO_x	nitrogen oxides
CO_2	carbon-di-oxide
HSU	hatridge smoke units
PAH	polycyclic aromatic hydrocarbons
DOC	diesel oxidation catalysts

References

1. Hatami, M.; Gannji, D.D.; Gorji-Bandpy, M. A review of different heat exchangers designs for increasing the diesel exhaust waste heat recovery. *Renew. Sustain. Energy Rev.* **2014**, *37*, 168–181. [CrossRef]
2. Zhang, Z.; Li, L. Investigation of In-Cylinder Steam Injection in a Turbocharged Diesel Engine for Waste Heat Recovery and NOx Emission Control. *Energies* **2018**, *11*, 936. [CrossRef]
3. Gu, W.; Weng, Y.; Wang, Y.; Zheng, B. Theoretical and experimental investigation of an organic Rankine cycle for a waste heat recovery system. *Proc. Inst. Mech. Eng. Part A J. Power Energy* **2009**, *223*, JPE725. [CrossRef]
4. Borsukiewicz-Gozdur, A.; Nowak, W. Maximizing the working fluid flow as a way of increasing power output of geothermal power plant. *Appl. Therm. Eng.* **2007**, *27*, 2074–2078. [CrossRef]
5. Chinese, D.; Meneghetti, A.; Nardin, G. Diffused introduction of organic Rankine cycle for biomass-based power generation in an industrial district: A systems analysis. *Int. J. Energy Res.* **2004**, *28*, 1003–1021. [CrossRef]
6. Drescher, U.; Brüggemann, D. Fluid selection for the organic Rankine cycle (ORC) in biomass power and heat plants. *Appl. Therm. Eng.* **2007**, *27*, 223–228. [CrossRef]
7. De Oliveira Neto, R.; Sotomonte, C.A.R.; Coronado, C.J.; Nascimento, M.A. Technical and economic analyses of waste heat energy recovery from internal combustion engines by the Organic Rankine Cycle. *Energy Convers. Manag.* **2016**, *129*, 168–179. [CrossRef]
8. Li, L.; Ge, Y.T.; Tassoua, S.A. Experimental study on a small-scale R245fa organic Rankine cycle system for low-grade thermal energy recovery. *Energy Procedia* **2017**, *105*, 1827–1832. [CrossRef]
9. Remeli, M.F.; Tan, L.; Date, A.; Singh, B.; Akbarzadeh, A. Simultaneous power generation and heat recovery using a heat pipe assisted thermoelectric generator system. *Energy Convers. Manag.* **2015**, *91*, 110–119. [CrossRef]

10. Mastrullo, R.; Mauro, A.W.; Revellin, R.; Viscito, L. Modeling and optimization of a shell and louvered fin mini-tubes heat exchanger in an ORC powered by an internal combustion engine. *Energy Convers. Manag.* **2015**, *10*, 697–712. [CrossRef]

11. Hatami, M.; Ganji, D.D.; Gorji-Bandpy, M. Numerical study of finned type heat exchangers for ICEs exhaust waste heat recovery. *Case Stud. Therm. Eng.* **2014**, *4*, 53–64. [CrossRef]

12. Pak, B.C.; Baek, B.J.; Groll, E.A. Impacts of fouling and cleaning on the performance of plate fin and spine fin heat exchangers. *KSME Int. J.* **2003**, *17*, 1801–1811. [CrossRef]

13. Bei, C.; Zhang, H.; Yang, F.; Song, S.; Wang, E.; Liu, H.; Chang, Y.; Wang, H.; Yang, K. Performance analysis of an evaporator for a diesel engine—Organic Rankine Cycle (ORC) combined system and influence of pressure drop on the diesel engine operating characteristics. *Energies* **2015**, *8*, 5488–5515. [CrossRef]

14. Zhang, H.G.; Wang, E.H.; Fan, B.Y. Heat transfer analysis of a finned-tube evaporator for an engine exhaust heat recovery. *Energy Convers. Manag.* **2013**, *65*, 438–447. [CrossRef]

15. Chen, H.; Wang, Y.; Zhao, Q.; Ma, H.; Li, Y.; Chen, Z. Experimental investigation of heat transfer and pressure drop characteristics of H-type finned tube banks. *Energies* **2014**, *7*, 7094–7104. [CrossRef]

16. Jin, Y.; Tang, G.-H.; He, Y.-L.; Tao, W.-Q. Parametric study and field synergy principle analysis of H-type finned tube bank with 10 rows. *Int. J. Heat Mass Transf.* **2013**, *60*, 241–251. [CrossRef]

17. Song, S.; Zhang, H.; Lou, Z.; Yang, F.; Yang, K.; Wang, H.; Bei, C.; Chang, Y.; Yao, B. Performance analysis of exhaust waste heat recovery system for stationary CNG engine based on organic Rankine cycle. *Appl. Therm. Eng.* **2015**, *76*, 301–309. [CrossRef]

18. Altosole, M.; Benvenuto, G.; Campora, U.; Laviola, M.; Trucco, A. Waste heat recovery from marine gas turbines and diesel engines. *Energies* **2017**, *10*, 718. [CrossRef]

19. Fernández-Yañez, P.; Armas, O.; Capetillo, A.; Martínez-Martínez, S. Thermal analysis of a thermoelectric generator for light-duty diesel engines. *Appl. Energy* **2018**, *226*, 690–702. [CrossRef]

20. Fernandez-Yanez, P.; Gomez, A.; García-Contreras, R.; Armas, O. Evaluating thermoelectric modules in diesel exhaust systems: Potential under urban and extra-urban driving conditions. *J. Clean. Prod.* **2018**, *182*, 107–1079. [CrossRef]

21. Fernández-Yáñez, P.; Armas, O.; Kiwan, R.; Stefanopoulou, A.G.; Boehman, A.L. A thermoelectric generator in exhaust systems of spark-ignition and compression-ignition engines. A comparison with an electric turbo-generator. *Appl. Energy* **2018**, *229*, 80–87. [CrossRef]

22. Raei, B.; Shahraki, F.; Jamialahmadi, M.; Peyghambarzadeh, S.M. Different methods to calculate heat transfer coefficient in a double-tube heat exchanger. A. comparative study. *Exp. Heat Transf.* **2018**, *31*, 32–46. [CrossRef]

23. Shah, R.K.; Sekulic, D.P. *Fundamentals of Heat Exchanger Design*; John Wiley & Sons, Inc.: Hoboken, NJ, USA, 2003. [CrossRef]

24. Navarro, H.A.; Cabezas-Gómez, L.C. Effectiveness-NTU computation with a mathematical model for cross-flow heat exchangers. *Braz. J. Chem. Eng.* **2007**, *24*, 509–521. [CrossRef]

25. Rajesh, R.; Senthilkumar, P.; Mohanraj, K. Design of heat exchanger for exhaust heat recovery of a single cylinder compression ignition engine. *J. Eng. Sci. Technol.* **2018**, *13*, 2153–2165.

26. Holman, J.B. *Experimental Techniques for Engineers*; McGraw Hill Publications: New York, NY, USA, 2003.

27. Lakshmanan, T.; Nagarajan, G. Experimental investigation on dual fuel operation of acetylene in a DI diesel engine. *Fuel Process. Technol.* **2010**, *91*, 496–503. [CrossRef]

28. Zeeshan, M.; Nath, S.; Bhanja, D. Numerical study to predict optimal configuration of fin and tube compact heat exchanger with various tube shapes and spatial arrangements. *Energy Convers. Manag.* **2017**, *148*, 737–752. [CrossRef]

29. Amanatidis, S.; Ntziachristos, L.; Giechaskiel, B.; Samaras, Z.; Bergmann, A. Evaluation of an oxidation catalyst ("catalytic stripper") in eliminating volatile material from combustion aerosol. *J. Aerosol Sci.* **2013**, *57*, 144–155. [CrossRef]

30. Hung, T.C.; Shai, T.Y.; Wang, S.K. A review of organic rankine cycles (ORCs) for the recovery of low-grade waste heat. *Energy* **1997**, *22*, 661–667. [CrossRef]

31. Muthiya, S.J.; Pachamuthu, S. Electrochemical NO$_x$ reduction and oxidation of HC and PM emissions from biodiesel fuelled diesel engines using electrochemically activated cell. *Int. J. Green Energy* **2018**, *15*, 314–324. [CrossRef]

32. Nanthagopal, K.; Ashok, B.; Varatharajan, V.; Anand, V.; Kumar, R.D. Study on the effect of exhaust gas-based fuel preheating device on ethanol–diesel blends operation in a compression ignition engine. *Clean Technol. Environ. Policy* **2017**, *19*, 2379–2392. [CrossRef]

33. Weidemann, E.; Buss, W.; Edo, M.; Mašek, O.; Jansson, S. Influence of pyrolysis temperature and production unit on formation of selected PAHs, oxy-PAHs, N.-PACs, PCDDs, and PCDFs in biochar—A screening study. *Environ. Sci. Pollut. Res.* **2018**, *25*, 3933–3940. [CrossRef] [PubMed]

energies

MDPI

Article

Desiccant-Based Air Handling Unit Alternatively Equipped with Three Hygroscopic Materials and Driven by Solar Energy

Piero Bareschino [1], Francesco Pepe [1], Carlo Roselli [1], Maurizio Sasso [1] and Francesco Tariello [2,*]

[1] Dipartimento di Ingegneria, Università degli Studi del Sannio, Piazza Roma 21, 82100 Benevento, Italy; piero.bareschino@unisannio.it (P.B.); francesco.pepe@unisannio.it (F.P.); carlo.roselli@unisannio.it (C.R.); sasso@unisannio.it (M.S.)
[2] Dipartimento di Medicina e Scienze della Salute "Vincenzo Tiberio", Università degli Studi del Molise, 86100 Campobasso, Italy
*   Correspondence: francesco.tariello@unimol.it; Tel.: +39-0874404957

Received: 31 March 2019; Accepted: 21 April 2019; Published: 24 April 2019

Abstract: The energy demand for the air-conditioning of buildings has shown a very significant growth trend in the last two decades. In this paper three alternative hygroscopic materials for desiccant wheels are compared considering the operation of the air handling unit they are installed in. The analyses are performed by means of the TRNSYS 17® software, simulating the plant with the desiccant wheel made of: silica-gel, i.e., the filling actually used in the experimental plant desiccant wheel of the University of Sannio Laboratory; MIL101@GO-6 (MILGO), a composite material, consisting of graphite oxide dispersed in a MIL101 metal organic framework structure; Campanian Ignimbrite, a naturally occurring tuff, rich in phillipsite and chabazite zeolites, widespread in the Campania region, in Southern Italy. The air-conditioning system analyzed serves a university classroom located in Benevento, and it is activated by the thermal energy of a solar field for which three surfaces are considered: about 20, 27 and 34 m^2. The results demonstrate that a primary energy saving of about 20%, 29%, 15% can be reached with silica-gel, MILGO and zeolite-rich tuff desiccant wheel based air handling units, respectively.

Keywords: desiccant wheel; solar heating and cooling; hygroscopic materials; dynamic simulations; energy and environmental analysis

1. Introduction

Air-conditioning in the tertiary sector is largely based on all-air or mixed air–water systems. In these plants most of the required energy is due to the removal of moisture from the air especially in hot and humid regions. Commonly the dehumidification process is conducted by cooling down the air below its dew point, at a temperature that is too low to supply this air to the conditioning space, therefore the dehumidified air is subsequently heated up. The cooling energy load is commonly satisfied by electric-driven chillers which have caused significant daytime electric peak loads in the developing and developed countries during the summer period, while also contributing to the increase in energy demands in the last decades. Desiccant-based HVAC (Heating Ventilation and Air-Conditioning) systems are a solution to satisfy the temperature and humidity levels required in buildings via decoupling latent and sensible loads, and thus significantly reducing the electric energy consumption. These systems operate with thermally-driven cycles that require thermal energy to regenerate the hygroscopic material. This heat can be supplied by several conventional means, such as waste heat, gas burner or electric heater. However, the best solution is often represented by the coupling with solar thermal collectors, since this approach allows huge primary energy savings and greenhouse gas emissions reductions.

The scientific literature in which hygroscopic materials are analyzed is as vast as the applications and materials studied, which are not only evaluated for the production of desiccant wheels (DWs) but, for example, also for the production of adsorption heat pumps or heat exchangers. The materials can be classified based on their composition and differ in their adsorption capacity and regeneration temperature [1]. In [2] the role and prospect of using agricultural waste material as "green" desiccants in desiccant-cooling systems has been evaluated.

Some of the experimental and numerical studies that consider desiccant material specifically for air handling unit (AHU) application are reported below.

In [3], a synthetic metal silicate-based DW is adopted in an AHU together with a sensible heat wheel and an electric heat pump. The coefficient of performance (COP) of varying air stream flow rates, regeneration temperatures and ambient conditions have been experimentally evaluated. The results are used to characterize a TRNSYS model that has been demonstrated to be capable of efficiently simulating the cooling system.

In [4] an experimental study has been carried out on an air handling unit using a DW made of a so called Functional Adsorbent Material Zeolite 01 (FAM-Z01). The authors investigated the effects of some parameters such as: the regeneration temperature, the process air stream's temperature and humidity, the desiccant wheel's rotational speed and the ventilation mass flow rate on the cycle performance. The FAM-Z01 shows low regeneration temperature and a maximum water removal capacity of 1.96 ± 0.12 kg/h in the tested conditions.

Kanoğlu et al. [5] considered an experimental innovative AHU with natural zeolite as the desiccant, paying particular attention to the energy and exergy analyses of the open-cycle realized in the plant. They measured a COP of 0.35, and an exergy efficiency of 11.1%. It is highlighted that the DW shows the greatest percentage of total exergy destruction (33.8%).

The performance of a DW, composed of metal silicate synthesized on inorganic fiber substrate, integrated in an air-conditioning system, have been analyzed and it has been compared with a conventional air-conditioning plant [6]. In this paper it is pointed out that the desiccant-assisted system's moisture removal capacity is about 15–30% greater in comparison to that of the conventional plant, and is capable of holding low humidity in the conditioned space.

Several parameters, such as the air humidity at the DW inlet, the temperature of the regeneration air, the air mass flow rate, etc., which may influence the performance of the DW, are evaluated and discussed for the composite desiccant material, which is a solid solution of LiCl and silica-gel. It behaves better than pure silica-gel in moisture adsorption, and a comparison between these two materials shows that the LiCl/silica-gel composites remove approximately 50% more moisture from air [7].

In [8] a zeolite-based DW, a superadsorbent polymer-made DW and a conventional silica-gel wheel have been compared. The results of the experimental tests reveal that the polymer desiccant wheel has a higher dehumidification capacity than the silica-gel wheel, when the temperature of the regeneration air is 50 °C and the relative humidity exceeds 60%, but it does not strongly improve with the increase of the regeneration temperature and furthermore it is more affected by the regeneration air velocity. Concerning the zeolite desiccant wheel, its dehumidification capacity decreases with decreasing supply air velocity and is not significantly affected by the regeneration air velocity. Finally the temperature of the dehumidified air exiting the silica-gel DW is significantly higher than that at the outlet of the superadsorbent polymer DW.

The behavior of four alternative desiccant wheels have been investigated and tested in two different laboratories [9]. The desiccant rotors are both commercial products and innovative ones. The materials they are made of are: titanium silicate, LiCl, silica-gel and a LiCl/silica-gel composite. The analyses of the main parameters affecting the performance of the wheels show that the best performance of silica-gel DW takes place in the range of 85–100 RPH, whereas the optimal rotational speed is lower for LiCl-based DW; the addition of LiCl to silica-gel increases the dehumidification capacity by about 3 g/kg; the dehumidification potential, in general, increases with the moisture content in the ambient air and the regeneration air temperature; the dehumidification performance improves when the regeneration air relative humidity decreases.

Some experimental results on a hybrid desiccant air-conditioning system equipped with a lithium chloride DW have been reported in [10] and the effects of the relevant operating parameters on the overall system performance have been analyzed. The Authors demonstrate that, with respect to a conventional vapor compression system, the hybrid desiccant cooling AHU reduces the electric power consumption by about 37.5% when the process air temperature and relative air humidity are held at 30 °C, and 55% respectively.

The impact of the features of adsorbent materials on the desiccant wheel performance has been studied through computer modeling in the work of Fong and Lee [11]. They compare a DW of a regular density silica-gel, one made of a synthetic zeolite, named AQSOA-Z02 [12], and another one based on a zeolite called CECA-3A [13]. With an absolute ambient air humidity of 16.04 g/kg, the three desiccant wheels achieve a moisture reduction at the regeneration temperature of 50 °C of 33.0%, 22.6% and 18.7% respectively; these percentages increase up to 65.2%, 64.5% and 51.1% respectively when the regeneration air temperature is increased up to 80 °C.

In [14] two alternative desiccant materials (silica-gel and titanium dioxide) have been compared. A solar-desiccant cooling system is numerically investigated through a validated TRNSYS model in three East Asian climatic conditions (temperate, subtropical and tropical). Titanium dioxide has been proven to be an interesting alternative material as it can reach a lower indoor humidity ratio and temperature with higher cooling performance than the silica-gel, considering the same specification of the solar thermal field and desiccant cooling plant. The system coefficient of performance is within the range 1.5–3, while the solar fraction is between 65% and 90%.

Also a composite desiccant material made of a biopolymer template, chitosan, in which nanoscale boehmite particles are embedded, has been proven to be an interesting renewable material, and a candidate to replace silica-gel due to its high moisture removal capacity. The results showed the formation of crystalline, nanostructured composite with moisture adsorption capacity that is higher by about 50% than the material weight [15].

In this paper an air-conditioning system driven by evacuated tube solar collectors and equipped with a DW is numerically investigated with a parametric approach considering:

- different solar thermal field configurations, three collecting surfaces (about 20, 27 and 34 m^2) and different tilt angles (20–55°);
- three desiccant rotor materials, that is the one which is actually installed in an available test plant (silica-gel), a composite material denominated MIL101@GO-6 (MILGO), made of graphite oxide dispersed in the MIL101 metal organic framework structure, and a naturally occurring zeolite-rich tuff, denominated Campanian Ignimbrite, which is rich in phillipsite and chabazite and is widespread in many areas of Campania region, in southern Italy.

Energy and environmental indexes have been assessed comparing the innovative materials with the conventional one in order to identify the best choice in the base case i.e., when the innovative system meets only the cooling and heating loads of a university classroom located in Benevento and when further low-temperature loads are taken into account.

2. Hygroscopic Materials: Modeling and Characterization

As alternative to the cooling dehumidification there is the moisture reduction by adsorption through hygroscopic materials. As already stated, silica-gel has been historically used as the material of choice for DWs, due to its good water adsorption capacity, relatively low cost and high mechanical resistance when exposed to repeated adsorption/desorption cycles. Notwithstanding, a significant interest has been shown in recent years toward either higher performing materials, or naturally abundant materials which may be obtained at a lower cost. In this context, in the present paper two alternative materials are considered as alternative to silica-gel: the first one is MIL101@GO-6 (or MILGO), a material having outstanding water adsorption properties [16], and Campanian ignimbrite, a naturally occurring zeolitic tuff, particularly rich in phillipsite and chabasite zeolites, which is abundant in many areas of Campania region, in southern Italy, well known for its water adsorption performances [17].

In order to describe the behavior of the desiccant wheel a mathematical model consisting of mass and energy balances for gas-side and solid-side was implemented in previous works [18,19]:

$$\rho_m \frac{\partial \omega}{\partial \theta} + \rho_m V \frac{\partial \omega}{\partial z} = \rho_m D_s \frac{\partial \omega}{\partial \theta} + \frac{\varepsilon_d \rho_d}{\varepsilon} \frac{\partial M}{\partial \theta} \tag{1}$$

$$\frac{\partial M}{\partial \theta} = K(M_e + M) \tag{2}$$

$$\left(\varepsilon \rho_m c_{p,m} + \varepsilon_d \rho_d c_{p,d} \right) \frac{\partial T}{\partial \theta} + \varepsilon \rho_m c_{p,m} V \frac{\partial T}{\partial z} = \varepsilon k_m \frac{\partial^2 T}{\partial z^2} + \frac{q_s}{M_w} \varepsilon_d \rho_d \frac{\partial M}{\partial \theta} \tag{3}$$

where ω is the air absolute humidity, M the moisture content of solid adsorbent (M_e at equilibrium condition), T the temperature, ρ the density, c_p the heat capacity, k the thermal conductivity, K the effective mass transfer coefficient, V the air superficial velocity, ε the void fraction, θ the time, z the axial coordinate, D_s the surface diffusion coefficient, q_s isosteric heat of adsorption, M_w the molecular weight of water, and the subscript m and d, where used, refer the above to moist air and solid adsorbent, respectively. Details about the equilibrium correlation and the absorption heat calculation, along with validation carried out with experimental data, as well as model implementation and results are reported in [18,19] for MILGO and tuff, respectively. The model was solved using the commercial software package Comsol Multiphisycs®. Wheel thickness L, discretized on the basis of a step size set to 1 mm, was chosen as spatial integration domain. Time intervals of lengths θ_{reg} and θ_{proc} for regeneration and dehumidification phases respectively, discretized on the basis of a step size set to 0.1 s, were chosen as temporal integration domains. A single simulation cycle consisted of a dehumidification phase followed by a regeneration one. For the dehumidification phase of the very first cycle the following first-run-only initial conditions were used:

$$M\ (z,0) = M_{in} \tag{4}$$

$$\omega\ (z,0) = \omega_{amb} \tag{5}$$

$$T\ (z,0) = T_{amb} \tag{6}$$

Every half a cycle, the last time values of all variables were taken as initial values for subsequent calculations. Computations were carried on for a sufficiently large number of cycles in order to approach a cyclic steady state profile in both adsorption and regeneration process. Analyses of the dehumidification performance were carried out considering the amplitude of the regeneration section, the DW rotation speed, the material porosity, etc., at different regeneration and process air conditions (T and ω) and for both the adsorbents. Under optimal parameter values, it was observed that the dehumidification effectiveness of MILGO DW was about 30% higher than that of the conventional silica-gel based desiccant wheel [18], while with zeolite-rich tuff rotor the moisture removal was better than with silica-gel DW, when relative humidity was low and the regeneration air temperature was very high.

A further model, the one of Maclaine–Cross and Banks [20], characterizes the combined mass and heat transfer processes taking place in a DW as a simple heat transfer process, by means of two independent characteristic potentials, F_1 and F_2, [21,22] that can be expressed for the specific pair silica-gel/air as [23]:

$$F_{1,j} = \frac{-2865}{\left(t_j + 273.15 \right)^{1.49}} + 4.344 \left(\omega_j / 1000 \right)^{0.8624} \tag{7}$$

$$F_{2,j} = \frac{\left(t_j + 273.15\right)^{1.49}}{6360} - 1.127\left(\omega_j/1000\right)^{0.07969} \tag{8}$$

where the subscript "*j*" refers to the generic thermo-hygrometric condition of the air at which the two potentials are evaluated, whereas ω and *t* are the humidity ratio (g/kg) and the air temperature (°C), respectively. The intersection of the isopotentials identifies the output conditions of the process air in the ideal case, when both the desorption and the adsorption processes are isoenthalpic. Jurinak's model [22] provides that the real output conditions are estimated using two efficiency indexes of the wheel, η_{F1} and η_{F2}, calculated similarly to the efficiency of a heat exchanger, as:

$$\eta_{F1} = (F_{1,2} - F_{1,1})/(F_{1,5} - F_{1,1}). \tag{9}$$

$$\eta_{F2} = (F_{2,2} - F_{2,1})/(F_{2,5} - F_{2,1}). \tag{10}$$

where potentials F_1 and F_2 must be evaluated in the states 1, 2 and 5 of Figure 1a. This model is adopted by the simulation software used in this work (see Section 3).

The parameters η_{F1} and η_{F2} for the silica-gel DW of the experimental plant were validated and calibrated in [24] and their values are 0.207 and 0.717 respectively. They are used below.

In order to continue using this model based on the characteristic potentials and efficiency indexes, Comsol Multiphysics® numerical model has been simulated for both alternative materials modifying one parameter a time in the subsequent ranges of interest for the case study under investigation: absolute humidity 0.010–0.020 kg/kg, with a step of 0.002 kg/kg; process air tperature 293–308 K, with a step of 5 K; regeneration air temperature 323–343 K, with a temperature step of 10 K. With these results the Jurinak's efficiency indexes have been characterized. The mean values of η_{F1} and η_{F2} that better reproduce the conditions at the outlet of the MILGO DWs have been found equal to 0.029 and 0.904 while those for zeolite-rich tuff are 0.219 and 0.634, respectively.

3. Methodology: Simulation Models, Plant Configuration and Analyses

The approach followed to elaborate the results consists of:

- numerical simulations, carried out to dynamically assess the energy flows in the considered plants;
- energy and environmental analyses based on seasonal and annual aggregated results.

3.1. Plants Simulation Model Characteristics and Operation

In this paper an innovative air-conditioning system (IS) and a conventional one (CS) are simulated by means of the software of dynamic simulation TRNSYS 17® [25] integrated with the additional components library TESS [26]. The simulations have been performed with a time-step of 1.5 min and considering the climatic conditions of Benevento, southern Italy. The user served by these plants is a university classroom with a floor surface of 63.5 m^2 and 30 seats, which is occupied in the weekdays from 9:00 to 18:00, whereas the air-conditioning plant is turned on at 8:30 in the morning, and it is switched off at 18:00, when the classroom is closed. The indoor air set-point temperatures in summer and winter operation are 26 °C and 20 °C, respectively, with a dead band of ±0.5 °C, while the air relative humidity is constantly held at 50 ± 10%. The model of the building was implemented through the "type 56" elementary unit of TRNSYS, with which a simulation project is made, using the envelope characteristics of Table 1.

Table 1. Classroom envelope characteristics [27].

Parameter	Opaque Components				Transparent Components		
	Roof	External Walls (N/S)	External Walls (E/W)	On the Ground Floor	North	South	East/West
U (W/m² K)	2.30	1.11	1.11	0.297	2.83	2.83	2.83
Area (m²)	63.5	36	15.87	63.5	8.53	9.40	0.976
g (-)	-	-	-	-	0.755	0.755	0.755

The innovative air-conditioning plant consists of a solar subsystem coupled with a desiccant wheel-based air handling unit (Figure 1). For the solar field the analyses developed later will consider evacuated tube solar thermal collectors (SC) with three aperture areas: about 20, 27 and 34 m². These collectors are connected to a 1000 L thermal energy storage tank (TS); to prevent the solar circuit from high temperature levels (>100 °C) a heat exchanger that dissipates thermal energy surplus or converts it in sanitary hot water or for other purposes (HW–HX, Hot Water Heat Exchanger) is considered.

The innovative AHU is configured as that installed at the University of Sannio laboratory [28]. It is a hybrid desiccant-based air handling unit because air cooling is also controlled with an electric chiller (CH). It has three air channels: one for process air, that is the air supplied to the conditioned space after dehumidification (1–2) and cooling (2–3–4); one for the cooling air, that is outdoor air cooled down by humidification (1–7); the last one for regeneration air, that is outdoor air heated up (1–5) by solar thermal energy or if necessary by the heat supplied by a natural gas fired boiler (B) to regenerate (5–6) the hygroscopic material of the desiccant wheel. The adsorption process realized on the surface of the lower part of the desiccant wheel allows a nearly isenthalpic dehumidification process, (see Figure 1a).

In heating mode the innovative AHU is arranged as a standard air handling unit that provides pre-heating (1–2–3), humidification (3–4) and post-heating (4–5), but thermal energy is supplied by the solar subsystem and if it is not sufficient by the boiler (Figure 1b).

The main components of the air-conditioning system are modeled by the types listed in Table 2. In this table are also reported the most important parameters used in the simulation model and the reference in which these mathematical models are described, validated and calibrated. Detailed information about the DWs considered in the analyses has been omitted as previously reported in Section 2. The TESS library type 1716 represents the DW, it has been characterized by the parameters η_{F1} and η_{F2} described previously.

To complete the description of the system operation logic implemented in the software, it is necessary to say that the circulation pump in the solar circuit starts working when the fluid temperature in the solar collectors is higher than that in the TS; thermal energy is taken from the tank to feed the heating coils (HC, HC2) when the AHU is on; the boiler operates as a back-up system if the temperature of the hot water is not high enough; the cooling coil is fed by the electric chiller during the cooling operation if the process air has not been sufficiently cooled after the cross flow heat exchanger (CF).

Table 2. Main TRNSYS submodels considered for the simulation and their main parameters.

Component (Reference)	Type	Library	Main Parameters	Value	Units
Cross flow heat exchanger [24]	91	Standard	Effectiveness	0.446	-
Humidifier [24]	506 c	TESS	Saturation efficiency	0.551	-
Natural gas boiler [24]	6	Standard	Nominal thermal power Efficiency	24.1 0.902	kW -
Air-cooled chiller [24]	655	TESS	Rated capacity Rated COP	8.50 2.98	kW -
Heating coil [24]	670	TESS	Liquid specific heat Effectiveness	4.190 0.864	kJ/(kg·K) -

Table 2. *Cont.*

Component (Reference)	Type	Library	Main Parameters	Value	Units
Cooling coil [24]	508	TESS	Liquid specific heat	4.190	kJ/(kg·K)
			Bypass fraction	0.177	-
Storage tank [29]	60 f	Standard	Volume	971	L
			Height	2.04	m
			Tank loss coefficient	1.37	W/(m²·K)
			Liquid specific heat	4.190	kJ/(kg·K)
Evacuated solar collectors	71	Standard	Tested flow rate	8.43×10^{-3}	kg/(s·m²)
			Intercept efficiency	0.676	-
			Efficiency slope	1.15	W/(m²·K)
			Efficiency curvature	0.004	W/(m²·K²)
			Fluid specific heat	3.85	kJ/(kg·K)

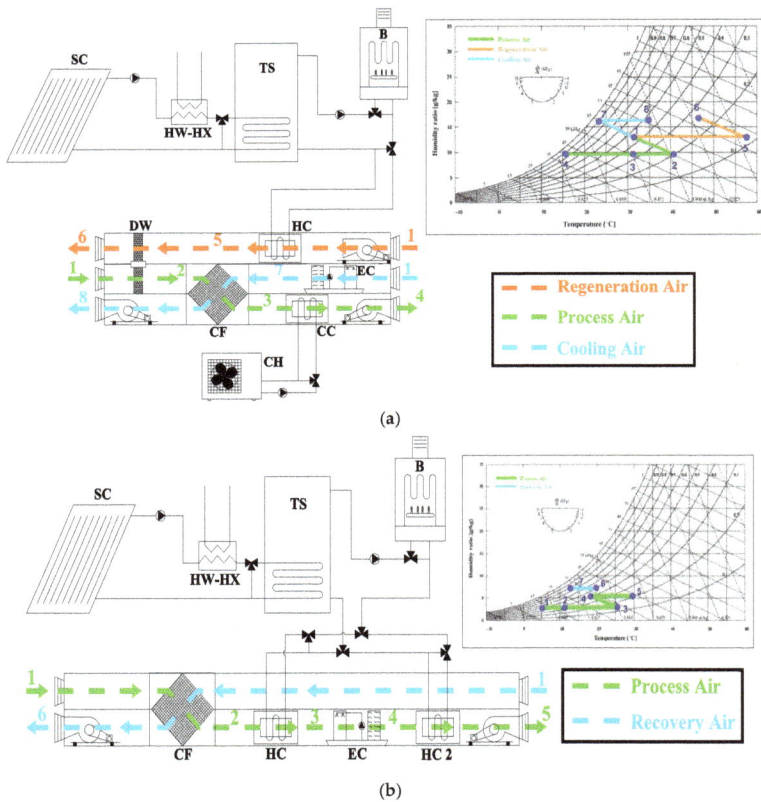

Figure 1. Alternative system layout in cooling mode (**a**) and heating mode (**b**).

In the CS AHU the outdoor air is dehumidified by its cooling below the dew point (cooling energy is removed by a 16 kW electric chiller), then it is heated up till the temperature is high enough to supply it to the conditioned space. In heating operation the AHU has a configuration similar to the innovative one but the boiler is the only heat source.

3.2. Energy and Environmental Indexes

The first two energy indexes, evaluated on the basis of the simulation results, are:

- the solar fraction (*SF*), that is the share of thermal enegy from the solar subsystem on the total thermal energy required by the AHU in the IS;
- the solar energy factor (*SEF*), that represents the ratio between the solar energy used in the AHU and that totally available.

The values of *SF* and *SEF* have been calculated on an annual and seasonal base. Concerning the solar fraction and the solar energy factor in cooling mode, *SF* and *SEF* are:

- the ratio between the solar energy used to regenerate the desiccnt wheel ($E_{th}^{TS-DWreg}$) and the total regeneration energy (E_{th}^{DWreg}):

$$SF_{Cooling} = \frac{E_{th}^{TS-DWreg}}{E_{th}^{DWreg}} \tag{11}$$

- the ratio between the solar energy used to regenerate the desiccant rotor and the total termal energy available from SC in summer ($E_{th,\,Cooling}^{SC}$),

$$SEF_{Cooling} = \frac{E_{th}^{TS-DWreg}}{E_{th,\,Cooling}^{SC}} \tag{12}$$

In heating mode *SF* is the ratio between the solar thermal energy supplied for pre-heating ($E_{th}^{TS-preheat}$) and post-heating ($E_{th}^{TS-postheat}$) and the total thermal energy for pre and post-heating ($E_{th}^{preheat} + E_{th}^{postheat}$), whereas the solar energy factor in the denominator has the total thermal energy from SC in the winter period ($E_{th,Heating}^{SC}$):

$$SF_{Heating} = \frac{E_{th}^{TS-preheat} + E_{th}^{TS-postheat}}{E_{th}^{preheat} + E_{th}^{postheat}} \tag{13}$$

$$SEF_{Heating} = \frac{E_{th}^{TS-preheat} + E_{th}^{TS-postheat}}{E_{th,Heating}^{SC}} \tag{14}$$

The total solar fraction and energy factor are, instead, evaluated as:

$$SF_{Total} = \frac{E_{th}^{TS-DWreg} + E_{th}^{TS-preheat} + E_{th}^{TS-postheat}}{E_{th}^{DWreg} + E_{th}^{preheat} + E_{th}^{postheat}} \tag{15}$$

$$SEF_{Total} = \frac{E_{th}^{TS-DWreg} + E_{th}^{TS-preheat} + E_{th}^{TS-postheat}}{E_{th,Total}^{SC}} \tag{16}$$

The following further two indexes provide a comparison between the IS and CS (Figure 2). They evalute respectively:

- the primary energy saving (*PES*) achieved by IS with respect to CS;
- the equivalent CO_2 emissions avoided by IS with respect to CS.

Concerning energy analysis, the comparison of IS and CS is performed considering only the primary energy demands related to fossil fuels, in fact the *PES* of non-renewable energy sources is:

$$PES = \left(1 - E_p^{IS} / E_p^{CS}\right) \times 100 \tag{17}$$

where:

$$E_p^{AS/CS} = \left(E_{el,CH}^{IS/CS} + E_{el,aux}^{IS/CS} + E_{el,non-HVAC}^{US}\right)/\eta_{EG} + E_{th,B}^{IS/CS}/\eta_B \qquad (18)$$

and:

$$E_{el,aux}^{IS} = E_{el}^{DC} + E_{el,pumps}^{IS} + E_{el,AHU}^{IS}, \qquad (19)$$

$$E_{el,aux}^{CS} = E_{el,pumps}^{CS} + E_{el,AHU}^{CS} \qquad (20)$$

The primary energy of the innovative and conventional system ($E_p^{IS/CS}$) is assessed taking into account that the Italian national electric system energy efficiency (η_{EG}), including transmission and distribution losses, is 42% [30], and using the boiler efficiency reported before (namely 90.2%). Furthermore, it is considered that solar energy does not determine a primary energy as it is a renewable energy source.

To assess the positive effects on the environment of the IS installation, equivalent CO_2 emissions of the two systems have been calculated and the equivalent avoided CO_2 emissions have been derived:

$$\Delta CO_2 = \left(1 - CO_2^{IS}/CO_2^{CS}\right) \times 100 \qquad (21)$$

where:

$$CO_2^{IS/CS} = \left(E_{el,CH}^{IS/CS} + E_{el,aux}^{IS/CS} + E_{el,non-HVAC}^{US}\right)\cdot\alpha + E_{th,B}^{IS/CS}\cdot\beta/\eta_B \qquad (22)$$

The specific emission factor of primary energy related to natural gas combustion, β, is equal to 0.207 kg CO_2/kW h$_p$, [30] while α, the specific emission factor of electricity drawn from the Italian grid, is equal to 0.573 kg CO_2/kW h$_{el}$ [30].

In addition to the results, already reported in [30] for the silica-gel DW, in this work the energy and environmental indexes are extended to the innovative plants with MILGO and zeolite-rich tuff DWs.

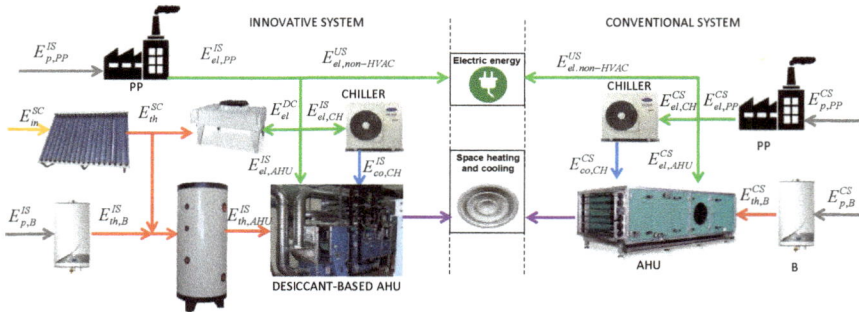

Figure 2. Scheme of the innovative and conventional systems.

4. Results

In the following subsections the results of the energy and environmental analyses will be shown grouping figures related to the same index, one subfigure for each hygroscopic material considered. The label "a" will refer to silica-gel, "b" to MILGO and "c" to zeolite-rich tuff. Each subfigure will contain data for different tilt angles (the analyses have been carried out considering tilt angles ranging from 20° to 55°), solar field collecting surfaces (three areas have been evaluated for the solar collectors field, namely 20, 27 and 34 m^2) or seasonal information (operation in cooling and heating mode).

4.1. Energy Analysis

On an annual basis the comparison of the proposed innovative system and the conventional one determines the results, in terms of *PES*, reported in Figure 3. The best performance was reached by the MILGO DW, for which *PES* was close to 29%, while in the case of silica-gel the maximum *PES* was slightly over 20%, and with the zeolite tuff DW it was about 14% in the best layout. For all three hygroscopic materials the widest collecting surface (about 34 m^2) determines the best energy results, but while with silica-gel and zeolite-rich tuff DW *PES* decreases when collector surface becomes smaller (Figure 3a,c), this does not happen with MILGO. The anomalous trends of Figure 3b are explained by the reduced amount of thermal energy needed to regenerate MILGO; the benefits in regeneration obtained with 27 m^2 of solar collectors, instead of 20 m^2, do not compensate the energy costs for dissipation; the situation changes at 34 m^2. The low regeneration energy of MILGO is further proven by the optimum tilt angle, which is about 50°, because the plant energy behavior is mainly influenced by the winter energy demand, in this situation solar energy is better exploited with a high tilt angle.

In order to assess the share of thermal energy supplied to the innovative AHU by the solar subsystem with respect to the total thermal energy demand, the *SF* has been evaluated in all the plant configurations and considering the operation in cooling mode (dotted line), heating mode (dashed line) and total (continuous line). Figure 4 shows the results for 20 m^2 of evacuated tube solar collectors and Figures 5 and 6 show the results for 27 and 34 m^2, respectively.

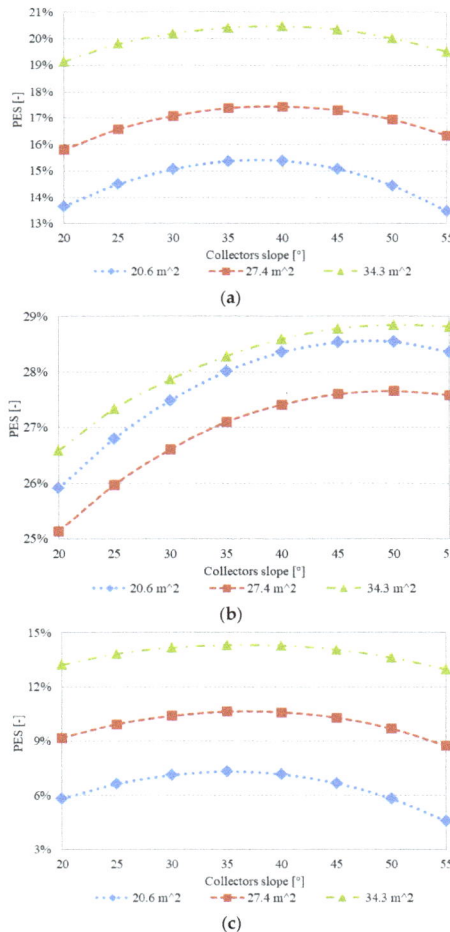

Figure 3. Primary energy savings as a function of the tilt angle and the solar field aperture area for: (**a**) silica-gel DW, (**b**) MILGO DW and (**c**) zeolite-rich tuff DW.

The *SF* during the cooling period shows a decreasing trend with the tilt angle in all the simulated cases and it is often higher than $SF_{Heating}$, that, instead has an opposite behavior. The total *SF* that takes into account both trends highlighted a maximum that is shifted more to the left or more to the right depending on the desiccant material considered. MILGO DW regeneration required low thermal energy, so the $SF_{Cooling}$ is higher than in the plants with silica-gel and zeolite-rich tuff DWs; it is equal to 100% with 34 m^2 of solar collectors. In this case, the corresponding SF_{Total} shows a maximum for high value of the tilt angle because it is affected mainly by winter operation. On the contrary, the zeolite-rich tuff DW-based plants needed more regeneration energy, therefore, $SF_{Cooling}$ is lower than for the AHU with silica-gel and MILGO DWs while the SF_{Total} shows a maximum at about 35°. The winter *SF* did not depend on the hygroscopic material considered and changed only with tilt angle and collecting surface.

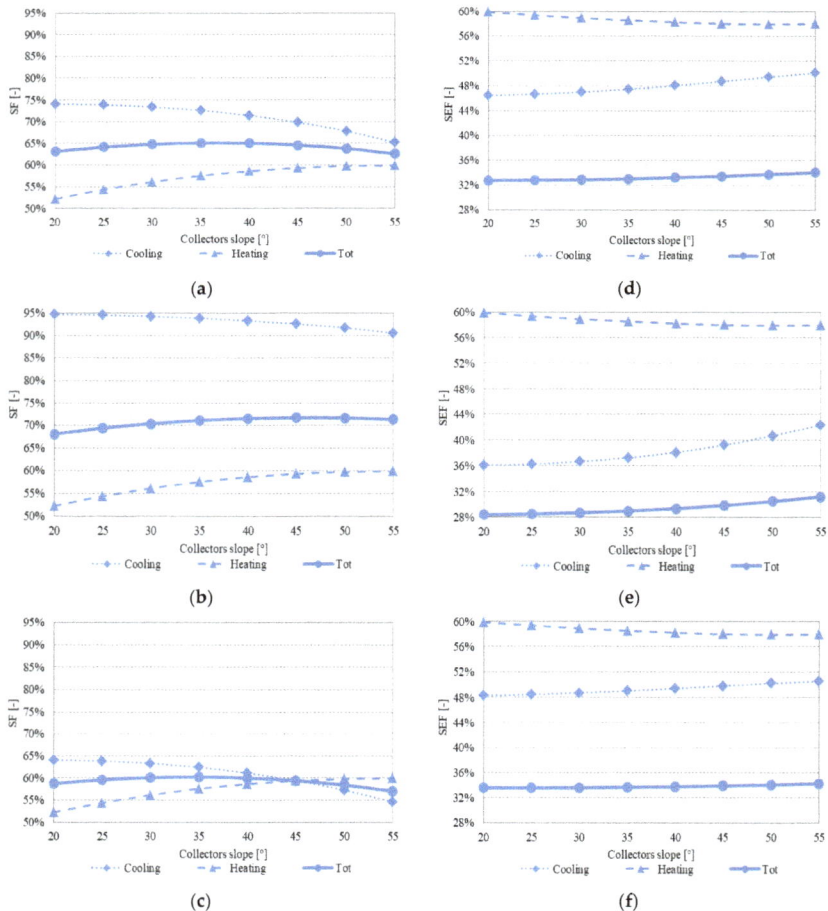

Figure 4. Solar fraction (left side) and solar energy factor (right side) as a function of the tilt angle in cooling operation, heating operation and total, for the plant with 20 m^2 of solar collectors and the innovative AHU equipped with: (**a**,**d**) silica-gel DW, (**b**,**e**) MILGO DW and (**c**,**f**) zeolite-rich tuff DW.

In order to have an idea of the amount of solar thermal energy exploited and of that dissipated, *SEF* is introduced. As for *SF* the $SEF_{Heating}$ index is independent from the DW material, it decreases with tilt angle and is lower when the collecting surface increases, because the dissipated solar thermal energy decreases. The solar thermal energy is used for a percentage in the range of 57–60% when 20 m^2 of evacuated tube solar collectors are considered, these percentages decrease to 52–56% and to 46–52% with the wider solar fields.

The solar energy factor for the summer period assumes lower values with respect to $SEF_{Heating}$ especially with MILGO DWs; in the worst case (34 m^2 solar collector and 20° tilt angle) it is equal to just over 23%, demonstrating a large amount of solar energy dissipated. This consideration is, in general, always true, $SEF_{Cooling}$ is lower than $SEF_{Heating}$, by about 50% in the best case. Unfortunately, the low seasonal SEF derived from the weekend days when the air-conditioning system is not working. All year round the solar energy factor is further reduced due to the long period in the intermediate season in which the HVAC system is switched off. In addition it can be noted that SF_{Total} trend is quite flat and it is not affected by the tilt angle.

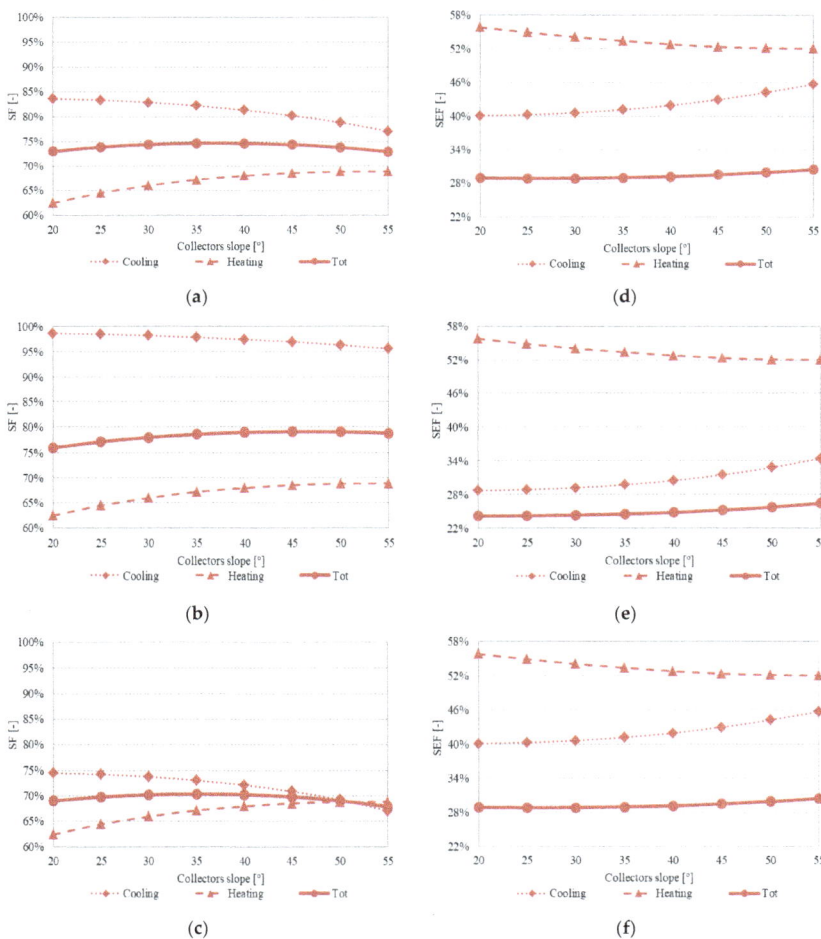

Figure 5. Solar fraction (left side) and solar energy factor (right side) as a function of the tilt angle in cooling operation, heating operation and total, for the plant with 27 m^2 of solar collectors and the innovative AHU equipped with: (**a,d**) silica-gel DW, (**b,e**) MILGO DW and (**c,f**) zeolite-rich tuff DW.

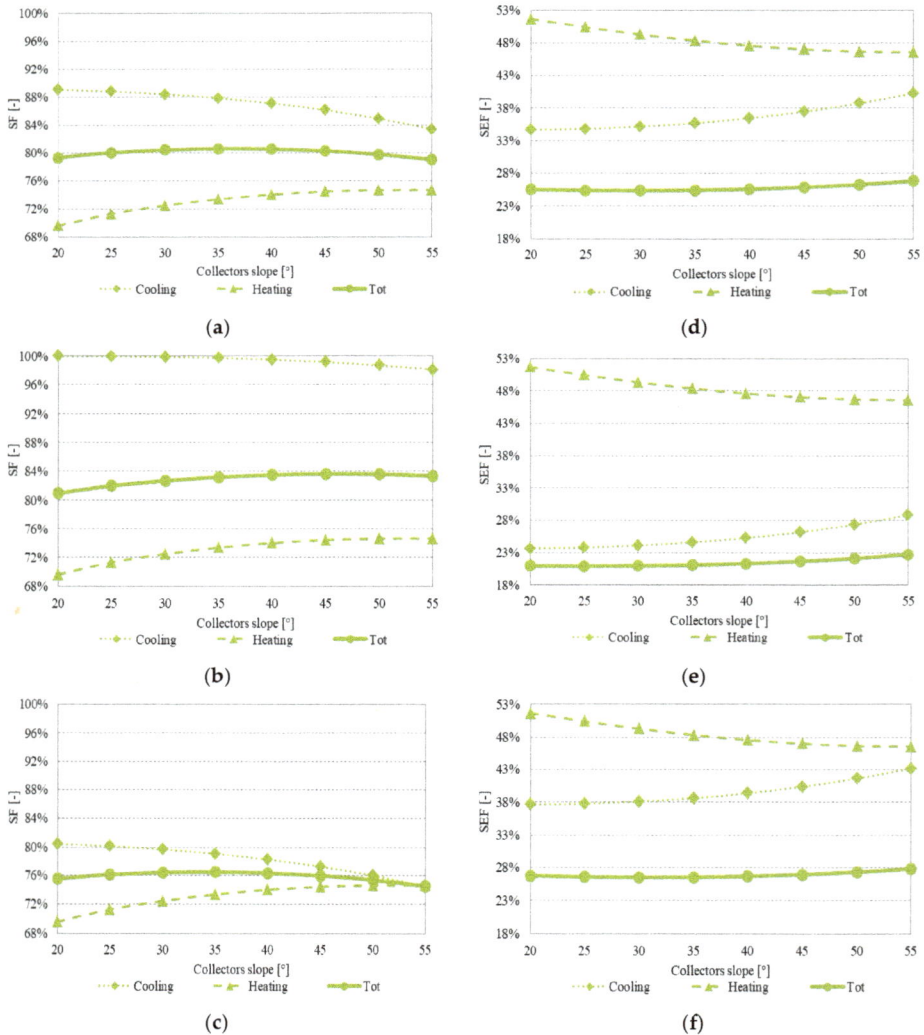

Figure 6. Solar fraction (left side) and solar energy factor (right side) as a function of the tilt angle in cooling operation, heating operation and total, for the plant with 34 m² of solar collectors and the innovative AHU equipped with: (**a,d**) silica-gel DW, (**b,e**) MILGO DW and (**c,f**) zeolite-rich tuff DW.

4.2. Environmental Analysis

In terms of avoided equivalent CO_2 emissions, the air-conditioning plants with silica-gel DW and those with zeolite-rich tuff DW have similar behaviors but the environmental performances with the conventional material are better than with tuff (Figure 7a,c). The parameter ΔCO_2 increases with the solar field aperture area and in the best case is about 17% and 11% with silica-gel and zeolite-rich tuff, respectively. The optimal tilt angle is close to 40° with the standard material while it moves towards 35° with tuff. MILGO DW-based plants use large amount of solar thermal energy and are responsible for low emissions, in fact, also in worst case higher ΔCO_2 are reached. As for *PES*, the largest solar collecting surface does not correspond with the best ΔCO_2 (Figure 7b). On the contrary, the lowest emissions take place with the 20 m^2 solar field because there is the minimum demand of electricity for dissipation (E_{el}^{DC}).

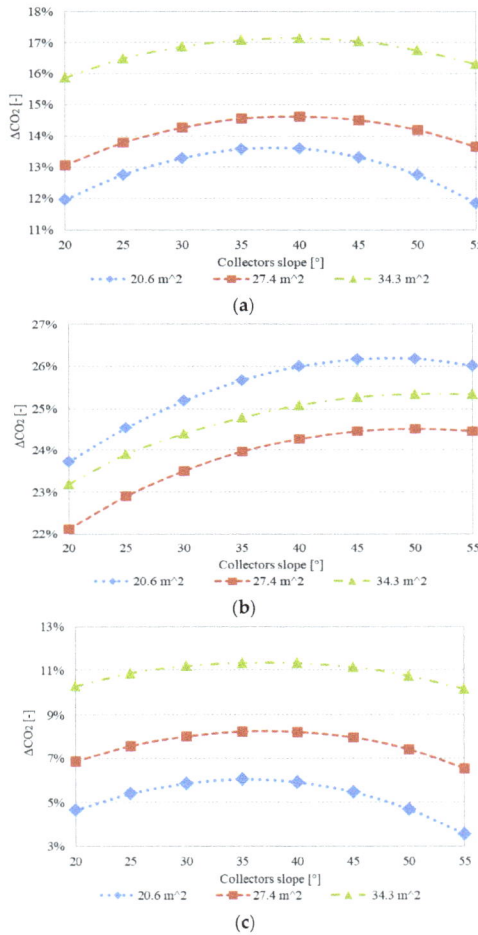

Figure 7. Equivalent CO_2 emissions avoided as a function of the tilt angle and the solar field aperture area for: (**a**) silica-gel DW, (**b**) MILGO DW and (**c**) zeolite-rich tuff DW.

5. Discussion

The indexes *SFs* and *SEFs* showed that the solar heating and cooling proposed plant is not well coupled with the user considered. A large share of solar thermal energy remains unused and needs to be dissipated, increasing the electricity demand. Consequently, if half this solar thermal energy in excess is exploited for other low temperature energy use, *PES* and ΔCO_2 increase significantly, approaching 63% and 60%, respectively.

The best results pertain to the configuration with MILGO DW, 34 m^2 of solar collectors, and a tilt angle of 35° (see Figures 8b and 9b). If the thermal energy surplus is completely considered for other applications in nearby users, the primary energy saving and the equivalent CO_2 emissions reach values over 65% under all the considered cases with a solar field of 34 m^2. Even when the smallest collecting surface (20 m²) is considered, the energy and environmental indexes reach and exceed 50% (see Figure 8 (right side), and Figure 9 (right side)).

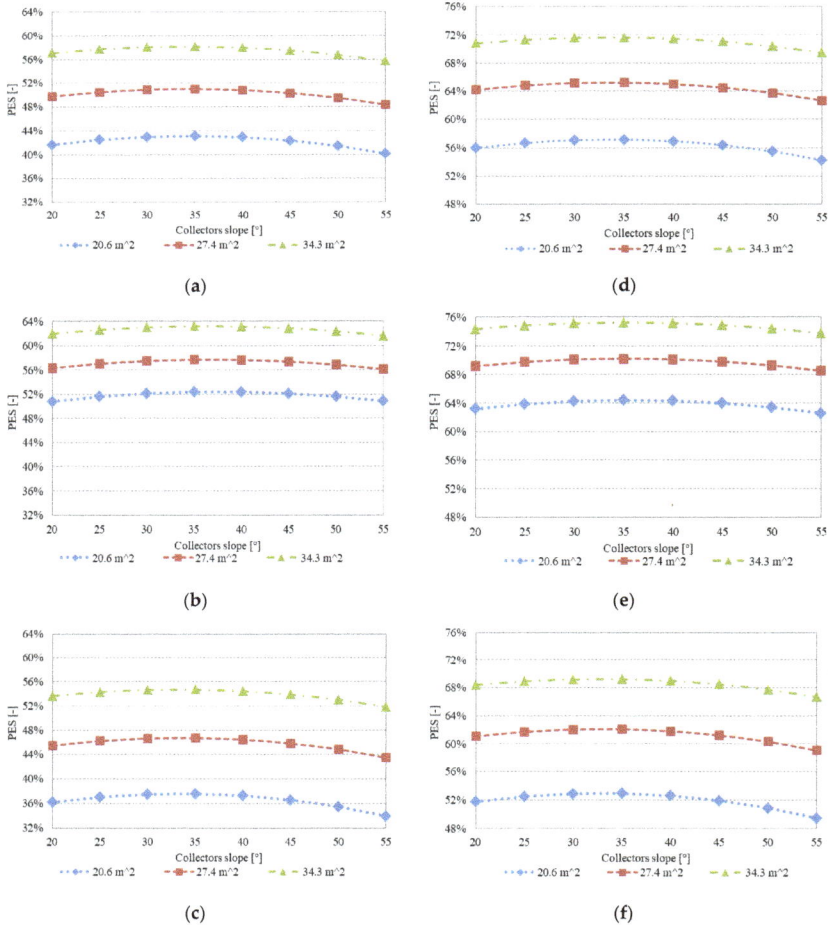

Figure 8. Primary energy savings as a function of the tilt angle and the solar field aperture area when half of the solar thermal energy surplus is used (left side) and when it is totally used (right side) for: (**a,d**) silica-gel DW, (**b,e**) MILGO DW and (**c,f**) zeolite-rich tuff DW.

The worst performances are observed using the tuff-based DW configuration, as a consequence of the climatic conditions chosen for carrying out this analysis. Campanian Ignimbrite-based DW, despite using a very cheap material largely available in the surroundings of Benevento, has interesting performances when compared to other hygroscopic materials only when relative humidity is low and the regeneration air temperature is very high, as reported by [19].

A further analysis based on economic considerations may be carried out in the future when there will be a clearer idea of the production costs of a DW based on tuff. It is considered likely that the initial cost of an AHU with a tuff desiccant wheel will be significantly lower with respect to the other two materials, therefore it will be necessary to see if the higher operating costs due to poorer performance will justify the adoption of this natural material.

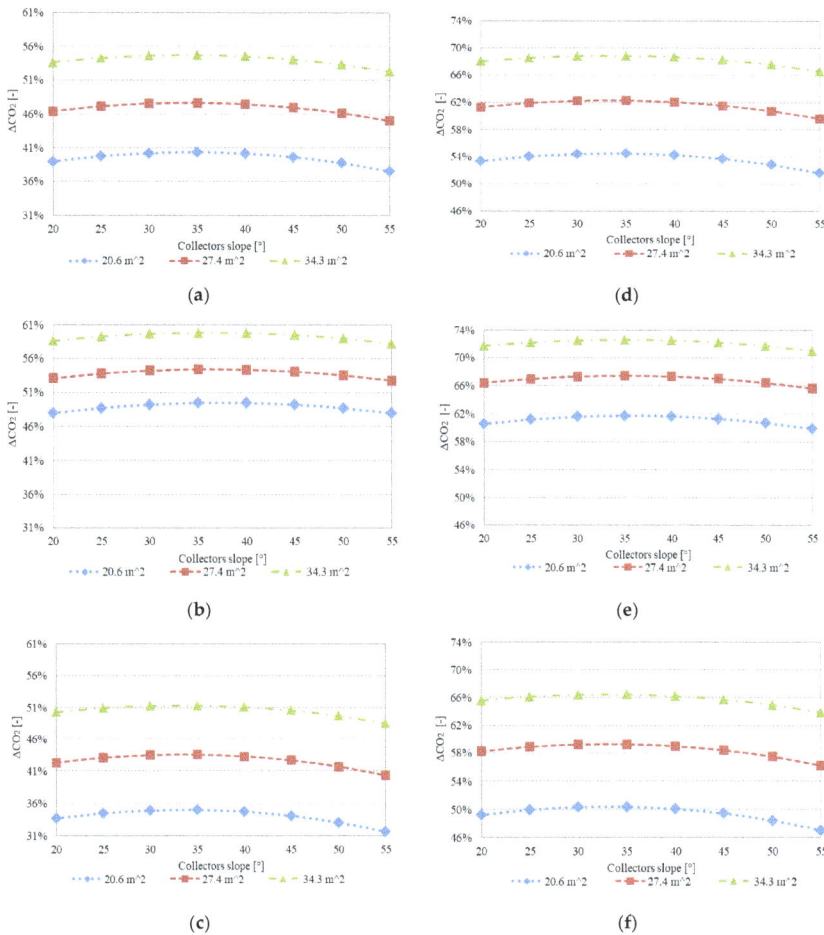

Figure 9. Equivalent avoided CO_2 emissions as a function of the tilt angle and the solar field aperture area when half of the solar thermal energy surplus is used (left side) and when it is totally used (right side) for: (**a,d**) silica-gel DW, (**b,e**) MILGO DW and (**c,f**) zeolite-rich tuff DW.

6. Conclusions

In this paper three DWs made of silica-gel, MILGO and zeolite-rich tuff have been considered in a desiccant based solar-driven air handling unit that operates for the air-conditioning of a university classroom located in Benevento. First of all, the performance of these innovative plants are compared with that of a conventional system; positive results are achieved under all the configurations considered, as demonstrated by the parametric study performed considering different evacuated tube solar collectors surfaces (about 20, 27 and 34 m^2) and tilt angle (20–55°). A comparison of the three innovative AHUs highlights that the best performances are demonstrated by the MILGO wheel system, while the poorest performances are those obtained by the tuff-based system. In the best case (MILGO DW, with a solar field of 34 m^2 and a tilt angle of about 50°) the primary energy saving approaches 29%. With respect to the equivalent CO_2 emissions, the optimal configuration does not overlap with the one characterized by the widest solar collector aperture area but, on the contrary, the best solution is the one with the smallest solar field. As demonstrated by the solar fraction and solar energy factor parameters, there is a large share of solar thermal energy surplus that needs to be dissipated, and this increases the electricity demand. Silica-gel DW-based plants show to be a more balanced solution for the solar thermal energy available in the studied location, while the tuff rotor operating far from its best operating conditions does not offer optimal performances.

When one takes in to account the possibility to further and totally use the solar energy in excess for other low temperature applications, for example in nearby users, the primary energy saving and the equivalent CO_2 emissions avoided boost up to 75% and 73%, respectively.

Author Contributions: The Authors have complementary expertise on the main topics of this paper and they jointly shared the structure and aims of the manuscript. More specifically, C.R. and F.T. dealt more with the air-conditioning plant modeling and simulation, P.B. and F.P. dealt more with the desiccant wheel materials, M.S. contributed more to the energy and environmental analysis. Finally, all the Authors equally contributed during the writing of the paper.

Funding: This research received no external funding.

Conflicts of Interest: The authors declare no conflict of interest.

Nomenclature

CO_2	Equivalent CO_2 emission (kg/year)
c_p	Specific heat (J/kgK)
D_s	Surface diffusion coefficient (m^2/s)
E	Energy (MWh/y)
F_1, F_2	Isopotential lines
K	Effective mass transfer coefficient (1/s)
k	Thermal conductivity (W/mK)
M	Moisture content of adsorbent material (kg$_{water}$/kg$_{adsorbent}$)
M_w	Molecular weight of water (kg/mol)
PES	Primary Energy Saving (%)
q_s	isosteric heat of adsorption (J/mol)
SEF	Solar Energy Factor (-)
SF	Solar Fraction (-)
T, t	Temperature (K), (°C)
V	air superficial velocity (m/s)
z	Axial coordinate (m)

Greek symbols

α	Specific emission factor of electricity drawn from the grid (kg CO_2/kW h$_{el}$)
β	Specific emission factor of primary energy related to natural gas combustion (kg CO_2/kW h$_{Ep}$)
ΔCO_2	Equivalent CO_2 avoided emission (%)
ε	Void fraction (-)
η	Efficiency (-)
θ	Time (s)
ρ	Density (kg/m^3)
ω	Air absolute humidity (kg$_{water}$/kg$_{dry air}$) or (g$_{water}$/kg$_{dry air}$)

Superscripts

CS	Conventional System
DC	Dry cooler
$DWreg$	Desiccant Wheel regeneration
IS	Innovative System
postheat	Post-heating phase
preheat	Pre-heating phase
TS	Thermal Storage
US	User

Subscripts

amb	Ambient
aux	Auxiliaries
B	Boiler
CH	Chiller
Co	Cooling
Cooling	Cooling mode
d	Adsorbent materia
e	At equilibrium condition
EG	Electric Grid
el	Electric
F_1, F_2	Isopotential lines
Heating	Heating mode
in	Initial
m	moist air
non-HVAC	not related to HVAC
p	Primary
PP	Power Plant
proc	Process
reg	Regeneration
SC	Solar thermal Collector
th	Thermal
tot, Total	Total

Acronyms

AHU	Air Handling Unit
B	Boiler
CC	Cooling Coil
CF	Cross-Flow heat exchanger
CH	Chiller
COP	Coefficient Of Performance
CS	Conventional System
DW	Desiccant Wheel
EC	Evaporative Cooler
HC, HC2	Heating Coils
HVAC	Heating, Ventilation and Air-Conditioning
HW-HX	Hot Water Heat exchanger
IS	Innovative System

MILGO Hygroscopic material, consisting graphite oxide dispersed in the MIL101 metal organic framework network structure

SC	Solar thermal Collector
SEF	Solar Energy Factor
SF	Solar Fraction
TS	Thermal Storage

References

1. Zheng, X.; Ge, T.S.; Wang, R.Z. Recent progress on desiccant materials for solid desiccant cooling systems. *Energy* **2014**, *74*, 280–294. [CrossRef]
2. Asim, N.; Emdadi, Z.; Mohammad, M.; Yarmo, M.A.; Sopian, K. Agricultural solid wastes for green desiccant applications: An overview of research achievements, opportunities and perspectives. *J. Clean. Prod.* **2015**, *91*, 26–35. [CrossRef]
3. Jani, D.B.; Mishra, M.; Sahoo, P.K. Performance analysis of a solid desiccant assisted hybrid space cooling system using TRNSYS. *J. Build. Eng.* **2018**, *19*, 26–35. [CrossRef]
4. Al-Alili, A.; Hwang, Y.; Radermacher, R. Performance of a desiccant wheel cycle utilizing new zeolite material: Experimental investigation. *Energy* **2015**, *81*, 137–145. [CrossRef]

5. Kanoğlu, M.; Çarpınlıoğlu, M.Ö.; Yıldırım, M. Energy and exergy analyses of an experimental open-cycle desiccant cooling system. *Appl. Therm. Eng.* **2004**, *24*, 919–932. [CrossRef]

6. Subramanyam, N.; Maiya, M.P.; Srinivasa Murthy, S. Application of desiccant wheel to control humidity in air-conditioning systems. *Appl. Therm. Eng.* **2004**, *24*, 2777–2788. [CrossRef]

7. Jia, C.X.; Dai, Y.J.; Wu, J.Y.; Wang, R.Z. Experimental comparison of two honeycombed desiccant wheels fabricated with silica gel and composite desiccant material. *Energy Convers. Manag.* **2006**, *47*, 2523–2534. [CrossRef]

8. White, S.D.; Goldsworthy, M.; Reece, R.; Spillmann, T.; Gorur, A.; Lee, D.Y. Characterization of desiccant wheels with alternative materials at low regeneration temperatures. *Int. J. Refrig.* **2011**, *34*, 1786–1791. [CrossRef]

9. Eicker, U.; Schürger, U.; Köhler, M.; Ge, T.; Dai, Y.; Li, H.; Wang, R. Experimental investigations on desiccant wheels. *Appl. Therm. Eng.* **2012**, *42*, 71–80. [CrossRef]

10. Jia, C.X.; Dai, Y.J.; Wu, J.Y.; Wang, R.Z. Analysis on a hybrid desiccant air-conditioning system. *Appl. Therm. Eng.* **2006**, *26*, 2393–2400. [CrossRef]

11. Fong, K.F.; Lee, C.K. Impact of adsorbent characteristics on performance of solid desiccant wheel. *Energy* **2018**, *144*, 1003–1012. [CrossRef]

12. Intini, M.; Goldsworthy, M.; White, S.; Joppolo, C.M. Experimental analysis and numerical modelling of an AQSOA zeolite desiccant wheel. *Appl. Therm. Eng.* **2015**, *80*, 20–30. [CrossRef]

13. Goldsworthy, M.J. Measurements of water vapour sorption isotherms for RD silica gel, AQSOA-Z01, AQSOA-Z02, AQSOA-Z05 and CECA zeolite 3A. *Microporous Mesoporous Mater.* **2014**, *196*, 59–67. [CrossRef]

14. Enteria, N.; Yoshino, H.; Mochida, A.; Satake, A.; Yoshie, R.; Takaki, R.; Yonekura, H.; Mitamura, T.; Tanaka, Y. Performance of solar-desiccant cooling system with silica–gel (SiO_2) and titanium dioxide (TiO_2) desiccant wheel applied in East Asian climates. *Sol. Energy* **2012**, *86*, 1261–1279. [CrossRef]

15. Rajamani, M.; Maliyekkal, S.M. Chitosan reinforced boehmite nanocomposite desiccant: A promising alternative to silica gel. *Carbohydr. Polym.* **2018**, *194*, 245–251. [CrossRef] [PubMed]

16. Yan, J.; Yu, Y.; Ma, C.; Xiao, J.; Xia, Q.; Li, Y. Adsorption isotherms and kinetics of water vapor on novel adsorbents MIL-101(Cr)@GO with super-high capacity. *Appl. Therm. Eng.* **2015**, *84*, 118–125. [CrossRef]

17. Caputo, D.; Iucolano, F.; Pepe, F.; Colella, C. Modeling of water and ethanol adsorption data on a commercial zeolite-rich tuff and prediction of the relevant binary isotherms. *Microporous Mesoporous Mater.* **2007**, *105*, 260–267. [CrossRef]

18. Bareschino, P.; Diglio, G.; Pepe, F.; Angrisani, G.; Roselli, C.; Sasso, M. Numerical study of a MIL101 metal organic framework based desiccant cooling system for air conditioning application. *Appl. Therm. Eng.* **2017**, *124*, 641–651. [CrossRef]

19. Bareschino, P.; Diglio, G.; Pepe, F. Modelling of a Zeolite-Rich Tuff Desiccant Wheel. *Adv. Sci. Lett.* **2017**, *23*, 6002–6006. [CrossRef]

20. Maclaine-Cross, I.L.; Banks, P.J. Coupled heat and mass transfer in regenerators– predictions using an analogy with heat transfer. *Int. J. Heat Mass Transf.* **1972**, *15*, 1225–1242. [CrossRef]

21. Howe, R.R. Model and Performance Characteristics of a Conditioning System Which Utilizes a Rotary Desiccant Dehumidifier. Master's Thesis, University of Wisconsin, Madison, WI, USA, 1983.

22. Jurinak, J.J. Open Cycle Solid Desiccant Cooling: Component Models and System Simulations. Ph.D. Thesis, University of Wisconsin, Madison, WI, USA, 1982.

23. Banks, P.J. Prediction of Heat and Mass Regenerator performance using nonlinear analogy method: Part 1—basis. *ASME J. Heat Transf.* **1985**, *107*, 222–229. [CrossRef]

24. Angrisani, G.; Roselli, C.; Sasso, M. Experimental validation of constant efficiency models for the subsystems of an unconventional desiccant-based Air Handling Unit and investigation of its performance. *Appl. Therm. Energy* **2012**, *33–34*, 100–108. [CrossRef]

25. Solar Energy Laboratory. *TRNSYS 17, a TRaNsient System Simulation Program*; University of Wisconsin: Madison, WI, USA, 2010.

26. *Thermal Energy System Specialists Components Library v. 17.01*; Thermal Energy System Specialists: Madison, WI, USA, 2004.

27. Angrisani, G.; Roselli, C.; Sasso, M.; Tariello, F.; Vanoli, G.P. Performance Assessment of a Solar-Assisted Desiccant-Based Air Handling Unit Considering Different Scenarios. *Energies* **2016**, *9*, 724. [CrossRef]

28. Angrisani, G.; Roselli, C.; Sasso, M.; Tariello, F. Dynamic performance assessment of a micro-trigeneration system with a desiccant-based air handling unit in Southern Italy climatic conditions. *Energy Convers. Manag.* **2014**, *80*, 188–201. [CrossRef]
29. Angrisani, G.; Canelli, M.; Roselli, C.; Sasso, M. Calibration and validation of a thermal energy storage model: Influence on simulation results. *Appl. Therm. Eng.* **2014**, *67*, 190–200. [CrossRef]
30. Angrisani, G.; Roselli, C.; Sasso, M.; Tariello, F. Dynamic performance assessment of a solar-assisted desiccant-based air handling unit in two Italian cities. *Energy Convers. Manag.* **2016**, *113*, 331–345. [CrossRef]

energies MDPI

Article

Sensitivity of Axial Velocity at the Air Gap Entrance to Flow Rate Distribution at Stator Radial Ventilation Ducts of Air-Cooled Turbo-Generator with Single-Channel Ventilation

Yong Li [1,2], Weili Li [1] and Ying Su [1,*]

[1]   School of Electrical Engineering, Beijing Jiaotong University, Beijing 100044, China
[2]   China North Vehicle Research Institute, Beijing 100072, China
*   Correspondence: 15117377@bjtu.edu.cn

Received: 17 July 2019; Accepted: 23 August 2019; Published: 6 September 2019

Abstract: In the design and calculation of a 330 MW water-water-air cooling turbo-generator, it was found that the flow direction of the fluid in the local stator radial ventilation duct is opposite to the design direction. In order to study what physical quantities are associated with the formation of this unusual fluid flow phenomenon, in this paper, a 100 MW air-cooled turbo-generator with the same ventilation structure as the abovementioned models is selected as the research object. The distribution law and pressure of the fluid in the stator radial ventilation duct and axial flow velocity at the air gap entrance are obtained by the test method. After the calculation method is proved correct by experimental results, this calculation method is used to calculate the flow velocity distribution of the outlets of multiple radial ventilation ducts at various flow velocities at air gap inlets. The relationship between the flow distribution law of the stator ventilation ducts and the inlet velocity of the air gap is studied. The phenomenon of backflow of fluid in the radial ventilation duct of the stator is found, and then the influence of backflow on the temperature distribution of stator core and winding is studied. It is found that the flow phenomenon can cause local overheating of the stator core.

Keywords: air-cooled steam turbine generator; single-channel ventilation; backflow; radial ventilation duct; fluid field

1. Introduction

Regarding the research on the temperature field of large turbo-generators, most scholars have mainly focused on the research on the fluid field and the temperature field [1–3]. Some research mainly focused on the heat transfer relationship between the temperature rise of structural parts such as the stator winding and fluid flow. There were also some studies on the internal fluid distribution law of generators, and the main research direction was the rational design of the ventilation system of the generator [4–6]. Unlike the above research contents, the main content of this paper is about which physical quantities of the fluid are related to the fluid flow distribution rules in the ventilation duct of the generator and which significant changes will occur with the changes of these physical quantities. The accurate measurement results of the fluid distribution in the ventilation duct of the stator core are the prerequisites for the accurate calculation of the above research contents.

In this paper, a 100 MW single-channel ventilation turbo-generator set is taken as the research object, and the experimental measurement method [7–9] of flow in stator radial duct distribution and air gap flow velocity of ventilation duct is proposed. The fluid pressure and flow velocity measuring sensors are buried at the generator stator ventilation duct and the air gap entrance. After measuring the pressure and flow speed at these positions, it is found that the fluid velocity at the local stator radial ventilation duct is significantly lower. Then the calculation results of fluid flow velocity in different ventilation ducts of the generator stator during the test conditions are obtained by the three-dimensional fluid field calculation method. The calculated results are in good agreement with the experimental results after comparison.

Though our analysis, it is found that the air gap inlet velocity is the main factor affecting the distribution of the fluid flow rate in the stator ventilation duct. Based on this calculation method, the results of velocity distribution in different radial ventilation ducts of generator stator are calculated at three different flow velocities at the air gap entrance. It is also found that with the increase of the velocity at the air gap entrance, the phenomenon of backflow in the stator iron core ventilation duct will occur locally.

2. Experimental Measurement and Result Analysis

2.1. Test and Measurement of the Velocities at Air Gap Inlet and Radial Ventilation Duct Outlets in the Generator

The schematic diagram of single ventilation is shown in Figure 1. There are 76 radial ventilation ducts in the stator of the test generator. Due to the large number of them, the test basically follows the principle of embedding a wind speed measuring component in every other wind duct. Due to the existence of the step structure on both sides of the stator core end, the fluid is blocked by the core step in this area, resulting in a large change in the fluid flow velocity in this area. Therefore, the measuring elements are embedded densely at the air gap entrance of the generator. The measuring component adopts the pitot tube, and the pitot tube is connected with two pressure measuring tubes at the tail. The measured full pressure and static pressure are led to the outside of the generator through the pressure tube, which is convenient for connecting the pressure collector and obtaining the test result. The pitot tubes are buried in the ventilation duct yokes of the stator core and care to ensure the same embedment depth and axial position. In the axial direction, the buried pitot tube is located in the axial center of the ventilation duct, and it is not affected by the boundary layer with low flow velocity at the wall surface. Each pitot tube in the radial direction is oriented toward the center of the circle to ensure the consistency between the measurement sensor and the direction angle of the flow velocity. In this way, the measuring point placed in the ventilation duct can clearly and effectively measure the wind speed and pressure in the entire wind channel, ensuring the consistency of data. The specific installation positions are shown in Figures 2 and 3.

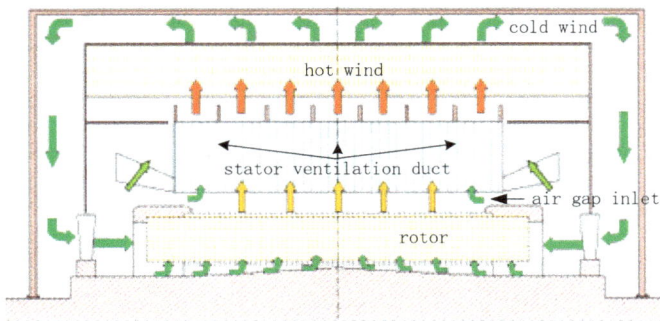

Figure 1. Single-channel ventilation air-cooled turbo-generator ventilation system.

Figure 2. Placement and measuring position of pitot tube in the stator ventilation ducts. (**a**) Placement position of pitot tube; (**b**) measuring position of ventilation duct.

Figure 3. The sensor placement at the air gap entrance.

Figure 3 shows the embedding position of the wind speed sensor at the air gap entrance. The sensor is located at the entrance of the generator air gap. Considering the influence of rotating air flow at the air gap entrance, the installation angle between the wind sensor direction and the axial direction is 40 degrees. Since the air volume passing through this position accounts for about 70% of the total air volume of the generator, it is very important to accurately measure the wind speed at this inlet position. At the same time, in single-channel ventilation of turbo-generator, the flow resistance at the air gap entrance is the biggest in the whole ventilation system. It is necessary to bury the pressure measuring point here, which is also of great guiding significance to verify the calculated value of the whole ventilation system design. This measurement will provide important inlet boundary conditions for the following finite element calculations.

2.2. Stator Radial Ventilation Duct Outlet Flow Rate Measurement Results

Table 1 shows that the static pressure at the measuring position of the stator core yoke ranges from 1000 to 1500 Pa. From the distribution of pressure difference, when the fluid is located at the end of both sides of the stator core, the pressure difference is small, and it is bigger close to the stator core center. According to the Bernoulli equation of fluid flow, it can be known that: This pressure difference is actually the dynamic pressure of the fluid, and this value can reflect the fluid velocity at the measurement position.

$$p_{total} - p_{static} = p_{Dynamic} = \frac{1}{2}\rho v^2 \tag{1}$$

Table 1. Pressure and flow velocity measurement results in stator radial ventilation ducts.

No. of Ventilation Ducts	Total Pressure Pa	Static Pressure Pa	Pressure Difference Pa	Calculated Wind Speed m/s
2	1514	1464	50	8.8
3	1324	1256	68	10.3
4	1188	1178	10	3.9
5	1122	1117	5	2.8
6	1153	1143	10	3.9
8	1205	1145	60	9.6
9	1178	1125	53	9.1
11	1173	1136	37	7.6
13	1193	1144	49	8.7
15	1238	1153	85	11.5
18	1231	1118	113	13.2
20	1191	1088	103	12.6
22	1201	1085	116	13.4
24	1231	1092	139	14.7
26	1211	1097	114	13.3
28	1270	1103	167	16.1
30	1257	1089	168	16.1
33	1317	1117	200	17.6
34	1367	1097	270	20.5
36	1286	1094	192	17.3
38	1259	1112	147	15.1
40	1359	1103	256	19.9
42	1304	1093	211	18.1
44	1263	1094	169	16.2
46	1353	1108	245	19.5
49	1142	1281	−139	14.7
51	1359	1152	207	17.9
53	1289	1144	145	15.0
55	1281	1134	147	15.1
57	1259	1140	119	13.6
59	1260	1124	136	14.5
61	1311	1162	149	15.2
64	1298	1206	92	11.9
66	1284	1209	75	10.8
68	1280	1216	64	10.0
70	1256	1203	53	9.1
71	1256	1212	44	8.3
73	1205	1191	14	4.7
74	1019	1172	−153	15.4
75	1259	1220	39	7.8
76	1576	1515	61	9.7

The calculated static pressure at the generator fan inlet is about 4500 Pa. It can be inferred from the static pressure measured by the sensor in Table 1 that the sum of fluid flow resistance at the generator end, air gap, core tooth, and other positions is 3000 Pa. The static pressure in the ventilation duct on both sides of the generator end is large, but the difference between the total pressure and the static pressure is small. The static pressure at the ventilation duct of the generator center is small, but the difference between the total pressure and the static pressure is large, indicating that the flow velocity in most ventilation ducts in the central area of the generator stator is larger than that at the end. Detailed analysis of the data in the table: It was found that the pressure difference in radial ventilation duct no. 49 and 74 was negative. After rechecking the measuring element, it was found that an installation error caused the negative measuring result.

In order to facilitate intuitive analysis, the measurement results are shown in Figure 4. The wind speed measurement results in the stator core ventilation duct have a total of 76 ventilation ducts throughout the core. The wind speed measurement components are densely packed at the ends and relatively less in the middle. It can be seen that the fluid radial velocities in the entire core ventilation ducts show a higher speed in the middle and a lower speed at both sides of the end. This indicates that the flow velocity from the two ends to the center gradually increases, and the maximum flow velocity is about 20 m/s.

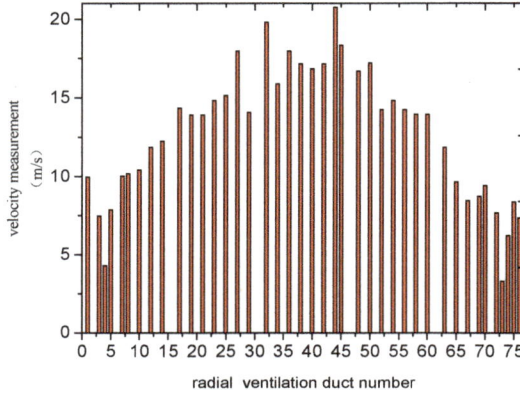

Figure 4. The distribution of wind speed values in the stator radial ventilation duct at 3000 r/min.

It can also be found that the flow velocity in the first three ventilation ducts is high on both sides of the stator, while the flow velocity in the fourth, fifth, and sixth ventilation ducts decreased rapidly, down to less than 4 m/s. It is not difficult to infer that the stator cores and winding near the fourth to the seventh ventilation ducts will have a higher temperature rise due to the lower wind speed.

The reason for the decrease of the fourth, fifth, and sixth ventilation duct wind speeds on both sides of the stator is that this position is the beginning of the normal stator ventilation duct. These ventilation duct structures are different from the first, second, and third which have the characteristics of the step structure, shown in Figure 5. The wind in the first three ventilation ducts, due to the iron core step structure block, has a higher wind speed in radial ventilation ducts. After passing through the first three ventilation ducts, the flow duct structure area is smooth in the axial direction and the flow becomes unobstructed. The wind has a high axial velocity when it flows in the air gap. This will cause the static pressure around the gap to drop. Thus, it results in a very low wind speed in radial ventilation ducts at this location. As the capacity of the generator increases, the fluid velocity at the entrance of the air gap increases, and this problem becomes more serious.

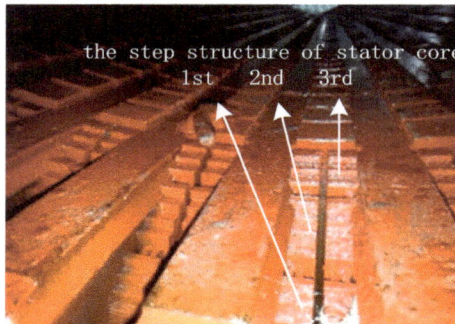

Figure 5. The step structure of stator end.

In order to study the influence of different axial velocities at the air gap entrance on the radial velocity of fluid in the adjacent ventilation duct, the finite element method is used to calculate the test condition.

3. Study on Sensitivity of Air Gap Inlet Flow Velocity to Stator Radial Ventilation Duct Flow Rate Distribution

3.1. Mathematical Model and Physical Model Description

3.1.1. Mathematical Model Description

Mass conservation equation [10–12]:

$$\mathrm{div}(\rho V) = 0 \tag{2}$$

Momentum conservation equation [10–12]:

$$\begin{cases} \frac{\partial(\rho u)}{\partial t} + \mathrm{div}(\rho V u) = \mathrm{div}(\mu \cdot \mathrm{grad} u) - \frac{\partial p}{\partial x} + S_u \\ \frac{\partial(\rho v)}{\partial t} + \mathrm{div}(\rho V v) = \mathrm{div}(\mu \cdot \mathrm{grad} v) - \frac{\partial p}{\partial x} + S_v \\ \frac{\partial(\rho w)}{\partial t} + \mathrm{div}(\rho V w) = \mathrm{div}(\mu \cdot \mathrm{grad} w) - \frac{\partial p}{\partial x} + S_w \end{cases} \tag{3}$$

Energy conservation equation [10–12]:

$$\frac{\partial(\rho T)}{\partial t} + \mathrm{div}(\rho V T) = \mathrm{div}\left(\frac{\lambda}{c} \mathrm{grad} T\right) + S_T \tag{4}$$

where ρ is fluid density; t is the time; V is the relative fluid velocity vector; u, v, and w are the components of V in x, y, and z axes; μ is viscosity coefficient; p is static pressure acting on a micro cell in air; S_u, S_v, and S_w are the source items of the momentum equation; λ is thermal conductivity; c is specific heat in constant pressure; S_T is the ratio of the heat generated in the unit volume and specific heat c.

Since the fluid in the stator calculation domain has a turbulent flow and the air in the radial ventilating duct of the generator can be regarded as incompressible, a standard k-ε model in the commercial solver "FLUENT" is used to solve the turbulence movement [10–12].

$$\begin{cases} \frac{\partial(\rho k)}{\partial t} + div(\rho k V) = div\left[\left(\mu + \frac{\mu_t}{\sigma_k}\right) grad k\right] + G_k - \rho\varepsilon \\ \frac{\partial(\rho\varepsilon)}{\partial t} + div(\rho V \varepsilon) = div\left[\left(\mu + \frac{\mu_t}{\sigma_\varepsilon}\right) grad\varepsilon\right] + G_{1\varepsilon}\frac{\varepsilon}{k}G_k - G_{2\varepsilon}\rho\frac{\varepsilon^2}{k} \end{cases} \tag{5}$$

where k is the turbulent kinetic energy, ε is the diffusion rate, G_k is turbulent generation rate, μ_t is turbulent viscosity coefficient, $G_{1\varepsilon}$ and $G_{2\varepsilon}$ are constant; σ_k and σ_ε are, respectively, the equation k and equation ε of Planck's constant turbulence.

3.1.2. Physical Model Description, Mesh Analysis, and the Boundary Conditions

a. Physical Model Description

The model of Figure 6a includes 76 air gap inlets and radial ventilation duct outlets, the stator core is near the ventilation ducts, and the fluid flow in the ventilation ducts will take away the loss in the core. The model of Figure 6b includes: The upper- and lower-layer bar of the generator stator slots and their main insulation, wedges, and ventilation ducts fluid area. The first three ventilation ducts of the iron core have the step structure shown in Figure 5, and their lengths in the radial direction are sequentially increased, and the corresponding fluid area here is sequentially reduced.

(a)

(b)

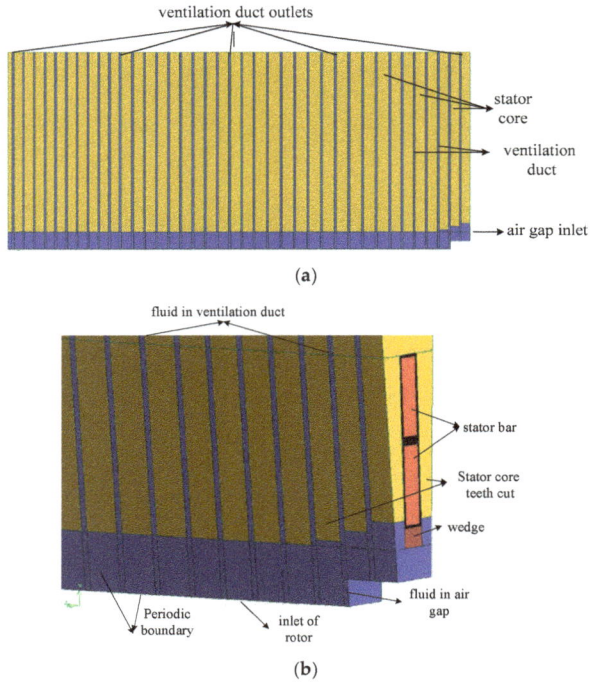

Figure 6. Physical model of stator ventilation system. (**a**) Calculation model of fluid field for 100 MW air-cooled turbo-generator; (**b**) partial display of calculation model.

b. Grid Mesh and Memory Usage

All the models adopted eight-node hexahedral mesh generation, as shown in Figure 7. The mesh generation was regular, and there were no over-sized or under-sized cells of the generation volume or area. The statistical results of the number of nodes and memory are shown in Table 2.

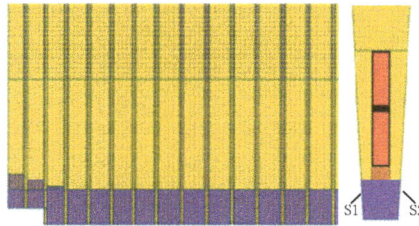

Figure 7. Model axial–radial and circumferential sections grid mesh.

Table 2. Statistical results of the number of nodes and memory.

	Cells	Faces	Nodes	Edges
Number Used	561,180	1,830,765	692,014	0
Mbytes Used	102	136	29	0
Number Allocated	561,180	1,830,765	692,014	0

In order to improve the computational accuracy of the fluid domain, the grid density in the fluid domain was dense, the grid density ratio between the fluid domain and the solid domain was about 10:1 in the axial direction. In the radial direction, the grid density was dense at the air gap and sparse at the yoke. While ensuring the accuracy of calculation, this effectively reduces the amount of computation and shortens the time. The grid mesh accuracy of this calculation method has been verified by the grid encryption method through previous work, which can better meet the calculation accuracy [13].

c. Boundary Conditions

The calculation boundary conditions are given as follows [14–16]:

(1) The air gap inlet is set as the velocity boundary condition, which is given according to the measurement results: The angle between the measuring element direction (Figure 3) and the axial direction is about 40 degrees, and the measured velocity is 74 m/s. Therefore, after triangle calculation, the inlet velocity of the air gap entrance is given as 56 m/s.
(2) The 38 radial ventilation outlets (half of generator) are basically balanced with the surrounding atmospheric pressure. Given the pressure boundary condition, the value is equivalent to one atmospheric pressure.
(3) The air gap exists in the whole circumferential direction, but the calculation model includes two additional surfaces, S1, S2 (shown in Figure 7), at the air gap that will increase the frictional resistance when fluid flows. In fact, this frictional resistance does not exist, so the periodic boundary conditions given for these two sides, eliminate the influence of frictional resistance on fluid flow.
(4) The velocity inlet of the rotor is given as the velocity boundary condition, and the velocity value is given according to the calculation results of the rotor ventilation calculation.

3.2. Comparison of Fluid Calculation Results and Measured Values at Test Conditions

The comparison of the test results and the calculation results of stator ventilation system are shown in Figure 8.

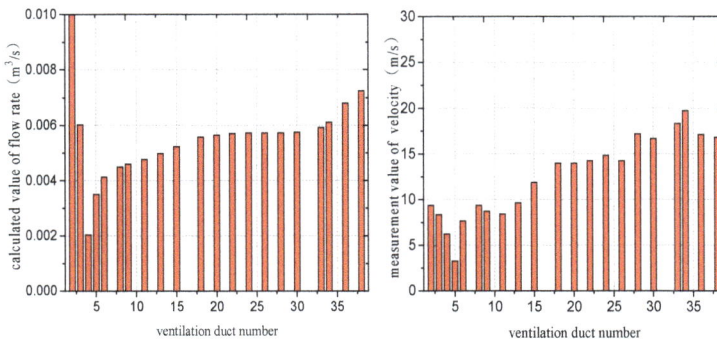

Figure 8. Comparison of calculated and measured values of fluid flow rate in the ventilation duct.

Because the fluid distribution is not uniform even in the same ventilation duct of the stator, the flow velocity value of one measuring point in the ventilation duct cannot completely reflect the distribution law of fluid velocity in this ventilation duct. Furthermore, the flow rate of the test result in this ventilation duct cannot be accurately obtained. However, since the fluid flow velocity measurement points are buried in the same position in each radial ventilation duct of the generator, the test velocity distribution law can indirectly reflect the flow rate distribution law in the stator radial ventilation ducts. The test results of flow velocities in different ventilation ducts and FEM calculation results of flow rates in different ventilation ducts are comparable.

It can be seen intuitively from Figure 8 that, in the fourth ventilation duct of the stator, the test results and calculated results both show a small flow rate at this position. In addition, the flow trend of fluid in different ventilation ducts is consistent. They all showed that the flow rate in the first three radial ventilation ducts is large, and then there is a trend of rapid increase after the flow rate decreases. The calculated flow distribution results are in agreement with the experimental measurements, which proves the accuracy of the calculation method.

3.3. Study on Sensitivity of Axial Velocity at the Air Gap Entrance to Flow Rate Distribution at Stator Radial Ventilation Ducts

In order to analyze the effect of different axial velocities at the air gap entrance on flow rate distribution at the stator radial ventilation ducts, this section gives four different wind speeds at the air gap entrance, which are 56, 70, 85, and 100 m/s respectively. Through the calculation of finite element theory, the distribution law of fluid flow in different radial ventilation ducts can be determined at the four different air gap inlet flow velocities.

When the air gap inlet velocity is 56 m/s, the flow velocity distribution of the fluid in the air gap and in each radial ventilation duct can be seen from Figure 9. After the fluid enters the air gap, the maximum flow velocity is 71 m/s, located below the second ventilation duct. According to the analysis of fluid continuity theory, the flow area in this place is the smallest position in the air gap, so the fluid velocity is the highest here. After fluid flowing through this position, the cross-sectional area in the air gap suddenly increases, but the fluid distribution does not rapidly expand as the air gap space increases, and the flow velocity in the air gap is still high. The velocity distribution trend shows the higher velocity in upper layer of the air gap, and the lower velocity in the lower layer. As the fluid in the air gap flows into the radial ventilation ducts sequentially, the flow velocity in the air gap becomes smaller and smaller. From the 38 radial flow velocity distributions, it can be seen directly that the velocity in the fourth and fifth ventilation ducts are lower, while the velocity of the first, second, 37th, and 38th are higher.

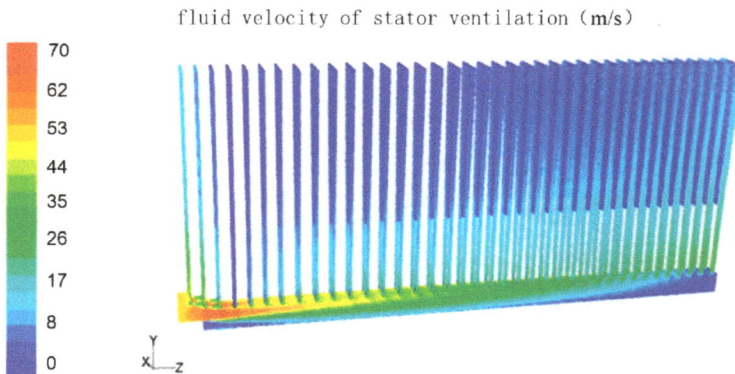

Figure 9. Fluid velocity distribution of stator air gap and the ventilation duct.

Similar to the above flow law, for the convenience of comparison, the fluid flow distribution in each radial ventilation duct is shown in Figure 10 under four different air gap entrance velocities.

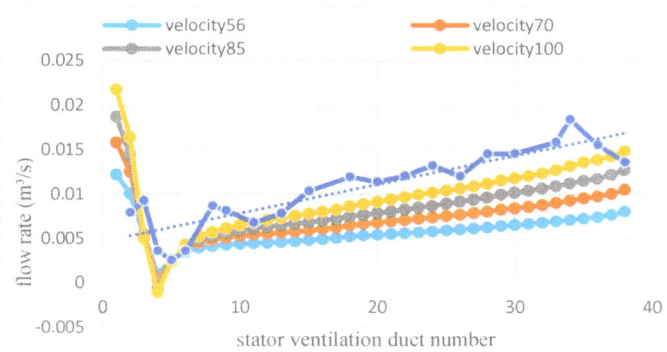

Figure 10. Flow rate distribution in the radial ventilation ducts at different air gap entrance velocities.

In Figure 10, the horizontal axis is the stator ventilation duct number. There is a total of six curves in the figure. Among them four curves are the four different inlet wind speed calculated values of the fluid velocity distribution in each ventilation duct of the stator; the other two curves are when the air gap inlet is 56 m/s, the flow rate test measured value and linear fit value at different radial ventilation ducts of the stator. From the calculated-value curves it can be seen that, with the increase of the flow velocities at the air gap entrance, the flow velocity in the other radial ventilation ducts except the fourth, fifth, and sixth ducts increases correspondingly; while the flow velocity in the fourth, fifth, and sixth ventilation ducts has the opposite distribution law, where the flow velocity does not increase but decreases.

After analyzing the calculation results of the fourth ventilation duct flow velocity, when the inlet speed is 56, 70, 85, and 100 m/s, the fluid average velocity of the radial ventilation duct outlets is 1.1, 0.24, −0.6, and −1.3 m/s. From the analysis of the calculation results, it can be seen that when the air gap inlet wind speed increases to a value between 70 and 85 m/s, the wind speed in the fourth ventilation duct will be close to zero. That is to say, there is no cooling wind blowing through the core and winding, the internal loss can only be taken away by the fluid in the adjacent ventilation duct; and when the speed at the air gap entrance is greater than the critical value above, the fluid velocity in the radial ventilation duct will form a reverse-direction reflux. At this time, the hot air from the other ventilation ducts will flow back into this ventilation duct. No matter which of any the above situations occurs, it is very unfavorable to the heat transfer of the generator stator structure and the safe operation of the generator.

3.4. The Influence of Backflow in the Radial Ventilation Duct on the Temperature of the Stator Winding and Core

Figure 11 shows the calculation results of the generator stator temperature field when the air gap inlet velocity is 56 m/s. It can be seen from the figure that the temperature of the stator core near the fourth radial ventilation duct is significantly higher than that of the other stator cores, but the temperature of the upper and lower windings of the stator is less affected by the low wind speed in this region.

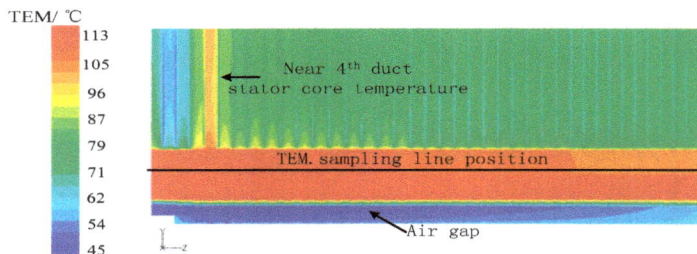

Figure 11. Temperature field of stator at air gap inlet velocity 56 m/s.

In order to analyze the axial distribution of the temperature of the stator core and winding in the case of backflow in the radial ventilation duct, Figure 12 shows the axial temperature distribution curve of the sampling line with the same radial height. The sampling line position is shown in Figure 11. When backflow occurs in the stator radial ventilation duct, the calculated temperatures of the stator core, winding, and cooling air along the axial direction are collected from the sampling line. At the same time, the corresponding air gap inlet velocity was 85 m/s.

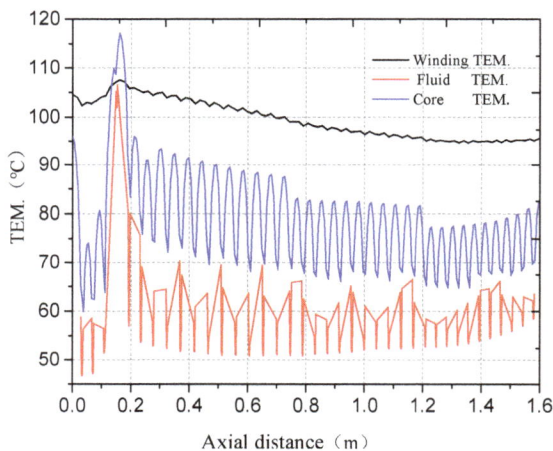

Figure 12. Winding and core temperature distribution at air gap inlet velocity 85 m/s.

It can be seen from Figure 12 that when backflow occurs in the local radial ventilation duct of the stator core, the negative pressure brings the hot wind at the back of the stator yoke into one or several radial ventilation ducts of the stator yoke to cool the stator core, resulting in a significant rise in the local stator core temperature, which reaches 115 °C. The core temperature is even higher than the average temperature of the stator windings. However, the average temperature of the stator winding is less affected by backflow. Therefore, the backflow of the fluid in the ventilation duct will have a direct impact on the temperature rise of the stator core, which is likely to produce local overheating of the core and affect the safe operation of the unit.

4. Conclusions

In this paper, a 100 MW air-cooled turbo-generator is taken as the research object. Through experimental measurement and finite element analysis of the model, the following conclusions are obtained:

1. Through the measurement of the fluid flow law in the stator radial ventilation duct, it is found that after the fluid passes through the first three radial ventilation ducts, the wind speed will decrease significantly in the back one or two radial ventilation ducts.
2. Through calculation and research, it is found that with the increase of the flow velocity at the air gap entrance, the flow velocity in several radial ventilation ducts after the step structure of the stator core end will gradually decrease to zero. At this time, if flow velocity at the air gap inlet continues to increase, the velocity direction of fluid in radial ventilation ducts will reverse, forming a backflow.
3. The backflow of the fluid in the ventilation duct will have a direct impact on the temperature rise of the local stator core, which is likely to produce local overheating of the core and affect the safe operation of the unit.

Author Contributions: Y.L. analyzed the data, contributed analysis tools and wrote the paper; W.L. conceived and designed the experiments; Y.S. performed the experiments.

Funding: We received National Natural Science Foundation for covering the costs to publish in open access, which name is "Cooling Mechanism of New Stator Tooth Inner Cooling and End Key-Type Shielding Structure Ventilation System for Air-Cooled Turbogenerator" and which number is 51477005.

Conflicts of Interest: The authors declare no conflict of interest.

References

1. Pickering, S.J.; Lampard, D.; Shanel, M. Modelling Ventilation and Cooling of The Rotors of Salient Pole Machines. In Proceedings of the IEEE International Electric Machines and Drives Conference (IEMDC), Cambridge, MA, USA, 17–20 June 2001; pp. 806–808. [CrossRef]
2. Shanel, M.; Pickering, S.J.; Lampard, D. Conjugate Heat Transfer Analysis of a Salient Pole Rotor in an Air-Cooled Synchronous Generator. In Proceedings of the International Electric Machines and Drives Conference, Madison, WI, USA, 1–4 June 2003; Volume 2, pp. 737–741. [CrossRef]
3. Shoulu, G.; Yichao, Y.; Yuzheng, L. Review on ventilation system of large turbine generator. *Energy Res. Inf.* **2004**, *4*, 195–200.
4. Shuang, L.; Zhihua, A.; Guangyu, Q. Research and performance verification of 150 MW air-cooled turbo-generator ventilation system. *Large Mot. Technol.* **2007**, *1*, 8–10.
5. Dongping, Z.; Rongshan, C. Calculation of ventilation and temperature rise of 100–200 MW air-cooled turbo-generator. *Power Gener. Equip.* **2006**, *3*, 193–195.
6. Guangde, L.; Weihong, Z. Ventilation system design of air-cooled turbo-generator. *Large Mot. Technol.* **1998**, *4*, 13–16.
7. Yamamoto, M.; Kimura, M. Ventilation and cooling technology of a new series of double-pole air-cooled steam turbine generator (Translated by Cai Qianhua). *Foreign Large Electr. Mot.* **2000**, *3*, 16–18.
8. Chauveau, E.; Zaim, E.H.; Trichet, D.; Fouladgar, J. A statistical of temperature calculation in electrical machines. *IEEE Trans. Mag.* **2000**, *36*, 1826–1829. [CrossRef]
9. Fujita, M.; Kabata, Y.; Tokumasu, T.; Kakiuchi, M.; Shiomi, H.; Nagano, S. Air-Cooled Large Turbine Generator with Multiple-Pitched Ventilation Ducts. In Proceedings of the IEEE International Conference on Electric Machines and Drives, San Antonio, TX, USA, 15 May 2005; pp. 910–917. [CrossRef]
10. Zixiong, Z.; Zengnan, D. *Viscous Fluid Mechanics*; Tsinghua University Press: Beijing, China, 1998; pp. 255–279.
11. Borishenko, A.N. *Aerodynamics and Heat Transfer in Motor*; Machinery Industry Press: Beijing, China, 1985.
12. Changming, Y. *Heat Conduction and Its Numerical Analysis*; Tsinghua University Press: Beijing, China, 1981; pp. 1–7.
13. Huo, F.; Li, Y.; Li, W. Calculation and analysis of stator ventilation structure optimization scheme of large air-cooled turbo-generator. *Chin. J. Electr. Eng.* **2010**, *6*, 95–103.

14. Li, H.J.W. Calculation and analysis of fluid velocity and fluid temperature in large air-cooled turbo-generator stator. *Proc. CSEE* **2006**, *26*, 168–173. [CrossRef]
15. Li, W.; Ding, S.; Zhou, F. Diagnostic numerical simulation of large hydro-generator with insulation aging. *Heat Transf. Eng.* **2008**, *29*, 902–909. [CrossRef]
16. Li, W.; Hou, Y. Heating analysis of stator strands of large hydro-generator based on numerical method. *Proc. CSEE* **2001**, *21*, 115–1186. [CrossRef]

energies

MDPI

Article

Model Verification and Justification Study of Spirally Corrugated Pipes in a Ground-Air Heat Exchanger Application

Kwang-Seob Lee [1,2], Eun-Chul Kang [2], Yu-Jin Kim [1,2] and Euy-Joon Lee [1,2,*]

[1]   Renewable Energy Engineering, University of Science and Technology, Korea, 217 Gajeong-ro, Yusung-gu, Daejeon 34113, Korea; kslee89@kier.re.kr (K.-S.L.); Yjin@kier.re.kr (Y.-J.K.)
[2]   Energy Efficiency and Materials Research Department, Korea Institute of Energy Research, 152 Gajeong-ro, Yusung-gu, Daejeon 34129, Korea; kec8008@kier.re.kr
*   Correspondence: ejlee@kier.re.kr; Tel.: +82-42-860-3514

Received: 26 September 2019; Accepted: 23 October 2019; Published: 24 October 2019

Abstract: Ground-air heat exchangers have become an important topic in recent years due to their contributions to the market growth of the ground source heat pump industry. This paper provides a comprehensive study and recommends suggestions on the selection process of a suitable pipe for an air-to-water heat pump (AWHP). Parametric studies including material, turbulent plate quantity, and pipe type were performed to identify an optimal pipe design for high-performance AWHP. Both numerical and experimental studies were carried out to validate current pipe models. Overall, there was good agreement between the numerical model and experimental results. It was determined that a spirally corrugated pipe exhibited excellent thermal power generation with little compromising pressure drop. Finally, a pipe selection example was demonstrated as a design guideline to size an optimal pipe for AWHP application.

Keywords: ground source heat pump; numerical and experimental studies; ground-air heat exchanger; geothermal energy; computational fluid dynamics; spirally corrugated pipe

---

1. Introduction

Ground-air heat exchangers (GAHX), commonly known as earth tubes, are underground heat exchangers that can capture or dissipate heat into the ground. At a certain depth, the ground temperature is nearly constant. Therefore, when outside air is moved by a blower through buried pipes, it is conditioned and then used for either partial or full home cooling in summer or heating in winter. Earth tubes are often a viable and economical alternative or supplement to conventional central heating or air-conditioning systems because there are no compressors, chemicals, or burners. In addition, only blowers are required to move the air. This nearly-free cooling technique can meet passive house standards and Leadership in Energy and Environmental Design (LEED) certification.

GAHX have been introduced in the literature during recent years. Yildiz et al. [1] analyzed a PV assisted closed-loop earth-to-air heat exchanger system, while Lee et al. [2] introduced a GAHX design for a heat pump system. Muehleisen [3] developed simple design tools for earth-air heat exchangers, while Hollmuller et al. [4] and Lee et al. [5] researched earth tube models using simulations. Vlad et al. [6] and De Paepe [7] introduced how to design GAHX. Rehau [8] produced an eco-air system based on GAHX technology. The Korea Institute of Energy Research [9] has investigated a trigeneration system with GAHX, while Vikas et al. [10] conducted GAHX performance analyses. Viorel et al. [11] developed a simple GAHX model and Huijun et al. [12] studied an evaluation model.

In recent research, Mroslaw et al. [13] conducted a comparison study of the thermal performance of GAHX, Cuny et al. [14] studied GAHX on different soil types, Amanowicz, Ł. [15] performed

GAHX CFD on different pipe paths, however, it did not include pipe shape. Zhao et al. [16] and Noorollahi et al. [17] have studied parametric design of GAHX, Agrawal et al. [18,19] discussed recent research trends on GAHX and GAHX effect on climate and soil conditions. Mahdavi et al. [20] evaluated photovoltaic-thermal(PVT)-GAHX system's potential. Ali et al. [21] has studied optimization of GAHX, and Li, H [22] has studied the feasibility of the GAHX system. Following the recent studies survey, GAXH are newly applied and coupled with renewable energy system.

GAHX also play an important role in geothermal technology. Ground source heat pumps rely on earth tubes to transfer heat to or from the ground. GAHX are more energy-efficient than air-source heat pumps because ground temperatures are more stable than air temperatures throughout the year.

The driving force for heat transfer is the difference between the ambient air and ground temperatures. This temperature difference (ΔT) changes little throughout a day, and the goal is to maximize ΔT. Optimization efforts must consider maximizing the pipe wall area to enhance convective heat transfer. It also requires pipe length to be long enough to obtain the desired temperature drop or rise in summer or winter, respectively, before reaching the house. However, pipe sizing is constrained to yield a more cost-effective design for users.

The heat transfer rate in GAHX applications is of paramount importance. If one can effectively control the rate at which the heat is transferred to or from the ground, pipe length and manufacturing cost can be reduced. The resistance level to heat flow of air cannot be controlled, but it can be influenced by the design. Specifically, turbulence generation within fluids can prevent the creation of a thermally resistant static "boundary layer" of fluid in contact with the transfer surface. An effective way to reduce this effect is to utilize a spirally corrugated pipe (i.e., deforming a tube with a continuous spiral indentation). Research has shown that by carefully choosing the depth, angle, and width of the indentation, boundary layer resistance can decrease faster than pressure.

Finally, different materials offer a variety of thermal conductivity, which is vital to the conductive heat transfer process to or from the ground via the pipe. This factor is also controlled by the designer along with the pipe boundary shape. The chosen material must be compatible with the process fluids, and it must have a low resistance to heat flow so that it does not become the overriding factor in the conductive heat transfer process.

In this study, GAHX produced the best design for maximum temperature variation and minimum pressure drop. This result showed a good match between the simulation results and the experimental results. Moreover, additional work has been done to explore various design alternatives to improve temperature reduction. As a result, the coefficient of performance (COP) of the AWHP system used in an office space was 3.0, calculated so the 111-m pipe length with a diameter of 250 mm could provide 6 kW of thermal power.

2. Computational Setup

Our simulation experiment consisted of two parts—validation and potential designs. The validation study focused on the two models—baseline and turbulent insert cases. The overall appearance of the baseline case is shown in Figure 1.

Only the pipe section beneath the ground was included in the computational domain. In other words, the temperatures at the pipe inlet and exit were assumed to be constant and the same at their nearest ground-level boundaries. The turbulent insert case featured the baseline case plus eight turbulent inserts, as shown in Figure 2.

The total computational pipe length for both cases was 88.6 ft (27 m) and the inner diameter of the pipes was 0.82 ft (0.250 m) with a thickness of 0.0279 ft (0.0085 m). In the turbulent insert pipe, the inserts were circular plates with angularly bent fins (Figure 3). These plates were placed 8.2 ft (2.5 m) from each other (Figure 2). There were three temperature sensors placed at 0, 10, and 20 m (Figures 2 and 3). However, only positions 0 and 20 m were monitored in the validation study.

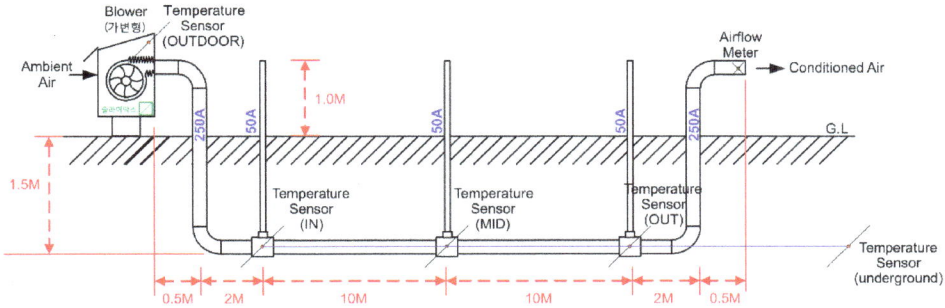

Figure 1. Baseline case.

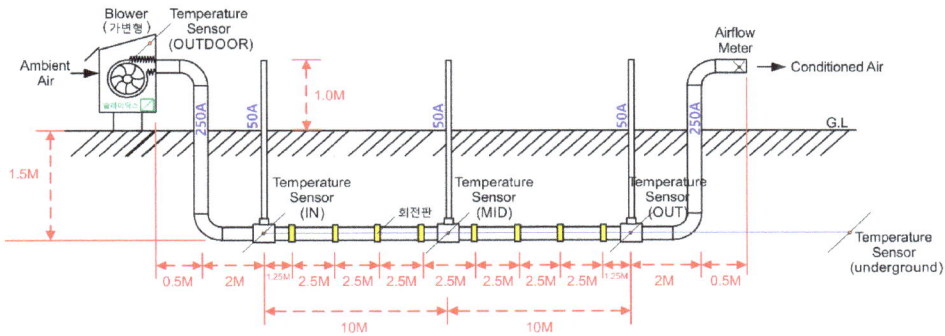

Figure 2. Turbulent insert case.

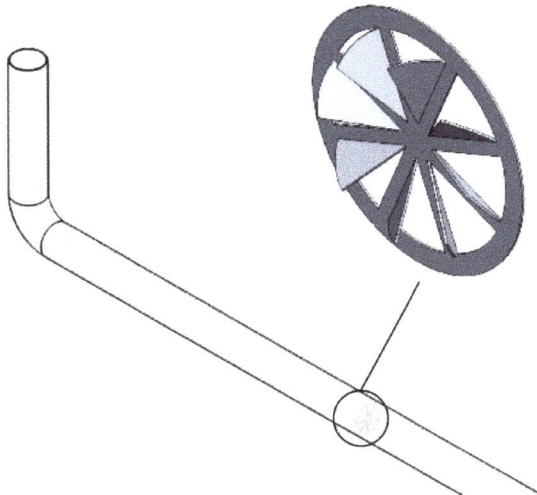

Figure 3. Turbulent insert pipe.

The simulation was performed for four consecutive days corresponding to the experimental data observed during summer in South Korea. The ground and outdoor temperatures are reported in Table 1. The highest and lowest temperature difference between the ground and outdoors was 12.8 and 9.2 °C, respectively.

Table 1. Environmental conditions for the reported data.

Date	Ground Temp (°C)	Outdoor Temp (°C)	Inlet Velocities (m/s)	
			Turbulent Insert	Baseline
Aug 13	22.5	33.4	3.53	3.71
Aug 14	22.6	35.4	3.55	3.74
Aug 15	22.7	34.5	3.55	3.72
Aug 16	22.8	32.0	3.67	3.86

A commercial simulation platform was used to perform numerical simulations. For the purpose of this study, a standard k-ε turbulence model with a standard wall function was chosen among the available models (Launder–Spalding wall function, scalable wall functions, non-equilibrium wall functions, enhanced wall treatment, etc.). The model was also chosen due to its stability and simplicity as the turbulent viscosity is calculated in a less complex way. The standard wall function has some weaknesses when applied to complex problems such as high-pressure gradients, buoyancy and complex strains, but in this case, its use is justified as the flow does not experience rapid changes near the pipe inside-walls. For boundary conditions, the velocity inlet and pressure outlet were prescribed at the pipe entrance, respectively. The air entering the pipe was assumed to have an outdoor temperature corresponding to each experimental day (see Table 1). The turbulent intensity was set at 10% as per Basse, N. T. [23] and the Reynolds number and the hydraulic diameter at the inlet was set at 0.82 ft (0.250 m). For the solution method, we used the SIMPLE scheme for pressure-velocity coupling with least-squares cells based on standard pressure. A first-order upwind scheme was applied for momentum, turbulent kinetic energy, and turbulent dissipation rate as its application is sufficient when applied to heat-flow applications with no presence of chemical reactions. However, a second-order upwind scheme was used for the pressure and energy equations. Convergence was reached when the residual was less than 10^{-6} for all parameters including continuity, energy, velocities, and the k and ε equations. Convergence also occurred when the pipe outlet temperature stabilized at a constant value.

3. Results

3.1. Grid Independence Study

A mesh independence study was performed for the turbulent insert case (Figure 2). Three different mesh sizes were investigated as shown in Table 2.

Table 2. Mesh independence study.

Size	Mesh Size	No. of Nodes	No. of Cells	ΔT (°C)
Coarse	1.21×10^{-2}–2.42	77,548	402,064	5.81
Medium	6.06×10^{-3}–1.21	123,945	672,063	5.94
Fine	3.55×10^{-3}–0.71	183,921	1,026,676	5.95

An unstructured tetrahedral mesh was used in all mesh generations (Figure 4). The value of Y+ depends on the quality of the mesh and was selected from the range of 30 < Y+ < 500, The temperature difference between the 0 and 20 m measurement points was investigated for changes in mesh size and results showed no significant differences. There was only a 0.13 °C temperature difference at these two points between the coarse and medium mesh sizes. In addition, the difference was only 0.04 °C between the coarse and fine mesh. With such a minimal change in temperature difference, the coarse mesh was selected to adequately represent the flow physics of the air through the pipe model. The pressure drop on each pipe was 755 pa (turbulent plate), 55 pa (spirally corrugated pipe), and 25 pa (baseline), and it was ignored from the point of the mesh independent study.

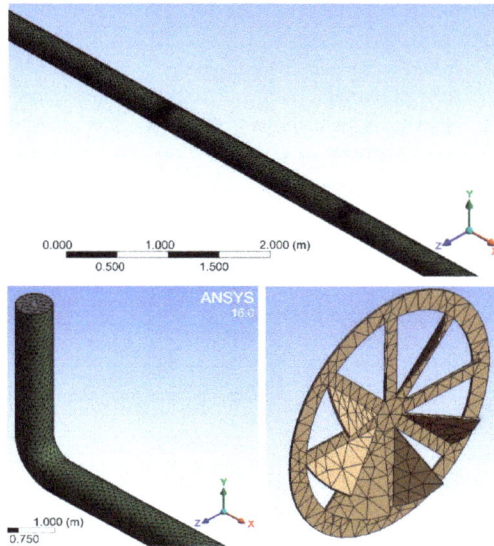

Figure 4. Mesh generation.

3.2. Validation Study

The validation study was performed for scenarios with and without turbulent plates. Experiment results were obtained for four days from 13–16 August 2016. The turbulent insert case exhibited slightly closer results to the experiments compared to the baseline case. However, ΔT errors between 0 and 20 m for both the baseline and the turbulent insert scenarios were less than 20%. These results were acceptable considering that all flows were turbulent. Moreover, the local errors at the two locations were less than 5 and 7% for the turbulent insert and baseline case, respectively (Tables 3 and 4)

Table 3. Baseline simulation results.

Date	Baseline Case								
	0 m			20 m			ΔT (20 m–0 m)		
	Sim.	Exp.	Diff.	Sim.	Exp.	Diff.	Sim.	Exp.	Diff.
Aug 13	31.5	32.4	3%	27.2	28.0	3%	4.3	4.4	2%
Aug 14	33.0	32.3	2%	28.1	28.2	0%	4.9	4.1	18%
Aug 15	32.4	31.9	2%	27.8	28.0	1%	4.7	3.9	20%
Aug 16	31.0	32.1	4%	26.1	28.0	7%	4.9	4.1	19%

Table 4. Turbulent insert simulation results.

Date	Turbulent Insert Case								
	0 m			20 m			ΔT (20 m–0 m)		
	Sim.	Exp.	Diff	Sim.	Exp.	Diff.	Sim.	Exp.	Diff.
Aug 13	32.0	32.7	2%	26.1	26.5	1%	5.81	6.20	6%
Aug 14	34.3	32.5	5%	27.8	27.0	3%	6.40	5.50	16%
Aug 15	33.6	32.2	4%	27.5	26.9	2%	6.06	5.30	14%
Aug 16	31.4	32.5	3%	26.6	27.0	1%	4.75	5.50	14%

Figure 5 shows the temperature profiles x = 0 m and x = 20 m. In the turbulent insert case, the temperature was more concentrically uniform at x = 0 m. The hot entering air stream remained at the center of the pipe, and air cooled radially as it moved closer to the pipe wall. Furthermore, the temperature was cooler near the pipe exit (x = 20 m). In the baseline case, temperature exhibited a non-uniform distribution within the cross-sectional profile at x = 0 m. A hotter air stream was located at the bottom of the pipe. The cool temperature layer was also thinner compared to the turbulent insert case. At x = 20 m, the temperature was concentrically distributed and remained higher in the center of the pipe.

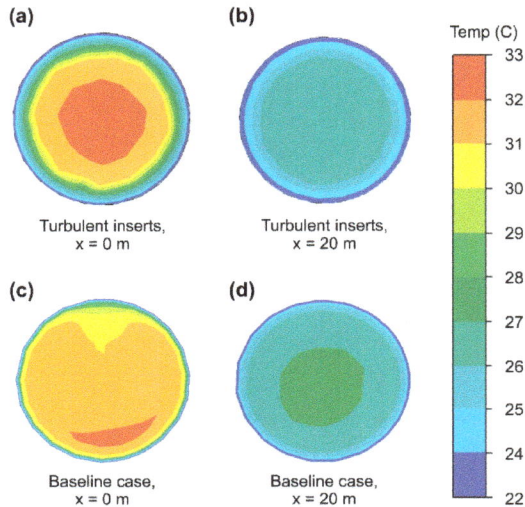

Figure 5. Temperature profiles at x = 0 m and x = 20 m for the baseline and turbulent insert cases.

The temperature profile in Figure 5 was explained by the following observations. In the turbulent insert case, the downstream plates acted as directional guides for the air to flow more uniformly upstream (Figure 6). Right after the inserted plates, the rotating flow became dominant, which enhanced the heat transfer process from the hot moving air to the cooler ground. In the baseline case, the hot air stream moved undisturbed along the pipe. Because of the downward momentum as air moved through the pipe, air sank towards the bottom of pipe after the first L-turn. This caused the temperature profile to be non-uniform near the inlet (x = 0 m). However, after traveling along the long pipe, the airflow became fully developed and uniform near the exit (x = 20 m, Figure 5).

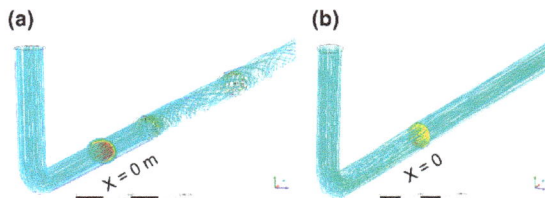

Figure 6. Streamlines originating from the pipe inlet (a) with turbulent insert, (b) baseline

Figure 7, shows the temperature profile in the middle plane along the pipe for the two scenarios. The turbulent insert case exhibited better heat transfer between the hot moving air inside the pipe and the cool ground. The turbulent plates tend to resist the airflow and create more rotational flows

(Figure 6). This kept the hot air moving in good contact with the cooler pipe wall. As a result, the downstream air cooled further as it traveled along the pipe. In contrast, the baseline case exhibited slower cooling as air moved along the pipe. At the same x position, one can see the distinct temperature difference between the two scenarios in Figure 7.

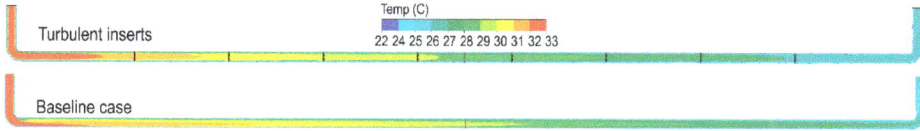

Figure 7. Temperature profile in the middle plane along the pipe for the two scenarios.

One major drawback of the turbulent insert case was the large pressure decrease. Figure 8 shows the pressure drop across the pipe for both scenarios. After each plate, pressure was significantly reduced. For the baseline case, the pressure drop was only 27 Pa across the pipe. In the next section, we will explore how the reduction in turbulent plates affects the desired temperature drop of the cooled air. Specifically, plate reductions can reduce the pressure drop, which can save power for the inlet fan.

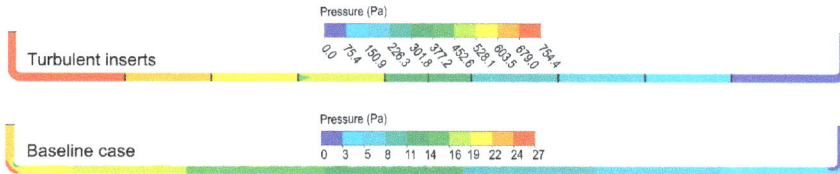

Figure 8. Pressure drop profile in the middle plane along the pipe for two scenarios.

3.3. Parametric Studies

3.3.1. Effects of Turbulent Plate Quantity

In this investigation, the plate quantity was reduced from 8 (original) to 6, 4, and 0 plates. The inlet velocity was kept constant at 3.53 m/s for all cases based on data from 13 August. For the first three cases with 8, 6, and 4 plates, the temperatures at x = 0 m were relatively similar. However, the temperature at x = 20 m increased as the number of plates decreased. As a result, the temperature differences between the two gauged positions also decreased as the number of plates decreased (Figure 10). Finally, in the zero-plate case, the temperature at both x = 0 m and x = 20 m was lower compared to cases with inserted plates (Figure 9).

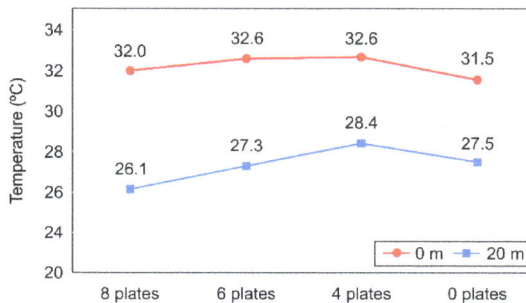

Figure 9. The effect of fins on the temperature of the two locations (0 and 20 m) at a v_{inlet} of 3.53 m/s.

The temperature differences between x = 0 m and x = 20 m for various plate quantities are plotted in Figure 10. We observed a decrease in cooling temperature as the number of plates decreased. In addition, we plotted the pressure drop for each case. By reducing two plates from eight to six, temperature difference decreased by only 0.63 °C. However, there was a big gap in pressure drop. Temperature differences between the four plates and zero plates cases were very similar. Thus, it is not recommended to use four plates in practice. Instead, either 6 or 0 plates is preferred depending on the amount of air cooling required before reaching the residential house.

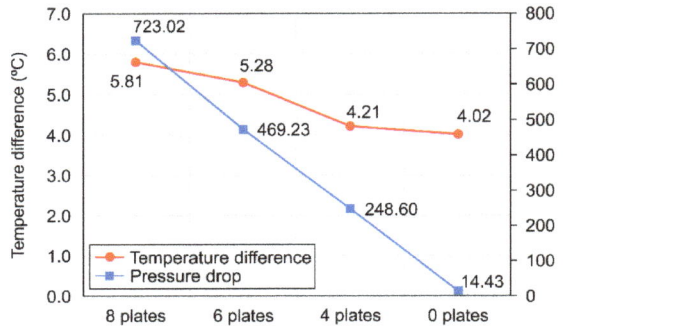

Figure 10. The temperature differences between x = 0 m and x = 20 m for various plate quantities.

3.3.2. Effects of Different Pipe Materials

In Figure 11, the effects of different materials on temperatures at two locations, x = 0 m and x = 20 m are shown. Table 5 shows the properties of the three materials investigated.

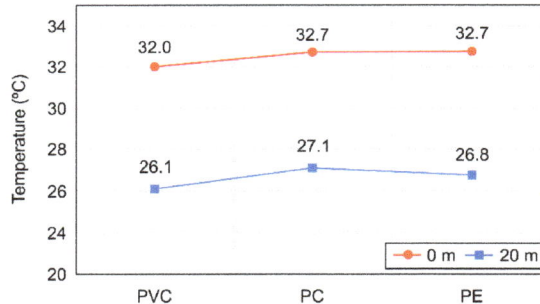

Figure 11. The effect of different materials on temperatures at 0 m and 20 m, at v_{inlet} = 3.53 m/s.

Table 5. Material properties.

	Material Properties			
	PVC	**PC**	**PE**	**Steel (for All Inserts)**
Density (kg/m^3)	1450	1200	960	8050
Thermal Conductivity (W/m-K)	0.28	0.25	0.33	16.27
Specific Heat (J/kg-K)	900	1300	1850	502.48

Overall, the temperatures at x = 0 m for all cases were nearly identical. However, the difference in the material had a small effect on downstream air temperature (x = 20 m). Figure 12 shows the temperature difference between the two locations for three different materials—polyvinyl chlorine

(PVC), polycarbonate (PC), and polyethylene (PE). The temperature difference was lowest in the PC case and highest in the PE case, though these differences were not significant.

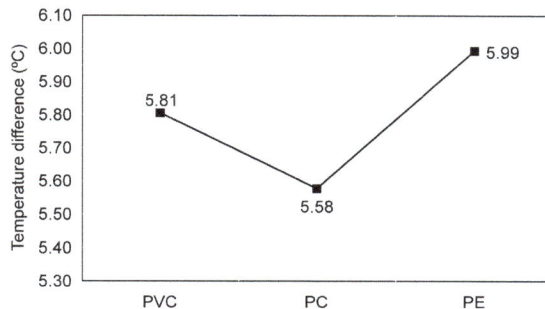

Figure 12. The temperature difference between x = 0 m and x = 20 m for different materials.

3.4. Pipe Type Exploration

In this section, three additional pipe types are detailed—corrugated, helical insert, and spirally corrugated pipes. The temperature drops between the two gauged positions are reported in Table 6. These results were obtained using the boundary conditions for 13 August 2016 (Table 1). The inlet velocity used for the corrugated, helical insert and spirally corrugated pipe cases was the same as the baseline case (3.71 m/s), while the turbulent insert case inlet velocity was kept at 3.53 m/s.

Table 6. Results for different pipe types using conditions from 13 August 2016.

Pipe Type	$T_{outdoor}$	$T_{x=0\,m}$ (°C)	$T_{x=20\,m}$ (°C)	$\Delta T_{20\,m-0\,m}$ (°C)	$\Delta T_{outdoor-20\,m}$ (°C)	V_{inlet} (m/s)
Baseline	33.4	32.0	27.4	4.6	6.0	3.71
Turbulent Insert	33.4	32.0	26.1	5.8	7.3	3.53
Corrugated	33.4	31.9	27.2	4.8	6.2	3.71
Helical Insert	33.4	31.5	26.5	5.0	6.9	3.71
Spirally Corrugated	33.4	30.9	25.7	5.2	7.7	3.71

From Table 6, the baseline case had the highest temperature at x = 20 m, resulting in the lowest temperature drop (6 °C). The turbulent insert case featured the best temperature drop at x = 20 m compared to the entering air temperature (7.3 °C) and the air temperature at x = 0 m (5.8 °C). The corrugated pipe case had a slightly better temperature decrease of 6.2 °C and 4.8 °C with respect to the entering and 0 m air temperature, respectively. The helical insert case had a second-best temperature drop when compared to the outside air (6.9 °C), although the temperature at 20 m was only 5 °C. Finally, the spirally corrugated pipe case exhibited the best improvement in temperature drop when compared to the outside air and second-best at 20 m. Figure 13 clearly reflects these temperature drop comparisons for the five investigated cases.

Figures 14 and 15 show the temperature and pressure profiles of the five studied cases. Although the turbulent insert case exhibited a reasonably good temperature drop along the linear pipe, its major drawback was the high-pressure drop (755 Pa), which caused extensive amounts of electrical energy to be consumed in the blower. The spirally corrugated pipe, however, demonstrated the best temperature drop (7.7 °C) without inducing a large pressure drop (only 55 Pa). The baseline case had the lowest pressure drop due to a lack of obstructions within the pipe. Taking temperature and pressure drops into consideration, the spirally corrugated pipe case stood out as the preferred option for practical use. However, more data is required for validation.

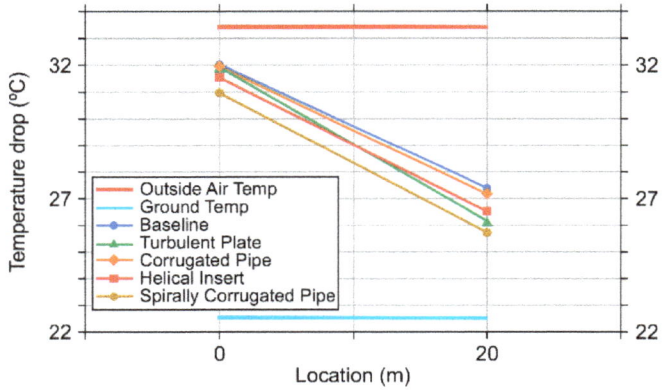

Figure 13. Temperature drop for the five investigated pipes.

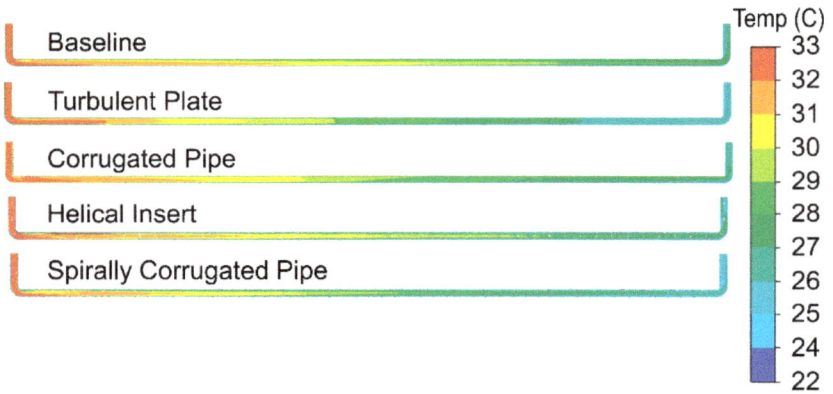

Figure 14. Temperature profile in the middle plane for five investigated pipes.

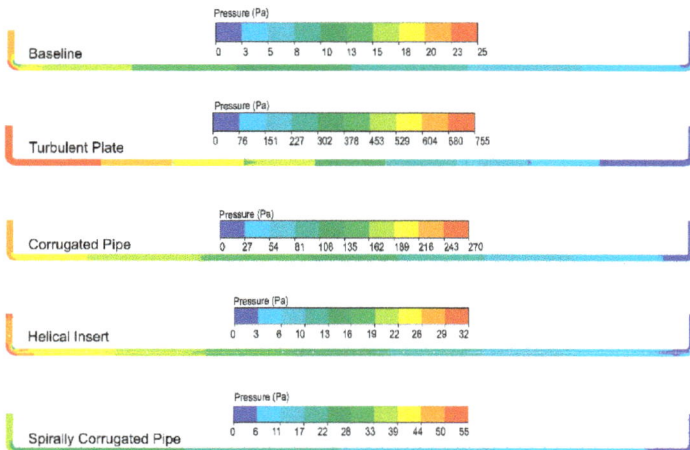

Figure 15. Pressure drop profile in the middle plane for five investigated pipes.

4. Design Guideline of a GAHX for an AWHP

In this section, we demonstrate how to select the most optimal pipe design in terms of type, material, and size. From the previous section, the spirally corrugated pipe was determined to be the most desirable. In the following example, a calculation is performed for a typical office space with a total floor area of 147 m² (1581 ft²). It is desirable for an AWHP system to achieve a coefficient of performance (COP) of approximately 3.0. According to Perko et al. [24], an average-sized space (186–223 m²) requires a thermal power of 10.6 kW to 11.4 kW, depending on climate conditions. In the worst-case scenario, a small area of 186 m² requires a heating power of 11.4 kW. The ratio (r) of floor area per kilowatt can then be calculated as:

$$r = \frac{A}{P} = \frac{186 \text{ m}^2}{11.4 \text{ kW}} = 16.32 \frac{\text{m}^2}{\text{kW}} \tag{1}$$

where, r is the floor area ratio per kW, A is floor area, and P is the required heating power.

The thermal power at the condenser was then calculated from the actual floor area of 147 m² using:

$$Q_{\text{condenser}} = \frac{A_{\text{office}}}{r} \approx 9 \text{ kW} \tag{2}$$

The target COP for the AWHP was set as 3.0. Therefore, the power required from the compressor was:

$$COP = \frac{Q_{\text{out}}}{W_{\text{e in}}} = \frac{Q_{\text{condenser}}}{Q_{\text{conpressor}}} \tag{3}$$

$$Q_{\text{compressor}} = \frac{Q_{\text{condenser}}}{COP} = \frac{9 \text{ kW}}{3} = 3 \text{ kW} \tag{4}$$

The required thermal power from air flowing through the buried pipe was therefore calculated as:

$$Q_{\text{geothermal}} = Q_{\text{condenser}} - Q_{\text{compressor}} = 9 \text{ kW} - 3 \text{ kW} = 6 \text{ kW} \tag{5}$$

In addition, the spirally corrugated pipe had a total length of 28 m. Therefore, heat extraction from the ground source was calculated as:

$$
\begin{aligned}
R_{\text{extract}} &= \frac{Q_{\text{pipe}}}{l_{\text{pipe, calculated}}} = \frac{C_{p \text{ air}} \times \dot{m}_{\text{air}} \times (T_{20m} - T_{0m})}{l_{\text{pipe, calculated}}} \\
&= \frac{1005 \frac{\text{J}}{\text{kg°C}} \times 0.223 \frac{\text{kg}}{\text{s}} \times 5.2°\text{C}}{20\text{m}} = 58.25 \text{ W/m}
\end{aligned} \tag{6}
$$

The target COP 3.0 resulted in a required thermal power from the underground pipe of 6 kW. From here, the total pipe length was calculated using:

$$l_{\text{pipe, calculated}} = \frac{Q_{\text{geothermal}}}{R_{\text{extract}}} = \frac{6000\text{W}}{58.25\text{W/m}} \approx 103 \text{ m} \tag{7}$$

Finally, the actual pipe length including the U-shape section to the ground surface was calculated by adding an extra 8 m:

$$l_{\text{pipe,final}} = 103 \text{ m} + 8 \text{ m} = 111 \text{ m} \tag{8}$$

The final optimal pipe design is summarized in Table 7. This design yielded a maximum 6 kW in thermal power on top of the 3-kW power required by the compressor. This allowed a COP of 3.0 for the AWHP system used to heat up an office space of 147 m² (1581 ft²) in winter.

Table 7. Design input and output of the office space calculation for a target COP of 3.0.

Design Input	Design Output
$A_{office} = 147 \text{ m}^2 \ (1581 \text{ ft}^2)$	$l_{pipe, \ calculated} = 103 \text{ m}$
$Q_{condenser} = 9 \text{ kW}$	$l_{pipe, \ final} = 111 \text{ m}$
$Q_{geothermal} = 6 \text{ kW}$	$\varphi_{pipe} = 0.25 \text{ m}$
$Q_{compressor} = 3 \text{ kW}$	
$COP = 3.0$	
$l_{pipe, \ total} = 28 \text{ m}$	
$l_{pipe, \ calculated} = 20 \text{ m}$	
$T_{20m} - T_{0m} = 5.2 \,°C$	
$v_{air} = 3.71 \text{ m/s}$	
$\dot{m} = 0.224 \text{ kg/s}$	
$T_{ground} = 22.5 \,°C$	
$T_{ambient} = 33.4 \,°C$	
$\rho_{air} = 1.225 \text{ kg/m}^3$	
$C_{p,air} = 1.005 \text{ kJ/kg.K}$	
$R_{extract} = 58.25 \text{ W/m}$	

5. Conclusions

In this paper, the GAHX was studied to yield the best design for a maximum temperature change and minimum pressure drop. The smooth pipe and turbulent insert case were validated against data obtained during the summer in South Korea. Results demonstrated good agreement between the simulated and experimental results. Additional work was performed to explore various design alternatives to improve temperature decrease. From the simulations, the turbulent insert case exhibited the best improvement in terms of temperature drop, yet had the highest pressure drop across the pipe. However, the spirally corrugated pipe demonstrated the best temperature drop (7.7 °C) with a small pressure drop of 55 Pa. Therefore, the spirally corrugated pipe was determined to be the optimal design. An example calculation for an average office space was then provided as a guideline to select the proper pipe size for a spirally corrugated pipe. It was calculated that a 111 m pipe length with a diameter of 250 mm can provide 6 kW of thermal power, which resulted in a COP of 3.0 for the AWHP system used in the office space. In our future work, we will conduct sensitivity analysis with turbulent intensity 5% to 15%. A mesh independent study on pressure drop and application of swept mesh will be considered on the meshing process to improve the model accuracy. We will conduct additional tests and experiments on the new design AWHP coupled with the GAHX based on the results of this study.

Author Contributions: Conceptualization, E.-J.L.; methodology, K.-S.L.; validation, E.-C.K.; formal analysis, K.-S.L.; investigation, E.-C.K. and Y.-J.K.; data curation, K.-S.L.; writing—original draft preparation, K.-S.L.; writing—review and editing, E.-J.L.

Funding: This research was funded by the Korea Institute of Energy Technology Evaluation and Planning (KETEP) and the Ministry of Trade, Industry & Energy (MOTIE) of the Republic of Korea, grant number 20188550000430.

Acknowledgments: This work was supported by the Korea Institute of Energy Technology Evaluation and Planning(KETEP) and the Ministry of Trade, Industry & Energy(MOTIE) of the Republic of Korea (No. 20188550000430).

Conflicts of Interest: The authors declare no conflict of interest.

References

1. Yildiz, A.; Ozgener, O.; Ozgener, L. Exergetic performance assessment of solar photovoltaic cell (PV) assisted earth to air heat exchanger (EAHE) system for solar greenhouse cooling. *Energy Build.* **2011**, *43*, 3154–3160. [CrossRef]
2. Lee, K.; Ryu, N.; Kang, E.; Lee, E. Ground air heat exchanger design and analysis for air source heat pump. *Korea Soc. Geotherm. Energy Eng.* **2016**, *12*, 1–6. [CrossRef]
3. Muehleisen, R.T. Simple design tools for earth-air heat exchangers. *Proc. SimBuild* **2012**, *5*, 723–730.

4. Hollmuller, P.; Lachal, B.M. *TRNSYS Compatible Moist Air Hypocaust Model*; Centre universitaire d'étude des problèmes de l'énergie, University of Geneve: Geneva, Switzerland, 1998.

5. Lee, K.H.; Strand, R.K. The cooling and heating potential of an earth tube system in buildings. *Energy Build.* **2008**, *40*, 486–494. [CrossRef]

6. Vlad, G.E.; Ionescu, C.; Necula, H.; Badea, A. Thermoeconomic design of an earth to air heat exchanger used to preheat ventilation air in low energy buildings. In Proceedings of the International Conference on Energy, Environment, Entrepreneurship, and Innovation, Lanzarote, Spain, 27–29 May 2011; pp. 11–16.

7. De Paepe, M.; Janssens, A. Thermo-hydraulic design of earth-air heat exchangers. *Energy Build.* **2003**, *35*, 389–397. [CrossRef]

8. Rehau Ecoair Ground-Air Heat Exchanger System. Available online: http://www.rehau.com (accessed on 20 May 2019).

9. Euyjoon, L.; Eunchul, K.; Kwangseob, L.; Eviguaniy, E.; Libing, Y.; Mohammed, G. *Kier-Canmet ENERGY Joint Study for CO2 free Trigenration System Development based on GAHX Report*; Korea Institute of Energy Research: Daejeon, South Korea, 2017.

10. Vikas, B.; Rohit, M.; Ghanshyam, D.A.; Jyotirmay, M. Performance analysis of earth–pipe–air heat exchanger for summer cooling. *Energy Build.* **2010**, *42*, 645–648.

11. Viorel, B. Simple and accurate model for the ground heat exchanger of a passive house. *Renew. Energy* **2007**, *32*, 845–855.

12. Huijun, W.; Shengwei, W.; Dongsheng, Z. Modelling and evaluation of cooling capacity of earth–air–pipe systems. *Energy Convers. Manag.* **2007**, *48*, 1462–1471.

13. Zukowski, M.; Topolanska, J. Comparison of thermal performance between tube and plate ground-air heat exchangers. *Renew. Energy* **2018**, *115*, 697–710. [CrossRef]

14. Cuny, M.; Lin, J.; Siroux, M.; Magnenet, V.; Fond, C. Influence of coating soil types on the energy of earth-air heat exchanger. *Energy Build.* **2018**, *158*, 1000–1012. [CrossRef]

15. Amanowicz, Ł. Influence of geometrical parameters on the flow characteristics of multi-pipe earth-to-air heat exchangers–experimental and CFD investigations. *Appl. Energy* **2018**, *226*, 849–861. [CrossRef]

16. Zhao, Y.; Li, R.; Ji, C.; Huan, C.; Zhang, B. Parametric study and design of an earth-air heat exchanger using model experiment for memorial heating and cooling. *Appl. Therm. Eng.* **2019**, *148*, 838–845. [CrossRef]

17. Noorollahi, Y.; Saeidi, R.; Mohammadi, M.; Amiri, A.; Hosseinzadeh, M. The effects of ground heat exchanger parameters changes on geothermal heat pump performance—A review. *Appl. Therm. Eng.* **2018**, *129*, 1645–1658. [CrossRef]

18. Agrawal, K.K.; Yadav, T.; Misra, R.; Agrawal, G.D. Effect of soil moisture contents on thermal performance of earth-air-pipe heat exchanger for winter heating in arid climate: In situ measurement. *Geothermics* **2019**, *77*, 12–23. [CrossRef]

19. Agrawal, K.K.; Yadav, T.; Misra, R.; Agrawal, G.D. The state of art on the applications, technology integration, and latest research trends of earth-air-heat exchanger system. *Geothermics* **2019**, *82*, 34–50. [CrossRef]

20. Mahdavi, S.; Sarhaddi, F.; Hedayatizadeh, M. Energy/exergy based-evaluation of heating/cooling potential of PV/T and earth-air heat exchanger integration into a solar greenhouse. *Appl. Therm. Eng.* **2019**, *149*, 996–1007. [CrossRef]

21. Ali, S.; Muhammad, N.; Amin, A.; Sohaib, M.; Basit, A.; Ahmad, T. Parametric Optimization of Earth to Air Heat Exchanger Using Response Surface Method. *Sustainability* **2019**, *11*, 3186.

22. Li, H.; Ni, L.; Liu, G.; Zhao, Z.; Yao, Y. Feasibility study on applications of an Earth-air Heat Exchanger (EAHE) for preheating fresh air in severe cold regions. *Renew. Energy* **2019**, *133*, 1268–1284. [CrossRef]

23. Basse, N.T. Turbulence intensity and the friction factor for smooth-and rough-wall pipe flow. *Fluids* **2017**, *2*, 30. [CrossRef]

24. Perko, J.; Dugec, V.; Topic, D.; Sljivac, D.; Kovac, Z. Calculation and design of the heat pumps. In Proceedings of the 2011 3rd International Youth Conference on Energetics (IYCE), Leiria, Portugal, 7–9 July 2011; pp. 1–7.

energies

MDPI

Article

Visualization Study of Startup Modes and Operating States of a Flat Two-Phase Micro Thermosyphon

Liangyu Wu [1], Yingying Chen [1], Suchen Wu [2], Mengchen Zhang [2], Weibo Yang [1,*] and Fangping Tang [1]

[1]   School of Hydraulic, Energy and Power Engineering, Yangzhou University, Yangzhou 225127, China; lywu@yzu.edu.cn (L.W.); yychen@microflows.net (Y.C.); tangfp@yzu.edu.cn (F.T.)
[2]   Key Laboratory of Energy Thermal Conversion and Control of Ministry of Education, School of Energy and Environment, Southeast University, Nanjing 210096, China; scwu@microflows.net (S.W.); 220130421@seu.edu.cn (M.Z.)
*   Correspondence: wbyang@yzu.edu.cn; Tel.: +86-514-8797-1315

Received: 5 August 2018; Accepted: 29 August 2018; Published: 30 August 2018

Abstract: The flat two-phase thermosyphon has been recognized as a promising technique to realize uniform heat dissipation for high-heat-flux electronic devices. In this paper, a visualization experiment is designed and conducted to study the startup modes and operating states in a flat two-phase thermosyphon. The dynamic wall temperatures and gas–liquid interface evolution are observed and analyzed. From the results, the sudden startup and gradual startup modes and three quasi-steady operating states are identified. As the heat load increases, the continuous large-amplitude pulsation, alternate pulsation, and continuous small-amplitude pulsation states are experienced in sequence for the evaporator wall temperature. The alternate pulsation state can be divided into two types of alternate pulsation: lengthy single-large-amplitude-pulsation alternated with short multiple-small-amplitude-pulsation, and short single-large-amplitude-pulsation alternated with lengthy multiple-small-amplitude alternate pulsation state. During the continuous large-amplitude pulsation state, the bubbles were generated intermittently and the wall temperature fluctuated cyclically with a continuous large amplitude. In the alternate pulsation state, the duration of boiling became longer compared to the continuous large-amplitude pulsation state, and the wall temperature of the evaporator section exhibited small fluctuations. In addition, there was no large-amplitude wall temperature pulsation in the continuous small-amplitude pulsation state, and the boiling occurred continuously. The thermal performance of the alternate pulsation state in a flat two-phase thermosyphon is inferior to the continuous small-amplitude pulsation state but superior to the continuous large-amplitude pulsation state.

Keywords: thermosyphon; two-phase flow; startup; phase change; operating state; visualization

1. Introduction

The rapid development of microelectronic technology poses an important challenge for high-heat-flux electronic cooling [1]. A number of efficient cooling technologies, including microchannels [2–5], heat pipes [6], boiling [7], solid–liquid change [8], fractal surface [9,10] and liquid cooling [11,12], have been proposed and are used for electronic component cooling [13], spacecraft thermal control [14], microfluidic engineering [15] and battery thermal management [16]. Of these, the flat two-phase thermosyphon has been regarded as the preferred heat removal technique in a confined space [17,18]. Differing from conventional heat pipes, the evaporator and condenser section of the flat two-phase thermosyphon are replaced by two plates. Therefore, the flat two-phase thermosyphon can expand one-dimensional heat transfer into two-dimensional heat transfer on a plane, resulting in an efficient heat transfer and satisfactory temperature uniformity [19]. Utilized in a solar collector system, the two-phase thermosyphon combines good energy behavior with simplicity of manufacture [14–17]. In addition, the flat two-phase thermosyphon can be integrated with electronic devices [20]. Therefore, the flat two-phase thermosyphon has been introduced as an effective way to meet the challenges of heat dissipation and temperature uniformity for high-heat-flux electronic devices.

Unlike in unconfined spaces, the vapor-liquid phase change inside a flat two-phase thermosyphon involves a direct interaction between boiling and condensation accompanied by complex gas–liquid two-phase flow behaviors because of the narrow space [21]. For example, when the liquid level in the cavity is high, the liquid surface may contact the condenser surface because of the two-phase flow fluctuation, which forms a "liquid bridge" between the liquid surface and the condenser surface because of the surface tension. The formation of a liquid bridge may increase the thermal resistance of the condenser surface as the liquid film thickness increases [22]. In addition, Wu et al. [23] reported that, instead of gravity and buoyancy, the surface tension and the shear force at the gas–liquid interface are the dominant forces affecting the gas–liquid two-phase flow during boiling and condensation in the flat two-phase thermosyphon. Therefore, the coupled boiling–condensation process has a non-negligible effect on the gas–liquid two-phase flow behavior and heat transfer process [24] in flat two-phase thermosyphons.

Available experimental and theoretical studies of flat two-phase thermosyphons focused primarily on the steady-state thermal performance, such as temperature uniformity [25,26], equivalent thermal conductivity [27], thermal resistance [28], and maximum heat transfer capacity. In addition, a number of studies have been conducted to investigate the coupled boiling–condensation heat transfer in a confined space [17,21,29]. However, few studies have given attention to the thermal response and corresponding gas–liquid two-phase flow in the flat two-phase thermosyphon, specifically the boiling and condensation behaviors under the startup and quasi-steady processes. Visual representation of vapor–liquid two-phase flow is of significance to understand the coupled boiling–condensation phase change heat transfer inside flat two-phase thermosyphons. Therefore, a visualization experiment was conducted to investigate the gas–liquid phase change heat transfer. The startup modes and operating state in the flat two-phase thermosyphon are investigated and analyzed by the observed dynamic temperature variations of the evaporator and condenser surface as well as the gas–liquid two-phase interface evolution.

2. Description of Experiment

In order to visualize the vapor–liquid two-phase flow, a flat two-phase thermosyphon was manufactured with transparent sidewalls. The visualization experimental setup, as shown in Figure 1, includes the flat two-phase thermosyphon, a heating unit, a cooling unit, and a data acquisition unit. The gas–liquid two-phase behavior and the coupled evaporator–condenser heat transfer process are observed in the flat two-phase thermosyphon with this experimental setup.

The flat two-phase thermosyphon primarily comprises a quartz glass tube, an evaporator plate, a condenser plate, a sealing plate of heat sink, and a charging pipe, as shown in Figure 2. The various components of the flat two-phase thermosyphon are illustrated in Figure 3. The glass tube is tightly clamped between the evaporator and condenser plates, and a close cavity is formed where the working medium is filled. The outer diameter of glass tube is 50 mm and the inner diameter is 44 mm. For clear observation of the vapor–liquid two-phase flow in the flat two-phase thermosyphon, a height of 15 mm was used for the glass tube during the experiment. Annular grooves were milled on the evaporator and condenser plates and were filled with fluorine rubber O-rings for sealing the flat two-phase thermosyphon. The evaporator and condenser sections are square brass plates with 45-mm sides, as shown in Figure 3. The thickness of the evaporator plate is 5 mm, while that of the condenser plate is 10 mm. A cooling channel is set on the back of the condenser plate; therefore, a sealing plate is needed for the sealing of the cooling water. The working medium is fed into the cavity through the charging pipe. In this study, de-ionized water was used as the working medium, as shown in Table 1.

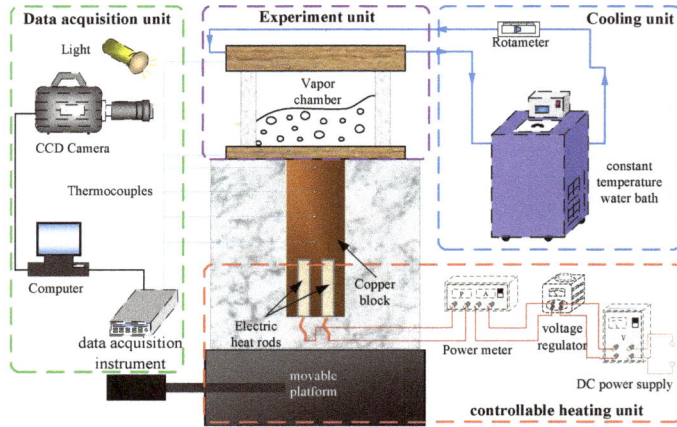

Figure 1. Schematic of experimental setup.

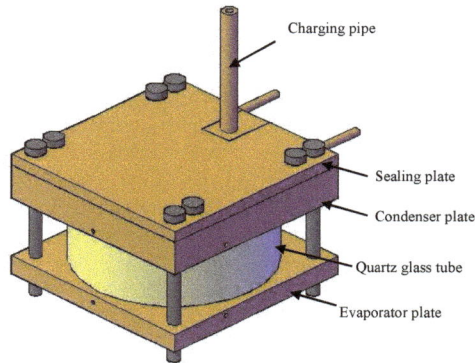

Figure 2. Schematic of flat two-phase thermosyphon.

Table 1. The thermophysical properties of de-ionized water (3.17 kPa, 25 °C).

Thermophysical Property	Value
Density (kg/m^3)	997
Enthalpy (kJ/kg)	104.67
latent heat (kJ/kg)	2435
thermal conductivity (W/K)	607
Specific isobar heat capacity (kJ/(kg·K)	4.182

The heating unit, which is used to provide and control a heat source for the flat two-phase thermosiphon, is supplied by a direct current power combined with a voltage regulator. Electric heating rods are embedded in the bottom of a copper block to heat the evaporator and a power meter is paralleled with the heating rods to measure the heating power, as shown in Figure 4. The electric heating rods are 6 mm in diameter and 50 mm in length. The 20-mm diameter copper block is clamped by a bracket comprising a number of polytetrafluoroethylene plates, so that the copper block can make close contact with the evaporator section of the flat two-phase thermosyphon. To obtain the accurate axial heat flux density of the copper block, four 0.5-mm diameter holes were drilled to a depth of 10 mm in the copper block. The holes were distributed along the axial direction of the copper block, and the distance from the holes to the top surface of the copper block are 5 mm, 20 mm, 35 mm, and 50 mm, as shown in Figure 5. During the experiment, heating power ranging from 20–90 W was applied to examine the effect of heat load on vapor–liquid two-phase flow.

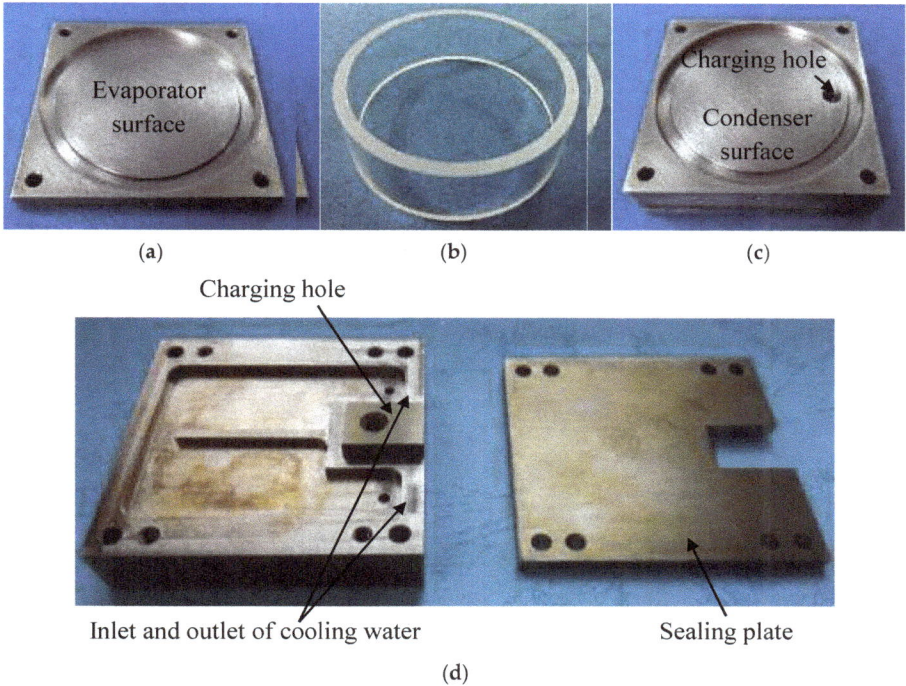

Figure 3. Structure of flat two-phase thermosyphon: (**a**) evaporator surface; (**b**) quartz glass tube; (**c**) condenser surface; and (**d**) cooling water tank at back of condenser section.

The cooling unit is connected to the condenser section of the flat two-phase thermosyphon, and comprises a constant temperature water bath and a glass rotameter. The constant temperature water bath provides a constant-temperature fluid circulation for the cooling of the condenser section, and the flow rate of the circulating water is measured by the glass rotameter. In the experiment, the water temperature was set to room temperature of 25 °C, and the flow rate of the circulating water was set to 80 mL/min. In order to measure the sensible heat gain of the circulating water, a number of thermocouples were installed at the inlet and outlet of the condenser section. The sensible heat gain is determined as the heat load of the flat two-phase thermosyphon. The cooling heat transfer rate is $Q_{cool} = \rho V C p \Delta T_{cool}$, where ρ and Cp are the density and specific heat of the cooling water, respectively, and ΔT_{cool} is the temperature difference between the cooling water at inlet and outlet.

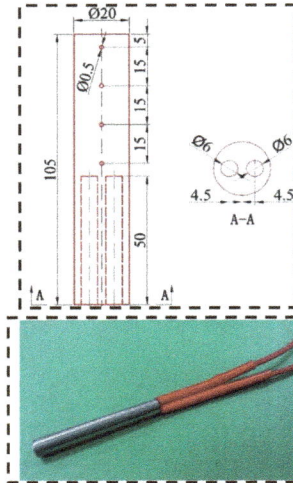

Figure 4. Structure of copper block and heating rods (unit: mm).

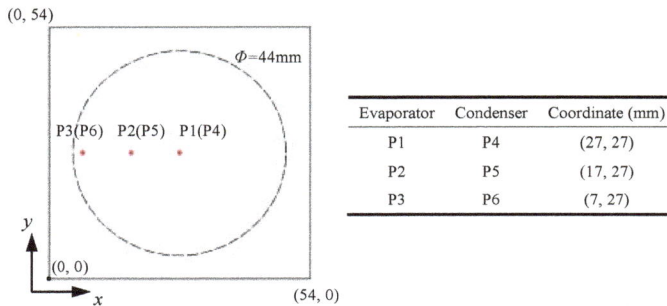

Evaporator	Condenser	Coordinate (mm)
P1	P4	(27, 27)
P2	P5	(17, 27)
P3	P6	(7, 27)

Figure 5. Position of measuring points on evaporator and condenser surface.

The data acquisition unit comprises a charge-coupled device (CCD) camera, a data collector (Agilent 34970A, Santa Clara, CA, USA), a computer, and a light. The vapor–liquid two-phase behavior in the confined cavity of the flat two-phase thermosyphon was monitored in real time and recording was started when the flat two-phase thermosyphon operated. The frame rate of CCD was set to 500 fps. To record the dynamic temperature variations of evaporator and condenser surfaces, three thermocouples (see Figure 5) were installed on the evaporator and condenser surfaces. The temperature was recorded by the data acquisition instrument. The sampling rate for the acquisition of the temperature is 2 Hz.

3. Results and Discussion

The vapor–liquid two-phase state inside the cavity is directly related to the wall temperature of the evaporator and condenser section, and, therefore, determines the phase change mechanisms and thermal performance of the flat two-phase thermosyphon [30]. An experiment was conducted to visually monitor the operating state of the working medium inside the cavity during the start-up and the quasi-steady processes. Based on the observed vapor–liquid two-phase state and the measured wall temperatures of the condenser and evaporator sections under different working conditions, it is possible to analyze the startup modes and operating states of a flat two-phase thermosyphon.

3.1. Startup Modes

When the evaporator section is heated, the flat two-phase thermosyphon initially goes through a start-up process and then attains a quasi-steady operating state. During the start-up process, the bubbles start to generate on the evaporator surface, and then gradually increase in size and rise from the surface. As a result, a complex two-phase vapor–liquid is formed close to the evaporator surface because of the dual effect of the natural convection and bubble disturbance. According to the dynamic temperature variations of the evaporator surface, two startup modes are identified for the flat two-phase thermosyphon: sudden startup mode and gradual startup mode.

The difference of two startup modes is mainly caused by the filling rate, i.e., the thermal response is largely dependent on the charging ratio. A larger degree of superheating is required to induce the startup of the flat two-phase thermosyphon when the charging ratio is higher. However, small superheating is required to induce the startup when the charging ratio is low. Therefore, different startup modes appear even though the heat input remains the same. The charging ratio φ is defined as the ration of the working fluid volume to the interior volume of thermosyphon.

3.1.1. Sudden Startup Mode

As shown in Figure 6a, in the sudden startup mode, the wall temperature of the evaporator section rises rapidly initially without temperature fluctuations, and the vapor–liquid two-phase working medium remains stationary in the early startup stage. Subsequently, the wall temperature continues to increase gradually, and no bubbles are generated inside the two-phase thermosyphon. During this process, a significant degree of superheating is required to induce the startup of the flat two-phase thermosyphon. The input heat flux is absorbed as sensible heat thorough the working medium and solid wall; therefore, the energy continues to be absorbed in the thermosyphon. As can be seen from the figure, at approximately 1000 s, the evaporator wall temperature suddenly decreases by approximately 10 °C (i.e., a temperature overshoot occurs in the startup process), and the condenser wall temperature suddenly rises by approximately 7 °C. Subsequently, there is a pronounced temperature pulsation for the condenser and evaporator walls.

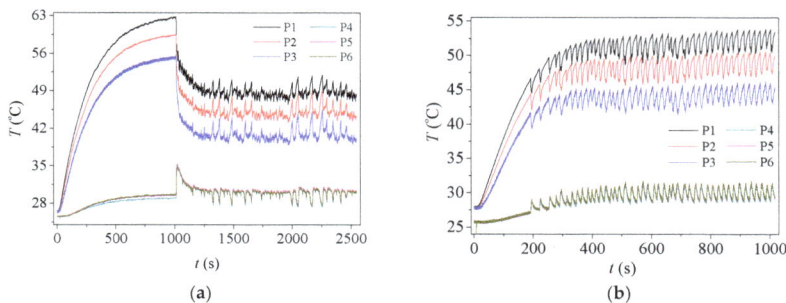

Figure 6. Dynamic wall temperature variations: (**a**) sudden startup mode (φ = 73%, q = 8.9 W/cm^2) and (**b**) gradual startup mode (φ = 47%, q = 8.5 W/cm^2).

This can be explained by the fact that, over time, the significant degree of superheating because of the continued accumulation of energy breaks the critical equilibrium state of the vapor–liquid two-phase working medium and bubbles begin to form on the evaporator surface and grow rapidly, and then leave the evaporator surface, causing a strong disturbance of the liquid surface [31]. When the vapor comes into contact with the condenser surface, it condenses to form condensate droplets and returns to the evaporation section because of the effect of gravity. The heat absorbed by the evaporator section is rapidly transmitted to the condensation section through the generation of bubbles and the condensation of vapor, and then the condenser and evaporator surface temperatures gradually attain an equilibrium state. Therefore, it can be concluded that the sudden startup is an overshoot startup process, in which the maximum surface temperature of evaporator section in the startup process is significantly greater than the value of steady-state wall temperature.

3.1.2. Gradual Startup Mode

In contrast to the sudden startup mode, there is no wall temperature overshoot in the gradual start-up mode, as shown in Figure 6b. The wall temperature of the evaporator section shows a significant pulsation after a short pre-heating period (approximately 200 s), and periodic variations of bubble formation, growth, and detachment from the evaporator surface can be observed inside the flat two-phase thermosyphon. Subsequently, increases in the departing frequency and the number of bubbles generated on the evaporator surface lead to a slow rise in the evaporator and condenser wall temperatures. Over time, the evaporator and condenser wall temperatures gradually become stable, and finally enter a steady state.

It can be seen from the figure that, during the entire startup process, the temperature change from the initial time to the steady state is gradual. This is because the bubbles increase in size shortly after the evaporator is heated in the gradual startup mode, i.e., a small degree of superheat is required for the formation of the critical bubble core. Therefore, the heat input into the evaporator section can be rapidly transmitted through the phase change process of the working fluid. As a result, a gradual increase of the evaporator section wall temperature (no overshoot) is observed in the gradual startup mode.

3.2. Operating States

The working fluid in the flat two-phase thermosyphon enters a quasi-steady state after the startup process. According to the wall temperature pulsation characteristics and visual images of the vapor–liquid two-phase fluid, three quasi-steady operating states are experienced in sequence with increasing heat load: continuous large-amplitude pulsation state (State A), alternate pulsation state (State B), and continuous small-amplitude pulsation state (State C).

3.2.1. Continuous Large-Amplitude Pulsation State (State A)

State A typically occurs when the evaporator section is supplied with a small heat load. As shown in Figure 7, during the first tens of seconds of a typical cycle, the liquid is superheated in the evaporator section and remains in the static status, while the vapor is cooled in the condenser section. When a critical degree of superheat is achieved, the liquid starts boiling and bubbles are generated. In the boiling process, the bubble generation only lasts for a few seconds, accompanied by a sudden large fluctuation in temperature; then the liquid returns to the static state and goes into the next cycle. The static status and boiling status can be clearly distinguished during this process. In this operating state, the wall temperature of the evaporator section fluctuates cyclically with a continuous large amplitude, where only a monotonic increase and decrease of the wall temperature are experienced in a typical cycle. Because of the small heat load, the energy absorption rate in the cavity is relatively slow, and bubbles are generated intermittently on the surface of the evaporator section. The following energy phases of the evaporation section are repeated cyclically until the end of the heat input: accumulation, release, re-accumulation, and re-release. In this process, the fluid inside a thermosyphon experiences boiling, stagnation, re-boiling, and re-stagnation.

In order to intuitively exhibit the features of a large-amplitude pulsation cycle, Figure 7b shows the wall temperature pulsation of the evaporator and condenser sections from 700 to 900 s, which includes four cycles of continuous large-amplitude pulsation. One pulsation cycle is from 700 to 769 s. Figure 7c shows the typical gas–liquid two-phase behavior for a pulsation cycle, for which the corresponding time points and wall temperatures are marked by point 1 to point 6 in

Figure 7b. During the cycle, the fluid remains in a static state from $t = 731$ s (point 1) to $t = 765$ s (point 2), and the wall temperature of the evaporator section gradually increases because of the energy accumulation. When the wall temperature increases to point 2, the degree of superheat reaches a critical value. As a result, the nucleation sites on the evaporation surface are activated and bubbles are generated continuously. Because of the bubble motion and vapor–liquid phase change, energy is rapidly transferred from the evaporator section to the condenser section, resulting in a significant decrease in the evaporator temperature and a rapid increase in the condenser temperature.

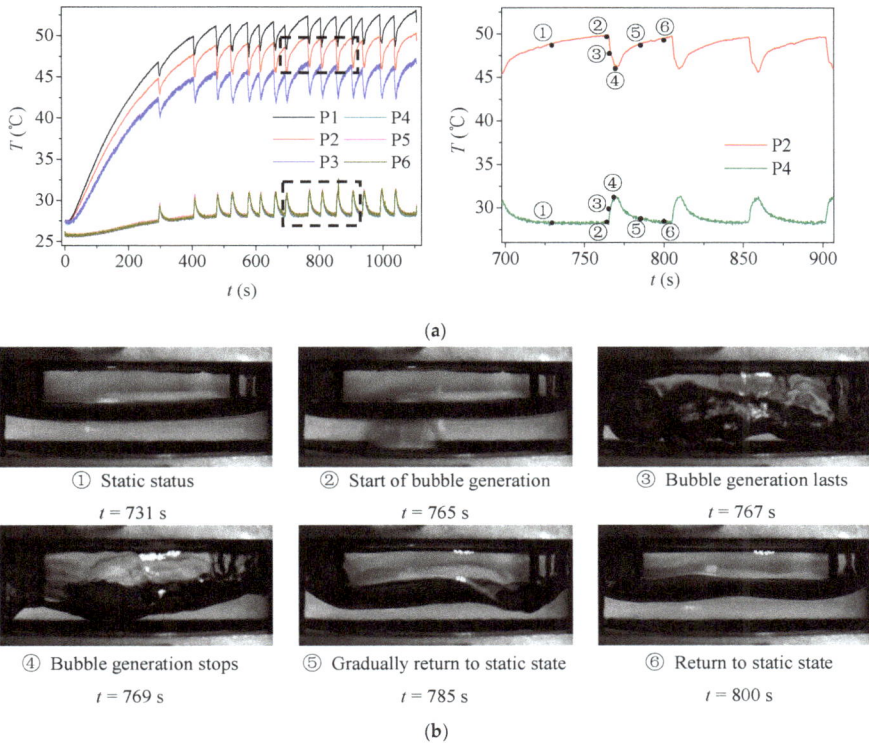

(a)

① Static status
$t = 731$ s

② Start of bubble generation
$t = 765$ s

③ Bubble generation lasts
$t = 767$ s

④ Bubble generation stops
$t = 769$ s

⑤ Gradually return to static state
$t = 785$ s

⑥ Return to static state
$t = 800$ s

(b)

Figure 7. Continuous large-amplitude pulsation state ($\varphi = 47\%$, $q = 6.4$ W/cm^2): (**a**) dynamic variation of wall temperature pulsation; (**b**) gas–liquid two-phase behavior in one cycle.

It should be noted that the continuous bubble generation (point 3) is of short duration and only lasts for approximately 4 s. After $t = 769$ s (point 4), the wall temperature of the evaporator section decreases; the degree of superheat is smaller than the critical value, causing bubble generation to cease, so the liquid gradually returns to the static state (point 5) and the energy absorption starts again in the next cycle (point 6). It can be seen that the heat accumulation in the evaporator section is of short duration in this operating state because of the small heat input, and the heat can be rapidly transferred to the condenser section through the motion of the vapor bubbles. Therefore, bubble generation is only observed for a short period.

3.2.2. Alternate Pulsation State (State B)

When compared with State A, the energy accumulation is accelerated with increasing heat input, so the duration of static status becomes shorter while the boiling status lasts longer in State B. As the bubble generation frequency increases, the pressure perturbation appears in the cavity, leading to the small amplitude fluctuation of wall temperature. In State B, the wall temperature of the evaporator section shows a fluctuation with small amplitude in addition to the large-amplitude fluctuations, and it exhibits a similar cyclical fluctuation. According to the pulsation amplitude and its duration for the wall temperature of evaporator section, State B can be divided into two alternate pulsation types. One is characterized by a long-duration single large-amplitude pulsation alternating with multiple short-duration small-amplitude pulsations (State B–1), and the other is characterized by a single short-duration large-amplitude pulsation alternating with multiple long-duration small-amplitude pulsations (State B–2).

Figure 8 shows the wall temperature pulsation and the corresponding gas–liquid two-phase flow behavior for State B–1 for the flat two-phase thermosyphon, in which three wall temperature cycles (from 590–690 s) are enlarged. As shown in the figure, the early stage of a cycle (from point 1 to point 2, 590–604 s) can be characterized by large-amplitude fluctuations, and the energy continues to be absorbed and the temperature slowly increases without fluctuations. Bubble generation then begins on the evaporator surface, and this process lasts for 28 s, significantly longer than for State A, as the heat input increases. It is noteworthy that after the bubble generation has commenced, the wall temperature decreases rapidly in the evaporator section and increases in the condenser section (from point 2 to point 3). Subsequently, the temperature exhibits multiple small-magnitude fluctuations (from point 3 to point 5). This can be attributed to the random generation and detachment of bubbles on the evaporator surface and the scouring effect of the gas–liquid two-phase fluid on the condenser surface. In this case, the wall temperatures of both the evaporator and condenser sections pulsate with small amplitudes close to the steady state. However, as the heat input is still not sufficient, the proportion of small-magnitude temperature pulsations to the entire pulsation cycle is small, i.e., the number of large-magnitude temperature pulsations dominates a cycle.

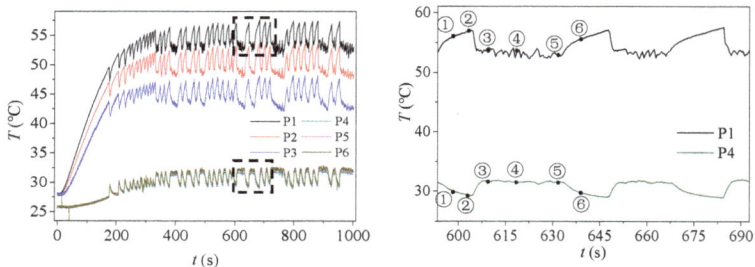

(a)

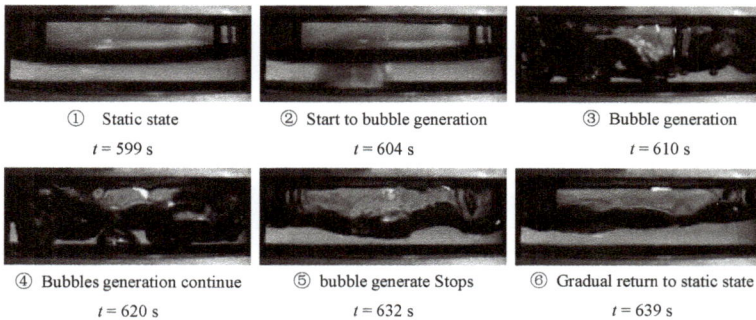

① Static state	② Start to bubble generation	③ Bubble generation
t = 599 s	*t* = 604 s	*t* = 610 s
④ Bubbles generation continue	⑤ bubble generate Stops	⑥ Gradual return to static state
t = 620 s	*t* = 632 s	*t* = 639 s

(b)

Figure 8. Alternate pulsation state (State B–1, φ = 47%, q = 10.9 W/cm^2): (**a**) temperature dynamic variation of evaporator wall; (**b**) gas–liquid two-phase behavior in one cycle.

With a further increase in the heat input, the operating state of State B–1 is transformed into State B–2. Figure 9 shows the wall temperature pulsations and the corresponding gas–liquid two-phase flow behavior in State B–2. Compared with State B–1, the duration of the small-magnitude pulsations is significantly greater in State B–2, and the duration of continuous bubble generation increases to approximately 100 s (from point 2 to point 6 in Figure 9a). The interval between bubbles decreased, and for the bulk of the time bubbles were generated continuously and rising from the evaporator surface, resulting in a significant fluid disturbance. According to the gas–liquid two-phase behavior images, it can also be seen that the disturbance of the working fluid amplifies with the heat input in the flat two-phase thermosyphon, which is favorable for the enhancement of heat transfer process. As a result, the heat input is efficiently transferred from the evaporator section to the condenser section through the intense gas–liquid phase change.

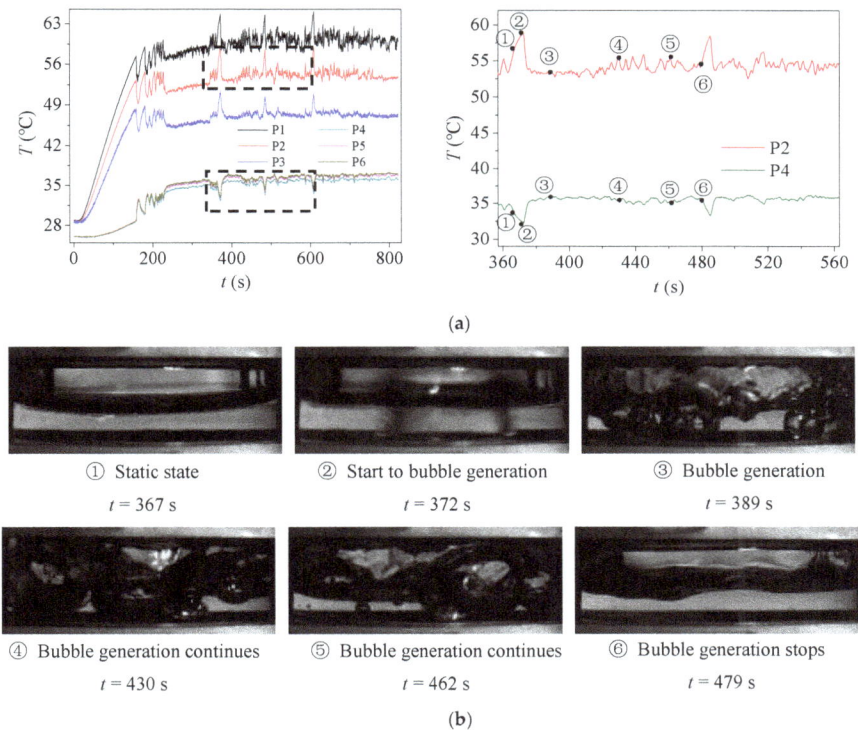

(a)

① Static state
$t = 367$ s

② Start to bubble generation
$t = 372$ s

③ Bubble generation
$t = 389$ s

④ Bubble generation continues
$t = 430$ s

⑤ Bubble generation continues
$t = 462$ s

⑥ Bubble generation stops
$t = 479$ s

(b)

Figure 9. Alternate pulsation state (State B-2, $\varphi = 47\%$, $q = 16.8$ W/cm^2): (**a**) characteristics of wall temperature pulsation; (**b**) gas–liquid two-phase behavior in one cycle.

3.2.3. Continuous Small-Amplitude Pulsation State (State C)

When the heat input is sufficiently great, the liquid is continuously boiling due to the intrinsically stochastic nature of liquid, resulting in continuous small-amplitude pulsation of wall temperature. In this case, the operating state enters a continuous small-amplitude pulsation state. Compared with the above operating states, there is no large-amplitude pulsation of wall temperature in State C as the boiling takes place continuously; there is no cyclical variation of wall temperature oscillation, and the wall temperature variation is random. Figure 10 shows the dynamic temperature variation of both the evaporator and condenser walls in State C. Once bubble generation begins in the cavity (point 2 in Figure 10), the bubbles are continuously generated on the evaporation wall. The wall temperature exhibits a small amplitude fluctuation, as opposed to a major change.

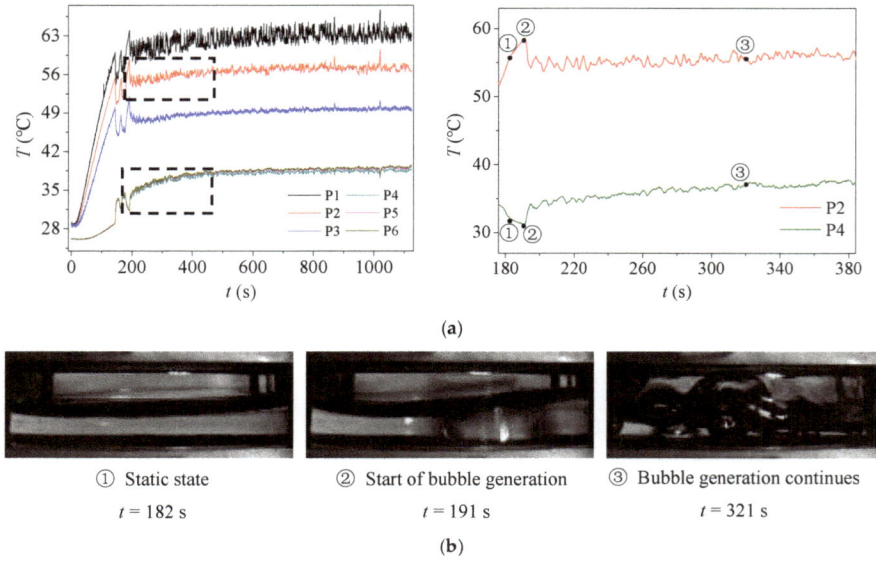

(a)

① Static state

$t = 182$ s

② Start of bubble generation

$t = 191$ s

③ Bubble generation continues

$t = 321$ s

(b)

Figure 10. Continuous small-amplitude pulsation operation ($\varphi = 47\%$, $q = 19.9$ W/cm^2): (**a**) characteristics of wall temperature pulsation; (**b**) gas–liquid two-phase behavior from the static state to continuous bubble generation.

During the continuous small-amplitude pulsation state, there is no obvious cycle. Once bubble generation begins in the cavity, the bubbles continuously generate on the evaporation wall. This phenomenon can be explained as follows: (1) the vapor–liquid phase-change heat transfer between the evaporator and the condenser becomes stronger in State C, where the stable boiling and condensation in the cavity ensure efficient heat transfer; (2) the phase change processes (including evaporation and boiling) on the evaporator surface are strengthened, and the nucleation sites on the evaporator surface are more easily activated and generate bubbles; (3) more bubbles are generated on the evaporator surface (see point 3 in Figure 10) than for the other operating states, and the liquid is more disturbed by the bubble motion. In summary, greater energy transfer intensity between the evaporator and condenser sections is achieved in the flat two-phase thermosyphon through efficient vapor–liquid phase changes.

3.3. Thermal Resistance

According to the above analysis, the heat load directly determines the operation state of vapor–liquid two phase flows inside a flat two-phase thermosyphon. This inevitably affects the thermal performance of evaporation and condensation phase change. In order to analyze the relationship between the heat transfer performance and heat load, the total thermal resistance of the flat two-phase thermosyphon R is introduced, defined by

$$R = \Delta\overline{T}/Q \tag{1}$$

where $\Delta\overline{T} = \overline{T}_e - \overline{T}_c$ is the difference between the average temperature of the evaporator surface, $\overline{T}_e = (T_1 + T_2 + T_3)/3$, and the average temperature of the condenser surface, $\overline{T}_c = (T_4 + T_5 + T_6)/3$; and Q is the actual heat load of the thermosyphon.

Figure 11 describes the effect of heat load on total thermal resistance of a flat two-phase thermosyphon. During the continuous large-amplitude pulsation state (State A), the heat load is small, and the bubbles are generated intermittently on the evaporator surface, so the heat transfer regime is the alternation of natural convection and nucleate boiling. In addition, the natural convection occupies most of time. This leads to a large thermal resistance of two-phase thermosyphon. When the operation state transforms from State A to the alternate pulsation state (State B), more heat is transferred through the bubble generation on the evaporator surface, i.e., the duration of nucleate boiling is longer. In addition, the motion of the bubbles causes the disturbance of the liquid near the evaporator surface, which also enhances heat transfer of the thermosyphon. The nucleate boiling occupies the most of time for State B. Since the good heat transfer performance of nucleate boiling, the total thermal resistance of State B is smaller than that of State A.

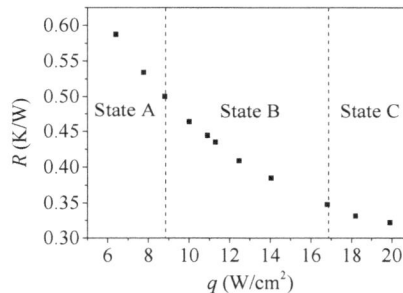

Figure 11. Effect of the heat load on total thermal resistance.

Unlike State A and B, during the continuous small-amplitude pulsation state (State C), the input heat is transferred wholly through nucleate boiling. Furthermore, the continuously generated bubbles cause a significant fluid disturbance, so the scouring effect of the gas–liquid two-phase fluid on the condenser surface make the condensate film thinner. This further reduces the thermal resistance as compared with State B. Therefore, as the thermal load increases, the thermal resistance of the flat two-phase thermosiphon decreases monotonously, thereby exhibiting better heat transfer performance. In other words, the thermal performance of alternate pulsation state in a flat two-phase thermosyphon is inferior to continuous small-amplitude pulsation state but superior to continuous large-amplitude pulsation state.

4. Conclusions

In this study, a flat two-phase thermosyphon with transparent sidewalls was manufactured for a visualization experiment. An experimental system was designed and conducted to investigate the phase-change heat transfer of the flat two-phase thermosyphon, with a particular focus on the startup mode and operating state. The dynamic temperature variations of the evaporator and condenser wall and the gas–liquid two-phase evolution in the flat two-phase thermosyphon were observed and analyzed. The primary conclusions were as follows:

1. Sudden startup and gradual startup were identified as the two types of startup modes in the flat two-phase thermosyphon, and the continuous large-amplitude pulsation state, alternate pulsation state, and continuous small-amplitude pulsation state were experienced in sequence with increasing heat load.

2. The continuous large-amplitude pulsation state occurred when a low heat load was applied to the evaporator section, in which the bubbles were generated intermittently.

3. The alternate pulsation state exhibited small fluctuations in addition to the large-amplitude fluctuations of the evaporator wall temperature because of the greater duration of boiling compared to the continuous large-amplitude pulsation state.

4. The continuous small-amplitude pulsation state occurred in the flat two-phase thermosyphon where the boiling occurred continuously and the wall temperature variation was random and exhibited no cyclical variations.

Author Contributions: W.Y. provided guidance and supervision. L.W. and Y.C. implemented the main research, discussed the results, and wrote the paper. S.W. and M.Z. collected the data. F.T. revised the manuscript. All authors read and approved the final manuscript.

Funding: This research was funded by the National Natural Science Foundation of China (grant number 51706194 and 51776037), the Joint Fund of Ministry of Education for Equipment Pre-research (grant number 6141A020225), and the Natural Science Foundation of Yangzhou City (grant number YZ2017103).

Conflicts of Interest: The authors declare no conflict of interest.

Nomenclature

Variables

C_p	specific heat (J/kg·K)
P	measuring points
q	heat flux (W/cm^2)
Q	heat input (W)
R	thermal resistance
t	time (s)
T	temperature (°C)
$\overline{T}$	average temperature (°C)
x	horizontal coordinate (mm)
y	vertical coordinate (mm)

Subscripts

c	condenser section
cool	cooling water
e	evaporator section
1~6	number of thermal couples

Greek symbols

ρ	density (kg/m^3)
φ	filling ratio
Φ	diameter (mm)

References

1. Murshed, S.M.S.; Nieto de Castro, C.A. A critical review of traditional and emerging techniques and fluids for electronics cooling. *Renew. Sustain. Energy Rev.* **2017**, *78*, 821–833. [CrossRef]
2. Chen, Y.P.; Zhang, C.B.; Shi, M.H.; Wu, J.F. Three-dimensional numerical simulation of heat and fluid flow in noncircular microchannel heat sinks. *Int. Commun. Heat Mass Transf.* **2009**, *36*, 917–920. [CrossRef]
3. Muszynski, T.; Andrzejczyk, R. Heat transfer characteristics of hybrid microjet–microchannel cooling module. *Appl. Therm. Eng.* **2016**, *93*, 1360–1366. [CrossRef]
4. Zhang, C.; Chen, Y.; Shi, M. Effects of roughness elements on laminar flow and heat transfer in microchannels. *Chem. Eng. Process.* **2010**, *49*, 1188–1192. [CrossRef]
5. Chen, Y.P.; Deng, Z.L. Hydrodynamics of a droplet passsing through a microfluidic t-junction. *J. Fluid Mech.* **2017**, *819*, 401–434. [CrossRef]
6. Chen, Y.P.; Zhang, C.B.; Shi, M.H.; Wu, J.F.; Peterson, G.P. Study on flow and heat transfer characteristics of heat pipe with axial "Ω"–shaped microgrooves. *Int. J. Heat Mass Transf.* **2009**, *52*, 636–643. [CrossRef]

7. Zhang, C.; Chen, Y.; Wu, R.; Shi, M. Flow boiling in constructal tree-shaped minichannel network. *Int. J. Heat Mass Transf.* **2011**, *54*, 202–209. [CrossRef]
8. Deng, Z.; Liu, X.; Zhang, C.; Huang, Y.; Chen, Y. Melting behaviors of pcm in porous metal foam characterized by fractal geometry. *Int. J. Heat Mass Transf.* **2017**, *113*, 1031–1042. [CrossRef]
9. Zhang, C.B.; Deng, Z.L.; Chen, Y.P. Temperature jump at rough gas-solid interface in couette flow with a rough surface described by cantor fractal. *Int. J. Heat Mass Transf.* **2014**, *70*, 322–329. [CrossRef]
10. Zhang, C.; Chen, Y.; Deng, Z.; Shi, M. Role of rough surface topography on gas slip flow in microchannels. *Phys. Rev. E* **2012**, *86*. [CrossRef] [PubMed]
11. Sharma, C.S.; Tiwari, M.K.; Zimmermann, S.; Brunschwiler, T.; Schlottig, G.; Michel, B.; Poulikakos, D. Energy efficient hotspot-targeted embedded liquid cooling of electronics. *Appl. Energy* **2015**, *138*, 414–422. [CrossRef]
12. Chen, Y.; Zhang, C.; Shi, M.; Yang, Y. Thermal and hydrodynamic characteristics of constructal tree-shaped minichannel heat sink. *AIChE J.* **2010**, *56*, 2018–2029. [CrossRef]
13. Weibel, J.A.; Garimella, S.V. Chapter four–recent advances in vapor chamber transport characterization for high-heat-flux applications. In *Advances in Heat Transfer*, 1st ed.; Sparrow, E.M., Cho, Y.I., Abraham, J.P., Gorman, J.M., Eds.; Elsevier: New York, NY, USA, 2013; Volume 45, pp. 209–301.
14. Zhang, C.; Shen, C.; Chen, Y. Experimental study on flow condensation of mixture in a hydrophobic microchannel. *Int. J. Heat Mass Transf.* **2017**, *104*, 1135–1144. [CrossRef]
15. Chen, Y.; Gao, W.; Zhang, C.; Zhao, Y. Three-dimensional splitting microfluidics. *Lab Chip* **2016**, *16*, 1332–1339. [CrossRef] [PubMed]
16. Chen, Y.P.; Wu, R.; Shi, M.H.; Wu, J.F.; Peterson, G.P. Visualization study of steam condensation in triangular microchannels. *Int. J. Heat Mass Transf.* **2009**, *52*, 5122–5129. [CrossRef]
17. Zhang, M.; Liu, Z.; Ma, G. The experimental investigation on thermal performance of a flat two-phase thermosyphon. *Int. J. Therm. Sci.* **2008**, *47*, 1195–1203. [CrossRef]
18. Liu, Z.; Zheng, F.; Liu, N.; Li, Y. Enhancing boiling and condensation co-existing heat transfer in a small and closed space by heat-conduction bridges. *Int. J. Heat Mass Transf.* **2017**, *114*, 891–902. [CrossRef]
19. Chen, X.; Ye, H.; Fan, X.; Ren, T.; Zhang, G. A review of small heat pipes for electronics. *Appl. Therm. Eng.* **2016**, *96*, 1–17. [CrossRef]
20. Jouhara, H.; Chauhan, A.; Nannou, T.; Almahmoud, S.; Delpech, B.; Wrobel, L.C. Heat pipe based systems–advances and applications. *Energy* **2017**, *128*, 729–754. [CrossRef]
21. Zhang, G.; Liu, Z.; Wang, C. An experimental study of boiling and condensation co-existing phase change heat transfer in small confined space. *Int. J. Heat Mass Transf.* **2013**, *64*, 1082–1090. [CrossRef]
22. Zhang, G.; Liu, Z.; Wang, C. A visualization study of the influences of liquid levels on boiling and condensation co-existing phase change heat transfer phenomenon in small confined spaces. *Int. J. Heat Mass Transf.* **2014**, *73*, 415–423. [CrossRef]
23. Wu, J.; Shi, M.; Chen, Y.; Li, X. Visualization study of steam condensation in wide rectangular silicon microchannels. *Int. J. Therm. Sci.* **2010**, *49*, 922–930. [CrossRef]
24. Lu, L.; Liao, H.; Liu, X.; Tang, Y. Numerical analysis on thermal hydraulic performance of a flat plate heat pipe with wick column. *Heat Mass Transf.* **2015**, *51*, 1051–1059. [CrossRef]
25. Blet, N.; Lips, S.; Sartre, V. Heats pipes for temperature homogenization: A literature review. *Appl. Therm. Eng.* **2017**, *118*, 490–509. [CrossRef]
26. Do, K.H.; Kim, S.J.; Garimella, S.V. A mathematical model for analyzing the thermal characteristics of a flat micro heat pipe with a grooved wick. *Int. J. Heat Mass Transf.* **2008**, *51*, 4637–4650. [CrossRef]
27. Kim, H.J.; Lee, S.-H.; Kim, S.B.; Jang, S.P. The effect of nanoparticle shape on the thermal resistance of a flat-plate heat pipe using acetone-based Al_2O_3 nanofluids. *Int. J. Heat Mass Transf.* **2016**, *92*, 572–577. [CrossRef]
28. Deng, Z.; Zheng, Y.; Liu, X.; Zhu, B.; Chen, Y. Experimental study on thermal performance of an anti-gravity pulsating heat pipe and its application on heat recovery utilization. *Appl. Therm. Eng.* **2017**, *125*, 1368–1378. [CrossRef]
29. Xia, G.D.; Wang, W.; Cheng, L.X.; Ma, D.D. Visualization study on the instabilities of phase-change heat transfer in a flat two-phase closed thermosyphon. *Appl. Therm. Eng.* **2017**, *116*, 392–405. [CrossRef]

30. Chen, Y.; Yu, F.; Zhang, C.; Liu, X. Experimental study on thermo-hydrodynamic behaviors in miniaturized two-phase thermosyphons. *Int. J. Heat Mass Transf.* **2016**, *100*, 550–558. [CrossRef]

31. Liu, X.; Chen, Y.; Shi, M. Dynamic performance analysis on start-up of closed-loop pulsating heat pipes (clphps). *Int. J. Therm. Sci.* **2013**, *65*, 224–233. [CrossRef]

Article

Analytical and Experimental Investigation of the Solar Chimney System

Zygmunt Lipnicki [1],*, Marta Gortych [1], Anna Staszczuk [1], Tadeusz Kuczyński [1] and Piotr Grabas [2]

[1]   Faculty of Civil and Environmental Engineering, University of Zielona Gora, Z. Szafrana St. 1,
      65-516 Zielona Góra, Poland; m.gortych@iis.uz.zgora.pl (M.G.); a.staszczuk@ib.uz.zgora.pl (A.S.);
      t.kuczynski@iis.uz.zgora.pl (T.K.)
[2]   Department of Research and Innovation in Economy, University of Zielona Gora, Licealna St. 9,
      65-417 Zielona Góra, Poland; piotrgrabas@poczta.fm
*   Correspondence: z.lipnicki@iis.uz.zgora.pl

Received: 29 March 2019; Accepted: 27 May 2019; Published: 29 May 2019

Abstract: In this, paper the authors propose a new simplified method of solving the problem of air flow through a solar chimney system using a classical system of equations for the principles of conservation (momentum, mass, and energy), as well as a general solution to research the problem using similarity theory. The method presented in this paper allows one to design a solar chimney. The theoretical analysis was compared with experimental studies on existing solar towers. The experimental and theoretical studies were satisfactorily consistent. For clarity, the phenomenon of heat flow in the solar chimney was described using dimensionless numbers, such as the Reynolds, Grashof, Galileo, Biot, and Prandtl numbers. In the equations for the dimensionless geometric parameters, the ratios of the collector radius to the thickness gap, height, and chimney radius were used. The method used to test the system of equations allows us to analyse various solar collectors easily. In the scientific literature, there is a lack of a simple calculation method to use in engineering practice, suitable for each type of solar chimney independent of dimensions and construction parameters.

Keywords: solar chimney; air flow; analytical and experimental solutions; method of calculation

---

1. Introduction

Solar chimneys offer considerable potential for energy supply in countries abundant in sunlight. Especially in the last few decades this technology has been studied widely [1–4]. Researchers have focused on developing the technology of thermal solar energy, with potential applications worldwide. Kasaeian et al. [5] presented a comprehensive review of solar chimney systems. Depending on the research directions, solar chimneys may serve various goals [6–9], such as natural ventilation in the buildings [2,5,10–13], small-scale power plants, generation in full-scale energy generation systems, and some applications in smart islands [14–18].

A power plant consisting of a solar collector and a chimney can work as a solar thermal power plant [6,7,18–25], which first converts solar energy into thermal energy in the solar collector to further convert it into kinetic energy in the chimney, with final electricity generation by applying a wind turbine and generator [14,16,26]. The construction of such a solar chimney may depend on the shape and dimensions of both the collector and chimneys and their canopy profiles [10,27].

Most researchers focus on solving the problem of heat flow in solar chimneys using numerical simulations instead of analytical solutions. Compared to the analytical method, fewer assumptions are used in numerical simulations. However, it is possible to obtain more detailed descriptions of the temperature and flow field. The number of studies on numerical methods adopting computational fluid dynamic programs (CFDs) to predict the efficiency of solar chimneys is constantly increasing. One of the first works using 2D numerical simulation to determine the temperature distribution and flow field in the collector was published by Pastohr et al. [28]. Similar numerical methods were used by Xu et al. [29,30]. Hamadan and Khashan [21] presented numerical results for a constant air flow inside the solar collector. The CFD analysis was used to determine the position of the turbine using the available power quantity. The CFD data showed that the height and size of the chimney are strongly related to each other, and the appropriate nozzle design at the entrance to the chimney is also important to reduce pressure losses. Koonsrisuk and Chitsomboon [31] developed a theoretical model that uses CFD analysis to conclude that the performance of a solar chimney with a sloping collector and divergent-top chimney is superior to a conventional chimney. In numerical 3D simulations, Guo et al. [32] validated experimental and numerical studies on the influence of solar radiation, turbine pressure drops, and ambient temperature on the solar collector's prototype in Manzanares, Spain. In the work of Abdelmohimen [33], the results of the collector tests from Manzanares were validated with the results of the numerical tests and proved to be consistent.

There are works discussing the problem of air flow through the solar collector system as an analytical issue. Pasumarthi and Sherif [34,35] developed an approximate model for studying the effect of various parameters on air temperature and velocity distribution in the system. Padki and Sherif [36] analyzed the solar collector system's efficiency, as influenced by the chimney, by developing an appropriate analytical model. This model allowed them to calculate the power output and efficiency of the solar chimney. Gannon and Von Backstrom [37] used a proposed mathematical model to apply a rotor for energy conversion in a solar chimney. They demonstrated that the use of blades to direct the air inlet resulted in improved system performance. Bernardes et al. [38] conducted a comprehensive analysis of analytical and numerical models describing the operation of a solar chimney power plant, which allowed them to estimate the output power of the solar chimneys. In addition, the influence of various environmental conditions and geometric dimensions on the power of solar chimneys was examined. The mathematical model has been used to predict the commercial performance of commercial solar chimneys on a large scale. It turned out that the height of the chimney, the pressure drop in the turbine, and the diameter of the chimney are of great importance [38]. Schlaich et al. [4] developed a mathematical model that is based on equations of momentum and energy conservation in solar chimney power plant. Zhou et al. [39] used a theoretical model to calculate the optimal chimney height for maximum output power. Guo et al. [40], in their model, considered the hourly changeability of solar radiation. It was shown that the output power increases almost linearly with the radius of the collector when the collector is small, assuming the use of an identical chimney. As the radius of the collector increases, this trend becomes slower. This process means that there is a limitation to the maximum radius of the collector above which there is no further increase in the output power. In the proposed analytical model of the solar chimney, Hamdan [41] obtained perfect compatibility with other published experimental and theoretical works. The obtained results indicate that the diameter of the collector and the diameter and height of the chimney and turbine are critical parameters for the construction of a solar plant. Kasaein et al. [5] reviewed experimental and theoretical studies of various solar collectors. They solved the system of equations for conservation of mass, momentum, and energy with appropriate boundary conditions for turbulent flows based on many experimentally determined parameters for air flowing through a chimney plant. While previous works show that the flow conditions inside the solar chimney can be well described theoretically, there is still a shortage of simple computational methods to be used in engineering practice, appropriate for each type of system, regardless of the dimensions and construction parameters. Therefore, this paper proposes a simplified method for solving the problem of air flow through the solar chimney. Again, this is based on the equations covering the principles of conservation (momentum, mass, and energy), and also the general solution of the problem using the theory of similarity. By using the original transformations of the equations describing the air flow in the solar chimney, combined with several additional assumptions, a system of equations was obtained, which allowed one to solve the problem analytically. This article is a continuation of the previous work [15]. Some of the theoretical considerations have been repeated and refined. The theoretical model has been strengthened by taking into account the local resistance of the air flow between the collector and the chimney. The work also included experimental research conducted on the collector–chimney model, and the results of theoretical and experimental research were compared. Experimental and theoretical studies have shown satisfactory consistency. For greater clarity, the phenomenon studied was described using dimensionless numbers, such as the numbers of Reynolds, Grashof, Galileo, Biot, and Prandtl. In the equations for dimensionless geometrical parameters of the solar chimney, the ratio of the radius of the collector to its height, as well as the height and radius of the chimney, were applied. Based on the original transformations and simplifying assumptions, a system of equations was obtained that solved the problem analytically. A universal procedure for solving a complex problem has been developed for use in engineering, for the design of a solar chimney. The proposed method allows for an easy analysis of various solar collectors.

2. Theoretical Model of the Solar Chimney System

Theoretical model of the solar chimney system is shown in Figure 1.

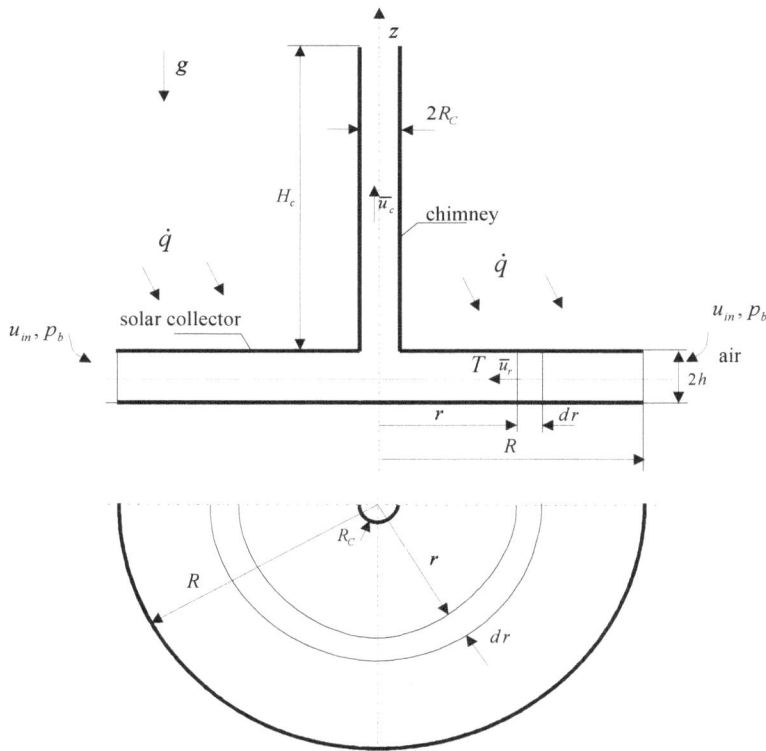

Figure 1. Solar chimney system.

The radius and height of the chimney are, respectively, represented by Rc and Hc. The distance between the top and bottom plates of the solar collector is constant and equal to $2h$, and the solar collector radius is represented by R. The absorbed solar radiation by the air flow in the collector is represented by $\dot{q}$. The velocity and pressure of the air flow at the entrance of the solar collector are, respectively, represented by u_{in} and p_b, where the velocity of the air in the chimney is constant and equal to u_c. The distributed temperature of the air inside the collector is represented by T.

2.1. The Analysis of the Air Flow in the Solar Collector

The incompressible flow of air (with a constant viscosity) inside the collector (see Figure 1), the stationary Navier-Stokes equation and principle of the conversation of mass flux crossing cylindrical surfaces have the forms [15]:

$$-\frac{1}{\rho}\frac{\partial p}{\partial r} = u_r\frac{\partial u_r}{\partial r} - v\left[\frac{1}{r}\frac{\partial}{\partial r}\left(r\frac{\partial u_r}{\partial r}\right) - \frac{u_r}{r^2} + \frac{\partial^2 u_r}{\partial z^2}\right] and\, u_r r = const, \tag{1}$$

where equation $u_r r = const$ results from the equation of conservation of mass.

After ignoring the small high-order parameters, the energy balance becomes

$$\dot{q}\cdot 2\pi\cdot r\cdot dr - 2\pi\cdot r\cdot dr\cdot\alpha(T - T_{ot}) = -2h\cdot\rho\cdot c_p\cdot 2\pi\cdot d(r\cdot\bar{u}_r\cdot T), \tag{2}$$

where $\dot{q}$, $\bar{u}_r$, T, T_{ot}, ρ, α, c_p, and h represent the heat flux density and average radial velocity in the gap of the collector for the laminar flow, density of air, temperature of air, ambient temperature, and convective heat transfer coefficient between the ambient air and plate of the collector, the specific heat of air, and the depth of the collector, respectively.

By applying the boundary conditions at the collector inlet, we get

$$p = p_b, u = u_{in} \text{ and } T = T_{ot} \text{ for } r = R, \tag{3}$$

The solutions to Equations (1) and (2) (in dimensionless form) in the collector can be expressed as the dimensionless pressure drop and dimensionless distribution of temperature, respectively [15]:

$$\Pi = \frac{6}{5}\left(\frac{1}{\tilde{r}^2}-1\right) + \frac{6}{n\mathrm{Re}}\ln\left(\frac{1}{\tilde{r}}\right); \theta = \tilde{q}\left\{1 - \exp\left[-\frac{Bi\tilde{\lambda}}{4n\mathrm{RePr}}\left(1-\tilde{r}^2\right)\right]\right\}, \tag{4}$$

where the following dimensionless variables were introduced

$$\tilde{r}=\frac{r}{R}; \Pi=\frac{2(p_b-p)}{\rho u_{in}^2}; \theta=\frac{T-T_{ot}}{T_{ot}}; \tilde{q}=\frac{\dot{q}}{\alpha T_{ot}};$$
$$Bi=\frac{\alpha h}{\lambda_s}; \tilde{\lambda}=\frac{\lambda_s}{\lambda}; \mathrm{Re}=\frac{u_{in}h}{v}; \mathrm{Pr}=\frac{v}{a}; n=\frac{h}{R}.$$

Equation (4) was obtained from the solution of Equations (1) and (2) for laminar flow between the collector plates.

$$\theta(\tilde{q}=3, Bi=0.005, \mathrm{Pr}=0.712, \tilde{\lambda}=2200, \text{ for n } = 0.04.)$$

As seen from the theoretical analysis (see Figure 2), the pressure drop and air temperature depend on the radial coordinates of the collector. As the Reynolds number increases, the air temperature in the collector increases more slowly. However, the influence of the Reynolds number on the pressure drop in the collector is negligible.

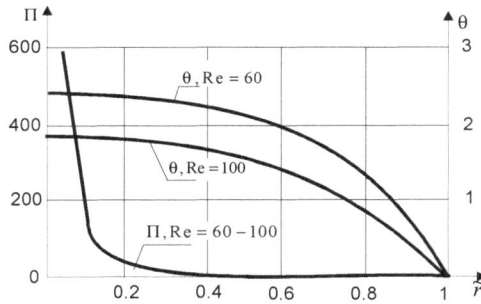

Figure 2. Distribution of dimensionless pressure drop, Π, and dimensionless temperature.

2.2. The Analysis of the Air Flow in the Solar Collector

For the chimney, the momentum equation of the flow of air inside the chimney (see Figure 1) in the cylindrical coordinate system (r, z), with a vertical axis (z) directed upwards, can be written in the form

$$0 = -\frac{1}{\rho_c}\frac{\partial p_c}{\partial z} - g + \nu\left[\frac{1}{r}\frac{\partial}{\partial r}\left(r\frac{\partial u_c}{\partial r}\right)\right]. \tag{5}$$

In Equation (5), we neglect the free convection in the solar collector. The density in the chimney describes the equation, and $\rho_c = \rho[1 - \beta(T_c - T_{ot})]$ represents the density of air in the chimney, which depends on the increase in temperature $(T_c - T_{ot})$. We assume linear dependence of density on temperature. The temperature T_c in the chimney is constant, the coefficient of the thermal expansion of air is represented by β, g represents the gravitational acceleration, and ν represents the kinematic viscosity of air. p_c and u_c represent the pressure and velocity inside the chimney, respectively.

We apply the boundary conditions

$$\text{for } z = 0, p_c = \hat{p}_{CO} \text{ and } z = H_C, p_c = p_H, \tag{6}$$

where the following correlations are used in the above equations:

$$\hat{p}_{CO} = p_{CO} - \Delta p, p_H = p_b - \rho g H_c, \tag{7}$$

In the theoretical model, the local pressure (Δp) in the flowing air between the solar collector and chimney is expressed as

$$\Delta p = \xi\frac{\rho_c \overline{u}_c^2}{2}, \tag{8}$$

where ξ represents the factor for the local loss in pressure in the air flow from the collector to the chimney, which results from the change of air flow direction and change of the channel cross-section between collector and chimney.

The velocity and average velocity in the chimney from the solution to the momentum equation (Equation (5)) are equal to

$$\widetilde{u}_c = \frac{1}{4\text{Re}\left(1 - \theta\frac{Gr}{Ga}\right)}\left[nGr\theta_c - \frac{1}{2}\widehat{\Pi}_{co}\frac{\text{Re}^2}{nm}\right](\hat{r}_c^2 - \hat{r}^2),$$

$$\widetilde{\overline{u}}_c = \frac{1}{8\text{Re}\left(1 - \theta_c\frac{Gr}{Ga}\right)}\left[nGr\theta_c - \frac{1}{2}\widehat{\Pi}_{co}\frac{\text{Re}^2}{nm}\right]\hat{r}_c^2. \tag{9}$$

The dimensionless pressure drop, $\widehat{\Pi}_{CO}$, at the beginning of the chimney is equal to

$$\widehat{\Pi}_{CO} = \Pi_{CO} + \xi \frac{16n^2}{\left(1 - \frac{Gr}{Ga}\theta_c\right)\tilde{r}_c^4},$$

where Π_{CO} represents the pressure drop calculated in the collector for $\tilde{r} = \tilde{r}_c$.

The velocity in the chimney, pressure drop at the beginning of the chimney, Reynolds number, Galileo number, Grashof number, and geometric parameter for the solar chimney can be presented in their dimensionless forms, respectively, as:

$$\tilde{u}_c = \frac{u_c}{u_{in}}; \Pi_{CO} = \frac{2(p_b - p_{CO})}{\rho u_{in}^2}; \text{Re} = \frac{u_{in}h}{\nu}; Ga = \frac{gR^3}{\nu^2}, Gr = \frac{g\beta T_{ot}R^3}{\nu^2}; m = \frac{H_c}{R}.$$

The condition for the air flow in the chimney is calculated with Equation (9):

$$Gr\theta_c > \frac{\widehat{\Pi}_{CO}\text{Re}^2}{2n^2m}. \tag{10}$$

From Equations (4) and (9), we achieve the coupled system below in Equation (11).

$$\frac{32n}{\tilde{r}_c^4} = \frac{nGr\theta_c - \frac{1}{2}\widehat{\Pi}_{CO}\frac{\text{Re}^2}{nm}}{\text{Re}}; \theta_c = \tilde{q}\left\{1 - \exp\left[-\frac{Bi\tilde{\lambda}}{4n\text{RePr}}\left(1 - \tilde{r}_c^2\right)\right]\right\};$$

$$\widehat{\Pi}_{CO} = \Pi_{CO} + \xi\frac{16n^2}{\left(1 - \theta_c\frac{Gr}{Ga}\right)\tilde{r}_c^4}; \Pi_{CO} = \frac{6}{5}\left(\frac{1}{\tilde{r}_c^2} - 1\right) + \frac{6}{n\text{Re}}\ln\frac{1}{\tilde{r}_c}. \tag{11}$$

Using the functions in the coupled system (Equation (11)), $\theta_c(\text{Re})$, and $\widehat{\Pi}_{CO}(\text{Re})$ enable us to calculate the Reynolds number (Re) from the following equation:

$$f(\text{Re}) = \frac{nGr\theta_c - \frac{1}{2}\widehat{\Pi}_{CO}\frac{\text{Re}^2}{nm}}{\text{Re}} - \frac{32n}{\tilde{r}_c^4} = 0. \tag{12}$$

The Reynolds number (Re) obtained from Equation (12) makes it possible to calculate the pressure drop $\widehat{\Pi}_{CO}$ at the beginning of the chimney and the dimensionless temperature parameter θ_c in the chimney.

$$To = 26\,^{\circ}\text{C}, h = 0.04\,\text{m}, R = 1\,\text{m}, Rc = 0.05\,\text{m}, \dot{q} = 800\,\text{W/m}^2, \alpha = 10\,\text{W/(m}^2\text{K}).$$

For example, in Figure 3, the calculation of the Reynolds number was performed for the sample solar chimney system. Subsequently, the dimensionless parameters, Π_{co} and θ_c, were calculated.

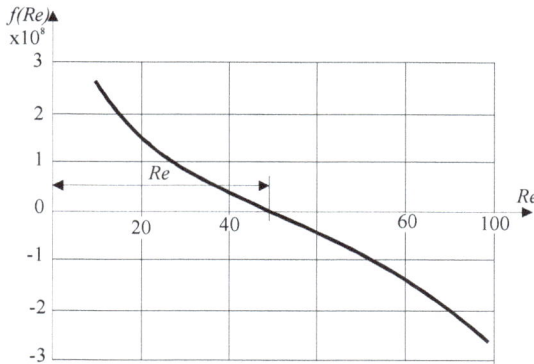

Figure 3. Example of the dependence of the function f(Re) on the Reynolds number (Re) for.

Table 1 and Figure 4 display the air velocity $\bar{u}_c$ and the air temperature T_C in the chimney.

Table 1. The air parameters in the chimney for $To = 26\,°C$, $h = 0.1$ m, $R = 10$ m, $Rc = 0.2$ m, $\dot{q} = 800$ W/m^2, $\alpha = 10$ W/(m^{2}K).

Hc, m	$p_b - p_{co}$, Pa	u_c, m/s	Tc, °C
5	2.39	3.75	106
10	4.65	5.23	105
15	6.59	6.23	104
20	8.85	7.22	103
30	12.40	8.55	100
40	17.65	10.20	97
50	20.15	10.91	96
100	37.64	14.92	88

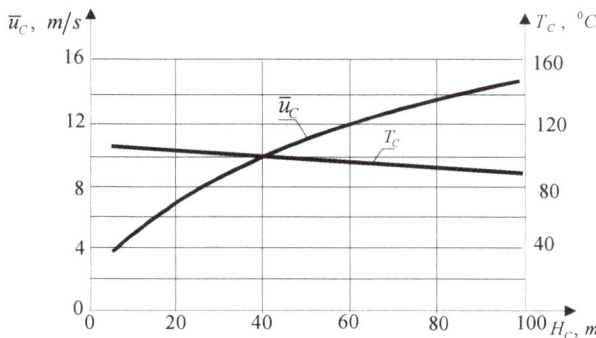

Figure 4. Dependence of air temperature (Tc) and velocity (uc) on the height of the chimney (H$_C$) for $To = 26\,°C$, $h = 0.1$ m, $R = 10$ m, $Rc = 0.2$ m, $\dot{q} = 800$ W/m^2, $\alpha = 10$ W/(m^{2}K).

The theoretical analysis on the turbulent flow of air is more challenging. In this case, the velocities of the air flows in the collector channel and chimney may be different. The theoretical model of air flow should be modified. Figure 4 presents the relationships of the height of the chimney with both the temperature (calculated according to Equation (4)) and velocity of the air inside the chimney (calculated according to Equation (9)). An increase in the chimney height causes an increase in air velocity and a decrease in air temperature. The increase in heat flux density increases both the temperature and air velocity in the chimney (see [15]).

Figure 5 shows the dependence of the energy efficiency of the solar chimney to the chimney height. Energy efficiency is the ratio of energy received to the energy supplied to the system. The energy obtained is the sum of the kinetic energy of the air and the increase in the enthalpy of the air. As shown in Figure 5, the energy efficiency of the solar chimney increases as the chimney height increases.

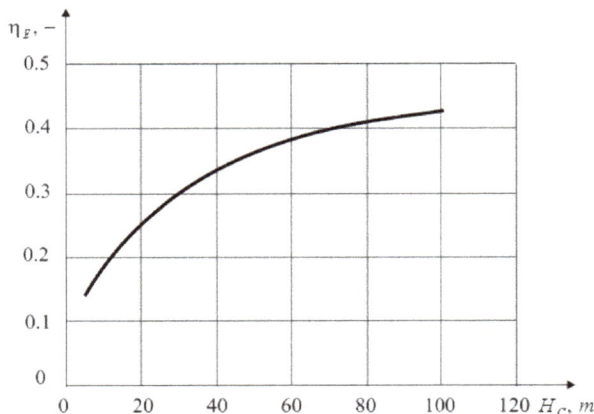

Figure 5. Energy efficiency η of the solar chimney as a function of chimney height, for $\dot{q} = 800$ W/m^2, $R = 10$ m, $Rc = 0.2$ m.

Designing the solar chimney based on the proposed theoretical flow characteristics allows one to easily determine the energy efficiency of the chimney. Particularly noteworthy is the possibility of optimizing the chimney height with the goal of maximizing its efficiency, while maintaining construction costs at an acceptable level. In this way, the profitability of the installation plant can be significantly improved.

3. Experimental Investigation

Figure 6 shows the experimental model of the solar chimney, which consists of two low cylindrical collectors and a vertical, thermally insulated pipe placed in the middle of the plate.

The solar collector consists of two horizontal, flat disks forming an air gap spaced from each other. The diameter of the upper metal plate is 2000 mm, and the width of the air gap is 80 mm. In the middle of the board there is a vertical, thermally insulated pipe with an inner diameter of 100 mm and a height of 2500 mm.

Collector heating combined with the effect of a thermal chimney forces the airflow in the chimney. The velocity was measured using a paddle anemometer and Pitot tube. The Testo 417 anemometer was used for the measurements, which allows quick and precise measurements of air velocity from the chimney. The accuracy of the measurements was ±0.1 m/s.

The results of the air velocity were measured with the paddle anemometer, which was mounted above the chimney. Then, the results of measurements were compared with the measurements using a Pitot tube, which was mounted on 2/3 of the chimney height, as shown in Figure 6. Moreover, the pressure drop at the collector's connection to the vertical pipe was measured using a differential inclined-tube manometer. The solar radiation flux was measured using a pyranometer located at a nearby meteorological station. The results of the measurements are presented in Table 2 and Figure 7.

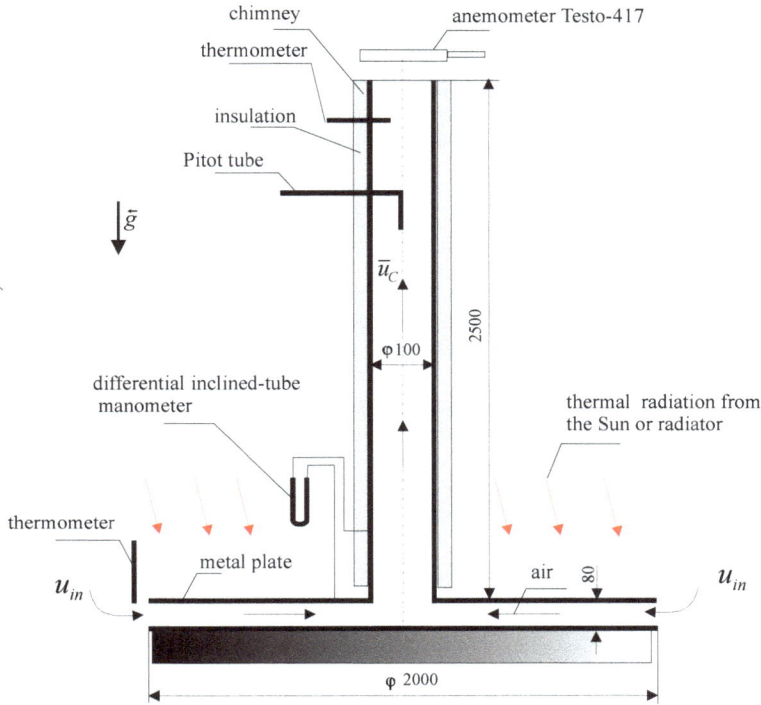

Figure 6. Experimental setup to validate the theoretical approach. Measured in mm.

Table 2. Results of experimental measurements.

No	1	22	3	4	5	6	7	8	9	10	11	12	13	14	15	16
$\dot{q}\,[W/m^2]$	145	150	400	550	600	610	680	710	715	730	790	800	800	840	845	920
$\dot{V}\,[m^3/h]$	19.8	21.2	19.8	25.2	17.2	17.0	22.3	21.2	20.9	20.9	28.3	28.3	31.1	28.3	25.4	19.8
$u_c\,[m/s]$	0.70	0.75	0.70	0.89	0.61	0.60	0.79	0.75	0.74	0.74	1.00	1.00	1.10	1.00	0.90	0.70

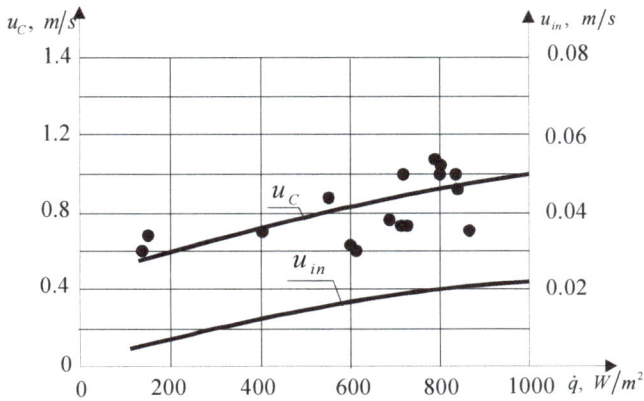

Figure 7. Comparison between the experimental and theoretical results.

Figure 7 presents a comparison of the results between theoretical studies (solid line) and experimental research (points) for the air velocity in the chimney, depending on the heat flux.

The results obtained in the analytical and experimental studies of the air velocity in the chimney are consistent and can be considered satisfactory. The scattering of results was mainly caused by disturbances caused by wind action between the collector plates on its perimeter. In this zone, the air flow velocities were very small (approximately 0.02 m/s), which was consistent with the theoretical model, and was, therefore, very sensitive to external gusts, which affected the operation of the solar chimney system.

4. Conclusions

In this work, the obtained results of the experimental investigation and the theoretical considerations are satisfactory. The presented method is particularly suitable for laminar flow conditions in a collector (for relatively small systems, not for electricity generation). Under turbulent flow conditions (for larger systems with higher internal air velocities, which can be used for small-scale electricity generation), the results are less accurate, although their accuracy still seems to be acceptable at first. The approximated results are sufficient for the purpose of a design evaluation of solar collector–chimney systems. The considerations observe both the first and second principles of thermodynamics.

The paper presents the dependence of the energy efficiency of the solar collector system to the height of the chimney. It was shown that the energy efficiency of the solar collector system increases as the chimney height increases.

The presented modelling allows one to mimic very small systems, which is a foundation for future work to look into larger systems, up to the size of power plants.

Author Contributions: Conceptualization, Z.L., M.G., A.S., T.K. and P.G.; Methodology, Z.L., M.G., A.S., T.K. and P.G.; Validation, Z.L., M.G., A.S., T.K. and P.G.; Formal Analysis, Z.L., M.G., A.S., T.K.; Investigation, Z.L., P.G.; Writing—Original Draft Preparation, Z.L., M.G., A.S., T.K.; Writing—Review & Editing, Z.L., M.G., A.S., T.K.; Visualization, Z.L., M.G.; Supervision, Z.L., T.K.

Funding: This research received no external funding.

Conflicts of Interest: The authors declare no conflict of interest.

Nomenclature

c_p	specific heat at a constant air pressure, J kg^{-1} K^{-1};
g	gravitational acceleration, m s^{-2};
H_c	chimney height, m;
$2h$	collector height, m;
m	ratio, $= Hc/R$;
n	ratio, $= h/R$;
p	pressure, Pa;
p_b	barometric pressure, Pa;
p_{co}	inlet pressure in the chimney, Pa;
$\dot{q}$	heat flux density, W m^{-2};
R	radius of the collector, m;
R_c	radius of the chimney, m;
T_{ot}	ambient temperature, °C;
T	temperature of air in the collector, °C;
T_c	temperature of air in the chimney, °C;
u_c	velocity in the chimney, m s^{-1};
u_{in}	velocity input into the collector, m s^{-1};
z, r	cylindrical coordinates, m;
α	convective heat transfer coefficient, W m^{-2}K^{-1};

λ	heat conductivity of air, W m^{-1}K^{-1};
λ_s	heat conductivity of the plate collector, W m^{-1}K^{-1};
ξ	local factor for pressure loss,
v	kinematic viscosity of air, m^{2}s^{-1};
ρ	density of air, kg m^{-3};
$\widetilde{u}_c$	dimensionless velocity in the chimney;
$\widetilde{r}_c$	dimensionless radius in the chimney;
$\widetilde{r}$	dimensionless radius in the collector;
$\widetilde{\dot{q}}$	dimensionless heat flux density;
Π	dimensionless pressure;
θ	dimensionless temperature;
Bi	Biot number;
Ga	Galileo number;
Gr	Grashof number;
Pr	Prandtl number;
Re	Reynolds number.

References

1. Haaf, W.; Friedrich, K.; Mayr, G.; Schlaich, J. Solar Chimneys—Part I: Principle and Construction of the Pilot Plant in Manzanares. *Int. J. Sol. Energy* **1983**, *2*, 3–20. [CrossRef]
2. Maia, C.B.; Ferreira, A.G.; Valle, R.M.; Cortez, M.F.B. Theoretical evaluation of the influence of geometric parameter and materials on the behavior of the airflow in a solar chimney. *Comput. Fluids* **2009**, *38*, 625–636. [CrossRef]
3. Schlaich, J. *The Solar Chimney. Electricity from the Sun*; Edition Axel Menges: Stuttgart, Germany, 1995.
4. Schlaich, J.; Bergermann, R.; Schiel, W.; Weinrebe, G. Sustainable electricity generation with solar updraft towers. *Struct. Eng. Int.* **2003**, *3*, 222–229. [CrossRef]
5. Kasaeian, A.B.; Molana, S.; Rahmani, K.; Wen, D. A review on solar chimney systems. *Renew. Sustain. Energy Rev.* **2017**, *67*, 954–987. [CrossRef]
6. Chergui, T.; Larbi, S.; Bouhdjar, A. Thermo-hydrodynamic aspect analysis of flows in solar chimney power plants-a case study. *Renew. Sustain. Energy Rev.* **2010**, *14*, 1410–1418. [CrossRef]
7. Maia, C.B.; Castro, J.O.; Cabezas-Gómezb, L.; Hanriota, S.M.; Ferreira, A.G. Energy and exergy analysis of the airflow inside a solar chimney. *Renew. Sustain. Energy Rev.* **2013**, *27*, 350–361. [CrossRef]
8. Naeeni, N.; Yaghoubi, M. Analysis of wind flow around a parabolic collector (1) fluid flow. *Renew. Energy* **2007**, *32*, 1898–1916. [CrossRef]
9. Naeeni, N.; Yaghoubi, M. Analysis of wind flow around a parabolic collector (2) heat transfer from receiver. *Renew. Energy* **2007**, *32*, 1259–1272. [CrossRef]
10. Bansal, N.K.; Mathur, J.; Mathur, S.; Jain, M. Modeling of window-sized solar chimneys for ventilation. *Build. Environ.* **2005**, *40*, 1302–1308. [CrossRef]
11. Khanal, R.; Lei, C. A numerical investigation of buoyancy induced turbulent air flow in an inclined passive wall solar chimney for natural ventilation. *Energy Build.* **2015**, *93*, 217–226. [CrossRef]
12. Khanal, R.; Lei, C. Ascalinginvestigation of the laminar convective flow in a solar chimney for natural ventilation. *Int. J. Heat Fluid Flow* **2014**, *45*, 98–108. [CrossRef]
13. Layeni, A.T.; Nwaokocha, C.N.; Giwa, S.O.; Olamide, O.O. Numerical and Analytical Modeling of Solar for Chimney Combined Ventilation and Power in Buildings. *Covenant J. Eng. Technol.* **2018**, *2*, 36–58.
14. Li, J.; Guo, P.; Wang, Y. Effects of collector radius and chimney height on power output of a solar chimney power plant with turbines. *Renew. Energy* **2012**, *47*, 21–28. [CrossRef]
15. Lipnicki, Z.; Gortych, M.; Staszczuk, A.; Kuczyński, T. The theoretical analysis of mass and energy flow through solar collector—Chimney system. *Civ. Environ. Eng. Rep.* **2017**, *24*, 117–131. [CrossRef]
16. Mekhail, T.; Rekaby, A.; Fathy, M.; Bassily, M.; Harte, R. Experimental and Theoretical Performance of Mini Solar Chimney Power Plant. *J. Clean Energy Technol.* **2017**, *5*, 294–298. [CrossRef]
17. Papageorgiou, C. Floating solar chimney with multi-pole generators. In Proceedings of the IASTED International Conference on Power and Energy System, Crete, Greece, 22–24 June 2011; pp. 60–65.

18. Cannistraro, G.; Cannistraro, M.; Trovato, G. Islands "Smart Energy" for Eco-Sustainable Energy a Case Study "Favignana Island". *Int. J. Heat Technol.* **2017**, *35*, S87–S95. [CrossRef]
19. Asnaghi, A.; Ladjevardi, S.M. Solar chimney power plant performance in Iran. *Renew. Sustain. Energy Rev.* **2012**, *16*, 3383–3390. [CrossRef]
20. Azizian, K.; Yaghoubi, M.; Niknia, I.; Kanan, P. Analysis of Shiraz Solar Thermal Power Plant Response Time. *J. Clean Energy Technol.* **2013**, *1*, 22–26. [CrossRef]
21. Hadman, M.O.; Khashan, S. Numerical Investigation of Solar Chimney Power Plant in UAE. In *ICREGA'14—Renewable Energy: Generation and Applications*; Springer Proceedings in Energy; Springer: Cham, Switzerland, 2014; pp. 513–524.
22. Hamdan, M.O. Analytical Thermal Analysis of Solar Chimney Power Plant. In Proceedings of the ASME 2010 4th International Conference on Energy, Phoenix, AZ, USA, 17–22 May 2010.
23. Sangi, R. Performance evaluation of solar chimney power plants in Iran. *Renew. Sustain. Energy Rev.* **2012**, *16*, 704–710. [CrossRef]
24. Tayebi, T.; Djezzar, M. Numerical Analysis of Flows in a Solar Chimney Power Plant with a Curved Junction. *Int. J. Energy Sci.* **2013**, *3*, 280–286.
25. Zhou, X.; Wang, F.; Ochieng, R.M. A review of solar chimney power technology. *Renew. Sustain. Energy Rev.* **2010**, *14*, 2315–2338. [CrossRef]
26. Hanna, M.B.; Mekhail, T.A.M.; Dahab, O.M.; Esmail, M.F.C.; Abdel-Rahman, A.R. Experimental and Numerical Investigation of the Solar Chimney Power Plant's Turbine. *Open J. Fluid Dyn.* **2016**, *6*, 332–342. [CrossRef]
27. Cottam, P.J.; Duffour, P.; Lindstrand, P.; Fromme, P. Effect of canopy profile on solar thermal chimney performance. *Sol. Energy* **2016**, *129*, 286–296. [CrossRef]
28. Pastohr, H.; Kornat, O.; Gurlebeck, K. Numerical and Analytical Calculations of the Temperature and Flow Field in the Upwind Power Plant. *Int. J. Energy Res.* **2004**, *28*, 495–510. [CrossRef]
29. Xu, G.L.; Ming, T.Z.; Pan, Y.; Meng, F.L.; Zhou, C. Numerical analysis on the performance of solar chimney power plant system. *Energy Convers. Manag.* **2011**, *52*, 876–883. [CrossRef]
30. Xu, H.; Karimi, F.; Yang, M. Numerical investigation of thermal characteristics in a solar chimney projects. *J. Sol. Energy Eng.* **2014**, *136*, 011008. [CrossRef]
31. Koonsrisuk, A.; Chitsomboon, T. Effects of flow are a changes on the potential of solar chimney power plants. *Energy* **2013**, *51*, 400–406. [CrossRef]
32. Guo, P.; Li, J.; Wang, Y. Numerical simulation of solar chimney power plant with radiation model. *Renew. Energy* **2014**, *62*, 24–30. [CrossRef]
33. Abdelmohimen, M.A.H.; Algarin, A.A. Numerical investigation of solar chimney power plants performance for Saudi Arabia weather conditions. *Sustain. Cities Soc.* **2018**, *38*, 1–8. [CrossRef]
34. Pasumarthi, N.; Sherif, S.A. Experimental and theoretical performance of a demonstration solar chimney model: Part I: Mathematical model development. *Int. J. Energy Res.* **1998**, *22*, 277–288. [CrossRef]
35. Pasumarthi, N.; Sherif, S.A. Experimental and theoretical performance of a demonstration solar chimney model: Part II: Experimental and theoretical results and economic analysis. *Int. J. Energy Res.* **1998**, *22*, 443–461. [CrossRef]
36. Padki, M.; Sherif, S. On a simple analytical model for solar chimneys. *Int. J. Energy Res.* **1999**, *23*, 345–349. [CrossRef]
37. Gannon, A.J.; Von Backström, T.W. Controlling and maximizing solar chimney power output. In Proceedings of the 1st International Conference on Heat Transfer, Fluid Mechanics and Thermodynamics, Kruger Park, South Africa, 8–10 April 2002.
38. Bernardes, M.A.; Weinrebe, A.V.G. Thermal and technical analyses of solar chimneys. *Sol. Energy* **2003**, *75*, 511–524. [CrossRef]
39. Zhou, X.; Yang, J.; Xiao, B.; Hou, G. Analysis of chimney height for solar chimney power plant. *Appl. Therm. Eng.* **2009**, *29*, 178–185. [CrossRef]

40. Guo, P.; Li, J.; Wang, Y. Annual performance analysis of the solar chimney power plant in Sinkiang, China. *Energy Convers. Manag.* **2014**, *87*, 392–399. [CrossRef]
41. Hamdan, M.O. Analysis of solar chimney power plant utilizing chimney discrete model. *Renew. Energy* **2013**, *56*, 50–54. [CrossRef]

energies

MDPI

Article

Comprehensive Electric Arc Furnace Electric Energy Consumption Modeling: A Pilot Study

Miha Kovačič [1,2], Klemen Stopar [1], Robert Vertnik [1,2] and Božidar Šarler [2,3,*]

[1] Štore Steel Ltd., Železarska cesta 3, SI-3220 Štore, Slovenia; miha.kovacic@store-steel.si (M.K.);
    klemen.stopar@store-steel.si (K.S.); robert.vertnik@store-steel.si (R.V.)
[2] Faculty of Mechanical Engineering, University in Ljubljana, Aškerčeva 6, SI-1000 Ljubljana, Slovenia
[3] Institute of Metals and Technology, Lepi pot 11, SI-1000 Ljubljana, Slovenia
* Correspondence: bozidar.sarler@fs.uni-lj.si; Tel.: +386-1-4771-403

Received: 28 February 2019; Accepted: 3 June 2019; Published: 4 June 2019

Abstract: The electric arc furnace operation at the Štore Steel company, one of the largest flat spring steel producers in Europe, consists of charging, melting, refining the chemical composition, adjusting the temperature, and tapping. Knowledge of the consumed energy within the individual electric arc operation steps is essential. The electric energy consumption during melting and refining was analyzed including the maintenance and technological delays. In modeling the electric energy consumption, 25 parameters were considered during melting (e.g., coke, dolomite, quantity), refining and tapping (e.g., injected oxygen, carbon, and limestone quantity) that were selected from 3248 consecutively produced batches in 2018. Two approaches were employed for the data analysis: linear regression and genetic programming model. The linear regression model was used in the first randomly generated generations of each of the 100 independent developed civilizations. More accurate models were subsequently obtained during the simulated evolution. The average relative deviation of the linear regression and the genetic programming model predictions from the experimental data were 3.60% and 3.31%, respectively. Both models were subsequently validated by using data from 278 batches produced in 2019, where the maintenance and the technological delays were below 20 minutes per batch. It was possible, based on the linear regression and the genetically developed model, to calculate that the average electric energy consumption could be reduced by up to 1.04% and 1.16%, respectively, in the case of maintenance and other technological delays.

Keywords: steelmaking; electric arc furnace; consumption; electric energy; melting; refining; tapping; modeling; linear regression; genetic programming

1. Introduction

The electric arc furnace (EAF) is a central element and the highest energy consumer in the recycled steel processing industry. The EAF contains electric energy, with a moderate addition of chemical energy, that is used for generating the required heat for the melting of recyclable scrap. The heat energy is primarily generated by the burning arc between the electrodes and the scrap, or its melt. The EAF consists of a shell (walls with water cooled panels and lower vessel), a heart (refractory material that covers lower vessel), and a roof with the electrodes. A scheme of the EAF is presented in Figure 1 [1–3].

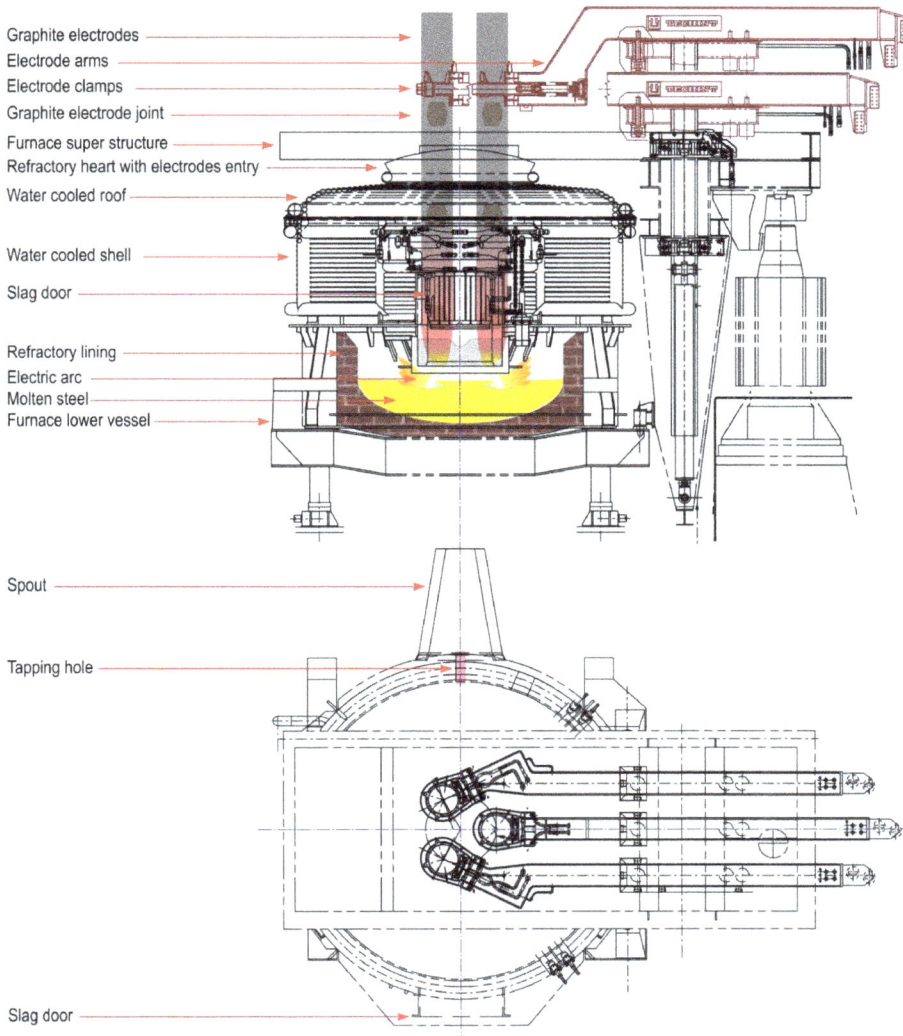

Figure 1. Scheme of the electric arc furnace.

The main EAF operation steps are as follows [1–3]: charging and melting, refining (oxidizing of the melt), chemical composition and temperature adjusting, and tapping (discharging of the furnace).

With respect to energy consumption, the contemporary research has mainly focused on the total (electric and chemical) consumed energy [2,4–6] and individual (electric or chemical) consumed energy [7,8] including other aspects of EAF operation such as transformer optimization [1,9–11], molten steel residue [12,13], scrap type [14,15], scrap management [14,16,17], electrode regulation [18–20], oxygen injectors [13,21–23], and slag cover [24].

The influences of maintenance on the power, steel, and cement industries were analyzed in [25]. The authors found that maintenance and rehabilitation were the key factors only when producing steel using the blast furnace. However, the influence of maintenance on producing steel from scrap through an EAF was not deduced due to insufficient data.

The concept of an adaptive hydraulic control system of the electrode positions was proposed in [26]. The underlying concept for adaptive control represents a simplified model of an EAF. The model also takes into account the influences of process disturbances such as scrap manipulation and its morphology. Several control algorithms are presented and critically assessed.

The dynamic control of an EAF is given in [27]. The electric arc model was divided into four parts by also considering the gas burners (natural gas, oxygen), slag, molten steel, and solid scrap. The developed model was used for predicting the chemical and electrical energy consumption while changing the scrap quantities during the gradual charging of the EAF. The research showed that a proper scrap charging strategy could reduce the energy consumption.

The decision support for the EAF operation was developed in [6] by using open source tools and took into account different EAF operator strategies. The designed decision support system could be integrated with complex EAF models.

The computationally reduced model of the EAF operation during only the refining stage was elaborated in [28]. The typical mass-energy influential parameters were employed including the equipment failures. The MATLAB software was used in the simulations. The authors stated that the model could be significantly improved with additional parameters (e.g., carbon concentration, temperature).

The energy consumption during the refining stage was modeled in [21] by using a comprehensive parameter analysis. The scrap melting evolution (i.e., quantities, timing) was also taken into account. The model was validated in practice on a 40 t EAF.

The paper in [29] focused on modeling the tapping temperature. The energy consumption could be optimized based on the consideration of the influential parameters. For modeling, an artificial neural network was used that combined the final fuzzy interference function. In addition, the operator strategies and experiences were taken into account.

A comprehensive approach toward the electric energy consumption of the EAF, used at the Štore Steel steelmaking company, is elaborated in this work. The entire set of influential parameters during all operation steps including maintenance and other technological delays in 2018 (3248 consecutively produced batches) were taken into account. To predict the electric energy consumption during the EAF operation, both linear regression and the genetic programming were used.

The rest of this paper is organized as follows. Typical processes related to the EAF used at the Štore Steel company including data collection are presented first. Afterward, the related process data from 3248 consecutive batches collected in 2018 were used to model the electric energy consumption with linear regression and genetic programming. The validation of the modeling results was conducted by using data from 278 batches (when the maintenance and other technological delays were below 20 minutes per batch), collected in 2019. The importance of the represented developments for the steel industry is given in the conclusions.

2. Materials and Methods

The Štore Steel company is one of the major flat spring steel producers in Europe. The company produces more than 1000 steel grades with different chemical compositions. The scrap is melted, ladle treated, and continuously cast in billets. The cooled-down billets are reheated and rolled in the continuous rolling plant. The rolled bars can be additionally straightened, examined, cut, sawn, chamfered, drilled, and peeled in the cold finishing plant. The Štore Steel company is known for its application of advanced artificial intelligence modeling tools [30] for better understanding and optimization of the processes.

The production process at the Štore Steel company starts with a 60 t EAF. The scrap is delivered in baskets by train from a scrapyard, located 300 m from the steel plant. The following types of scrap steel are used for melting: E1 (old thick steel scrap); E2 (old thin steel scrap); E3 (thick new production steel scrap); E8 (thin new production steel scrap); E40 (shredded steel scrap); scrapped non-alloyed steel; low-alloyed steel (moderate content of Cr); and pig iron.

The electric arc furnace is typically charged with three baskets. The first, second, and third baskets have the capacity of 22–30 t, 15–20 t, and 6–15 t, respectively. Each individual charging lasts approximately three minutes. The melting of the scrap after charging with the first, second, and third baskets lasts approximately 20 min, 15 min, and 10 min, respectively.

The following activities are conducted before charging with the first basket: examination, cleaning and reparation of the slag door and tapping spout with its refractory material; examination of the EAF refractory linings and reparation of the linings with the dolomite or magnesite; examination of the water-cooled panels; examination of the mast arm (which holds the electrodes); and the changing and settings of the electrodes.

For the slag formation, coke, lime, and dolomite are also used, which are deposited before melting the first basket. The slag insulation and protective ability expands the lifespan of the refractory material, preventing the EAF roof from exposure, and shielding the cooling panels from the intensive heat radiation.

Melting is conducted after swinging back the furnace roof. After lowering the electrodes, the burning arc between the graphite electrodes and the scrap or the molten steel is established. After the last basket has been melted, the EAF roof is swung off, and the remaining non-melted scrap is pushed into the melt bath.

In order to speed up the melting process, oxygen and natural gas from wall-mounted combined burners (natural gas) and injectors (oxygen, coke) are also used, in addition to the electric arc, to generate the complementary chemical heat. After melting the last basket during the refining process, the oxygen jets from the lances penetrate the slag and react with the liquid bath. In particular, the oxidation with the carbon, phosphorous, and sulfur is important. The oxidized products are trapped by the slag, which is removed through the slag doors by tilting the EAF backward. Afterward, the chemical composition analysis is conducted. After the chemical composition changes, the tapping (i.e., tilting the EAF forward) is conducted. The molten steel is charged into the ladle and consequently, the ladle treatment is conducted (e.g., slag formation, chemical composition adjustments, melt stirring). Typical delays during the refining process are connected with the chemical and temperature analysis, oxygen blowing, changing of the steel grade (especially Ca-treated steels for its improved machinability), and waiting for the lower electricity tariff.

In the present research, 26 process parameters including the electric energy consumption were considered. The data were taken from 3248 consecutively produced batches at the Štore Steel company during 2018. The dataset was composed of:

- Melting:

    ○ the considered process parameters were:

        ■ coke (kg): used for protective slag formation,
        ■ lime (kg): used for protective slag formation,
        ■ dolomite [kg]: used for protective slag formation,
        ■ E-type scrap (kg),
        ■ low-alloyed steel (moderate content of Cr) (kg),
        ■ packets of scrap (kg),
        ■ oxygen consumption (Nm^3) used for cutting the scrap and its combustion and forming the slag (important component of slag is FeO), and
        ■ natural gas consumption (Nm^3) used for heating the scrap.

    ○ The considered maintenance and other technological delays are:

        ■ lime addition (min): the additional time needed for lime addition,
        ■ scrap charging (min): the additional time needed for charging of the electric arc furnace with scrap,

- reparation of the linings with the dolomite or magnesite (min): the additional time needed for reparation of the refractory linings of the heart of the electric arc furnace,
- electrode settings (min): the additional time needed for electrode settings and replacing,
- other technological delays (min): the additional delays due to, for example, the maintenance of a dust collector, water cooling system, or overhead cranes,

- Refining and tapping:

  ○ the considered process parameters are:

    - oxygen consumption (Nm^3),which is used for uniform melt temperature distribution for removing the unwanted chemical elements such as sulfur or phosphorus,
    - limestone (kg), which is used for slag creation,
    - carbon content obtained by the first chemical composition analysis (%),
    - nominal final carbon content (%) where the melt can be used for producing several different grades of steel in further processing steps; the possibilities are determined from the first chemical composition analysis, and
    - carbon powder (kg), which is used for carbonizing and additional slag formation,

  ○ the considered maintenance and other technological delays:

    - chemical analysis delay (min): there can be problems with the sampling or the chemical analysis has to be repeated,
    - temperature and oxygen analysis delay (min): there can be problems with the sampling or the automatic lance used for the analysis,
    - extended refining (min): due to the chemical analysis and the temperature adjustments, the refining process needs to be extended in order to achieve a proper chemical composition and a proper temperature before tapping,
    - delay due to Ca-treated steel production (min): to produce Ca-treated steel, proper oxygen content is needed before tapping; in addition, the spout wear and geometry are important,
    - delay due to waiting for a lower electricity tariff (min): during the higher electricity tariff period (from 6:00 to 8:00 a.m.), the production in the steel plant stops,
    - delay due to steel grade changing (min): based on the first chemical analysis, the steel grade can be changed according to the foreseen planned production,
    - delay during tapping (min): delays can occur due to spout maintenance or spout blocking, ladle treatment and casting coordination and management, and, last but not least,

- Electric energy consumption (MWh).

The average values and the standard deviation of the individual parameters are presented in Table 1.

Table 1. The average values and the standard deviation of the individual parameters from 3248 consecutively produced batches at the Štore Steel company in 2018.

Parameter	Abbreviation	Average	Standard Deviation
Coke (kg)	COKE	814.27	89.35
Lime (kg)	CAO	998.16	90.20
Dolomite (kg)	CAOMGO	703.74	123.23
E-type scrap (kg)	E_SCRAP	42.54	5.32
Low-alloyed steel (moderate content of cr) (kg)	SCRAP_BLUE	6.19	5.17
Packets of scrap (kg)	SCRAP_PACK	7.03	3.99
Oxygen consumption (Nm^3)	OXYGEN_MELTING	1220.50	117.67
Natural gas consumption (Nm^3)	GAS	442.01	61.36
Lime addition (min)	CACO3_T	0.13	0.82
Scrap charging (min)	SCRAP_MANIPULATION_T	0.93	1.75
Reparation of the linings with the dolomite or magnesite (min)	REPARATION_MAINT	1.23	7.03
Electrode settings (min)	ELECTRODE_MANIPULATION_T	1.99	6.58
Other technological delays (min)	OTHER_T	5.48	42.44
Oxygen consumption (Mm^3)	OXYGEN_REFINING	459.00	115.81
Limestone (kg)	CACO3	72.75	185.92
Carbon content obtained by the first chemical composition analysis (%)	C_1	0.23	0.14
Required, final carbon content (%)	C_REQUIRED	0.41	0.16
Carbon powder (kg)	C	175.11	103.09
Chemical analysis delay (min)	CHEMICAL_ANALYSIS_T	4.02	3.48
Temperature and oxygen analysis delay (min)	OXYGEN_TIME_ANALYSIS_T	1.00	3.42
Extended refining (min)	REFINING_T	1.28	2.75
Delay due to Ca-treated steel production (min)	CA_TREATMENT_T	1.84	9.04
Delay due to waiting for lower electricity tariff (min)	PEAK_TARIFFE_T	5.20	27.76
Delay due to steel grade changing (min)	GRADE_CHANGING_T	2.87	9.13
Delay during tapping (min)	TAPPING_T	0.97	3.95

3. EAF Electric Energy Consumption Modeling

Based on the collected data (Table 1), the prediction of the EAF electric energy consumption was conducted by using linear regression and genetic programming. For the fitness function, the average relative deviation between the predicted and the experimental data was selected. The fitness function is defined as:

$$\Delta = \frac{\sum_{i=1}^{n} \frac{|Q_i - Q'_i|}{Q_i}}{n}, \tag{1}$$

where n is the size of the collected data and Q_i and Q'_i stand for the actual and the predicted electric energy consumption, respectively.

3.1. Linear Regression Modeling

The linear regression analysis results demonstrated that the model significantly predicted the electric energy consumption ($p < 0.05$, ANOVA) and that 63.60% of the total variances could be explained by independent variables variances (R-square). Out of the 25 independent parameters considered, only the following were not significantly influential ($p > 0.05$): lime, dolomite, scrap charging, chemical analysis delay, temperature and oxygen analysis delay, and delay during tapping.

The deduced linear regression model is:

$$\begin{aligned}
COKE \cdot & \ 0.002 + E\_SCRAP \cdot 0.152 + SCRAP\_BLUE \cdot 0.198 + SCRAP\_PACK \cdot 0.195 \\
& + OXYGEN\_MELTING \cdot 0.003 + GAS \cdot 0.005 + CACO3\_T \cdot 0.075 \\
& + SCRAP\_MANIPULATION\_T \cdot 0.003 + REPARATION\_MAINT \cdot 0.015 \\
& + ELECTRODE\_MANIPULATION\_T \cdot 0.015 + OTHER\_T \cdot 0.004 \\
& + OXYGEN\_REFINING \cdot (-0.003) + CACO3 \cdot 0.001 + C\_1 \cdot 0.73 + C\_REQUIRED \\
& \cdot (-0.45) + C \cdot 0.007 + OXYGEN\_TIME\_ANALYSIS\_T \cdot 0.007 + REFINING\_T \\
& \cdot 0.041 + CA\_TREATMENT\_T \cdot 0.013 + PEAK\_TARIFFE\_T \cdot 0.011 \\
& + GRADE\_CHANGING\_T \cdot 0.012 + TAPPING\_T \cdot 0.005 + 8.2872.
\end{aligned} \tag{2}$$

The average and maximal relative deviation from the experimental data was 3.60% and 36.75%, respectively. The calculated influences of the individual parameters (individual variables) on the electric energy consumption are presented in Figure 2. It is possible to conclude that E-type scrap, low-alloyed steel (moderate content of Cr), packets of scrap, oxygen consumption during melting, natural gas consumption, limestone, other technological delays, and coke injection during refining were the most influential factors. Based on the linear regression model, it was possible to calculate that the average electric energy consumption could be reduced by up to 1.04% in the case of the maintenance and other technological delays that we wanted to avoid. On the other hand, the time savings represented 24.89% of the average tapping time. As above-mentioned, during the higher electricity tariff period from 6:00 to 8:00 a.m., the production in the steel plant stopped.

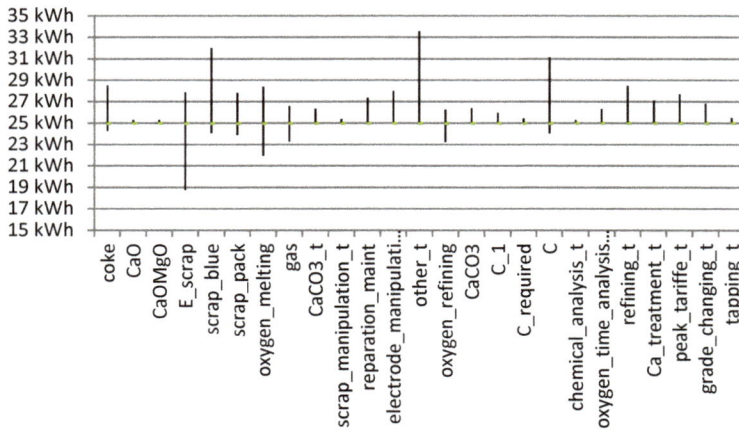

Figure 2. The calculated influences of the individual parameters on the electric energy consumption using the linear regression model.

3.2. Genetic Programing Modeling

Genetic programming is probably the most general evolutionary optimization method [31,32]. The organisms that undergo adaptation are in fact the mathematical expressions (models) for predicting the ratio between the material with the surface defects and the examined material. The models, i.e., the computer programs, consist of the selected function (i.e., basic arithmetical functions) and terminal genes (i.e., independent input parameters and random floating-point constants). Typical function genes are: addition (+), subtraction (−), multiplication (*), and division (/), and terminal genes (e.g., x, y, z). Random computer programs (Figure 3) for calculating various forms and lengths are generated by means of the selected genes at the beginning of the simulated evolution.

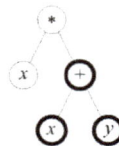

Figure 3. Random computer program as mathematical expression $x(x + y)$.

The varying of the computer programs is carried out by means of genetic operations (e.g., crossover, mutation) during several iterations, called generations. The crossover operation is presented in Figure 4. After the completion of the variation of the computer programs, a new generation is obtained. Each result, obtained from an individual program from a generation, is compared with the experimental data. The process of changing and evaluating the organisms is repeated until the termination criterion of the process is fulfilled.

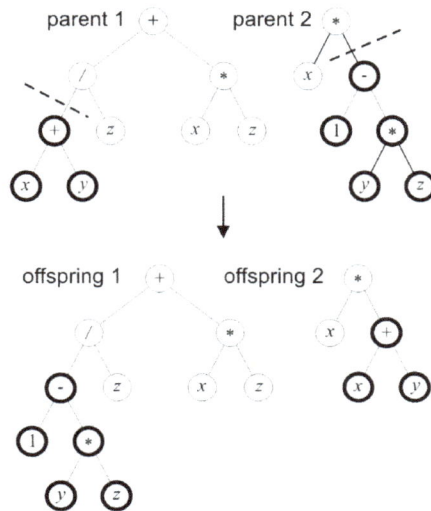

Figure 4. Crossover operation (out of two parental organisms, the offspring with randomly distributed genetic material are evolved).

An in-house genetic programming system, coded in the AutoLISP programming language, which is integrated into AutoCAD (i.e., commercial computer-aided design software), was used [33–35]. Its settings included the size of the population of organisms: 500; the maximum number of generations: 100; reproduction probability: 40%; crossover probability: 60%; maximum permissible depth in the creation of the population: 6; maximum permissible depth after the operation of crossover of two organisms: 10, and the smallest permissible depth of organisms in generating new organisms: 2.

The genetic operations of the reproduction and the crossover were used. To select the organisms, the tournament method with a tournament size 7 was used.

The in-house genetic programming system was run 100 times in order to develop 100 models for the prediction of electric energy consumption. Each run lasted approximately two and a half hours on an I7 Intel processor and 8 GB of RAM.

It must be emphasized that during the random generation of the computer programs (models for electric energy consumption), the already developed linear regression model (Equation (2)) was employed. The population size was 500. Out of these 500 organisms (computer programs), 50 were the same linear regression model, and the remaining 450 organisms were randomly generated at the beginning of the simulated evolution. Afterward, the population was changed with the genetic operations (e.g., crossover) without introducing any additionally developed linear regression models.

The best mathematical model obtained from 100 runs of genetic programming system was:

$$
\begin{aligned}
&8.39083 + 0.001 \cdot CACO3 + 0.00133 \cdot CAOMGO + 0.013176 \cdot \\
&CA\_TREATMENT\_T - 0.449208 \cdot C\_REQUIRED + 0.17427 \cdot E\_SCRAP + \\
&0.005 \cdot GAS + 0.0241847 \cdot GRADE\_CHANGING\_T + 0.003 \cdot \\
&OXYGEN\_MELTING - 0.003858 \cdot OXYGEN\_REFINING + 0.011 \cdot \\
&PEAK\_TARIFFE\_T + 0.056 \cdot REFINING\_T + 0.198 \cdot SCRAP\_BLUE + 0.195 \cdot \\
&SCRAP\_PACK + 0.00297 \cdot C_1^3 \cdot E\_SCRAP \cdot SCRAP\_PACK + C\_1(0.738316 + \\
&0.000198 \cdot CAOMGO + 0.000792 \cdot GAS + 0.007 \cdot OTHER\_T - 0.000594 \cdot \\
&OXYGEN\_REFININIG + 0.004 \cdot OTHER_T \cdot SCRAP\_PACK) + C(0.004954 + \\
&0.000792 \cdot C_1^2 \cdot C\_REQUIRED \cdot SCRAP\_PACK + C\_1(0.002376 + 0.000044 \cdot \\
&C\_REQUIRED \cdot SCRAP\_PACK)).
\end{aligned}
\tag{3}
$$

The average and the maximal relative deviation from the experimental data was 3.31% and 41.21%, respectively. The calculated influences of the individual parameters (individual variables) on the electric energy consumption are presented in Figure 5. It is possible to conclude that the dolomite, E-type scrap, low-alloyed steel (moderate content of Cr), other technological delays, and coke injection during refining were the most influential factors. Note that the coke, lime, limestone, scrap charging, reparation of the linings with the dolomite or magnesite, electrode settings, chemical analysis delay, oxygen and temperature analysis delay, and the delay during tapping were not considered in the model (Equation (3)). Additionally, based on the genetically developed model, it was possible to calculate that the average electric energy consumption could be reduced by up to 1.16% in the case of the maintenance and other technological delays.

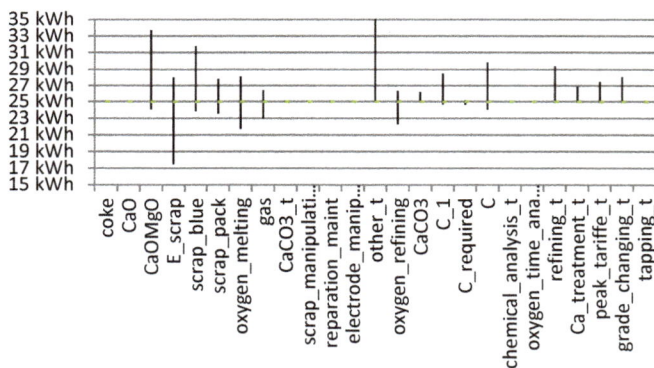

Figure 5. The calculated influences of the individual parameters on the electric energy consumption using a linear genetically developed model.

4. Validation of the Modeling Results

Additional data were gathered in January 2019 (278 batches) when the maintenance and technological delays were below 20 minutes. The average values and the standard deviation of the individual parameters are summarized in Table 2.

Table 2. The average values and the standard deviation of individual parameters from 3248 consecutively produced batches at the Štore Steel Ltd. in 2018.

Parameter	Average	Standard Deviation
Coke (kg)	800.58	68.56
Lime (kg)	989.06	105.04
Dolomite (kg)	696.31	59.42
E-type scrap (kg)	40.50	5.22
Low-alloyed steel (moderate content of cr) (kg)	7.41	4.95
Packets of scrap (kg)	7.67	2.96
Oxygen consumption (Nm^3)	1211.83	128.30
Natural gas consumption (Nm^3)	505.48	69.25
Lime addition (min)	0.13	1.01
Scrap charging (min)	0.21	0.42
Reparation of the linings with the dolomite or magnesite (min)	0.00	0.00
Electrode settings (min)	0.52	1.82
Other technological delays (min)	0.05	0.35
Oxygen consumption (Mm^3)	495.44	106.34
Limestone (kg)	96.40	210.69
Carbon content obtained by the first chemical composition analysis (%)	0.32	0.13
Required, final carbon content (%)	0.42	0.13
Coke (kg)	87.54	65.20
Chemical analysis delay (min)	4.22	2.62
Temperature and oxygen analysis delay (min)	0.51	1.44
Extended refining (min)	1.76	2.76
Delay due to Ca-treated steel production (min)	0.09	1.03
Delay due to waiting for lower electricity tariff (min)	0.01	0.08
Delay due to steel grade changing (min)	1.17	3.46
Delay during tapping (min)	1.25	1.65

The average relative deviation between the experimental data and the linear regression model was 3.65%, and that between the experimental data and the genetic programming model was 3.49%. This is in accordance with the average relative deviation from the data obtained in 2018. Consequently, we can conclude that the represented approach can be used as a precise EAF energy consumption tool that also considers the maintenance and technological delays.

5. Conclusions

The prediction of the electric energy consumption of the EAF operation at the Štore Steel company was presented. Twenty-five selected parameters from the individual production process steps in 2018 (3248 consecutively produced batches) were used for modeling. Two models were considered: the first was based on linear regression, and the second was based on the more accurate genetic programming. The average relative deviation of the models from the experimental data was 3.60% with the linear regression model, and 3.31% with the genetic programming model, respectively.

Based on the linear regression results, it was possible to conclude that 63.60% of the total variances could be explained by the variances of the independent variables. Based on the linear regression model, it was possible to calculate that the average electric energy consumption could be reduced by up to 1.04% in the case of maintenance and other technological delays, while on the other hand the time savings represented 24.89% of the average tapping time. Out of the 25 independent parameters, only lime, dolomite, scrap charging, chemical analysis delay, temperature and oxygen analysis delay, and delay during tapping were not significantly influential ($p > 0.05$).

An in-house genetic programming system, coded in AutoLISP, which is integrated into AutoCAD, was used to obtain 100 independent models for the prediction of the electric energy consumption during the EAF operation. A population size of 500 organisms was chosen. Out of these 500 organisms (computer programs), 50 were from the same developed linear regression model, and the remaining 450 organisms were randomly generated at the beginning of the simulated evolution. Afterward, the population was changed with the genetic operations (e.g., crossover) without introducing additionally developed linear regression models. Only the best ones were used for analysis. The most influential parameters (based on calculation) were dolomite, E-type scrap, low- alloyed steel (moderate content of Cr), other technological delays, and coke injection during refining. It must be emphasized that coke, lime, limestone, scrap charging, reparation of the linings with the dolomite or magnesite, electrode settings, chemical analysis delay, oxygen and temperature analysis delay, and delay during tapping were not considered in the genetically developed model.

Both models were also validated by using the data from 278 batches produced in 2019, when the maintenance and the technological delays were below 20 minutes per batch. The average relative deviation of the linear regression and genetic programming model prediction from the experimental data were 3.56% and 3.49%, respectively. This was in accordance with the average relative deviations from the data obtained in 2018.

The following points represent the highlights of our work:

- For modeling the EAF electric energy consumption, 25 parameters were used.
- Parameters involved melting (e.g., coke, dolomite, quantity), refining and tapping (e.g., injected oxygen, carbon, and limestone quantity), maintenance, and technological delays.
- The data from 3248 consecutively produced batches in 2018 were used.
- For modeling, linear regression and genetic programming were used.
- Both developed models were validated by using the data from 278 batches produced in 2019.
- Both models showed that the electric energy consumption could be reduced by up to 1.16% with the reduction of the maintenance and other technological delays.

In the future, a detailed analysis of charging and melting operation steps will be conducted including the time-dependent electric energy, natural gas, oxygen, and coke consumption. The represented approach is, with only slight modifications, practically applicable in a spectra of different EAFs as well as in other steelmaking process steps.

Author Contributions: M.K.: conceptualization, methodology, investigation, data analysis, software, writing, visualization; K.S.: conceptualization, investigation, data analysis, writing, editing, visualization; R.V.: software, data mining, data analysis, review and editing; B.Š.: project management, data analysis, review and editing.

Funding: This research was funded by the Slovenian Grant Agency, grant numbers P2-0162 and J2-7197 and the Štore-Steel Company (www.store-steel.si).

Conflicts of Interest: The authors declare no conflict of interest.

References

1. Stopar, K.; Kovačič, M.; Kitak, P.; Pihler, J. Electric arc modeling of the EAF using differential evolution algorithm. *Mater. Manuf. Process.* **2017**, *32*, 1189–1200. [CrossRef]
2. Toulouevski, Y.N.; Zinurov, I.Y. Modern Steelmaking in Electric Arc Furnaces: History and Development. In *Innovation in Electric Arc Furnaces*; Springer: Berlin/Heidelberg, Germany, 2013; pp. 1–24.
3. Toulouevski, Y.N.; Zinurov, I.Y. EAF in Global Steel Production. In *Energy and Productivity Problems*; Springer: Berlin, Germany, 2017; pp. 1–6.
4. Tunc, M.; Camdali, U.; Arasil, G. Energy Analysis of the Operation of an Electric-Arc Furnace at a Steel Company in Turkey. *Metallurgist* **2015**, *59*, 489–497. [CrossRef]
5. Damiani, L.; Revetria, R.; Giribone, P.; Schenone, M. Energy Requirements Estimation Models for Iron and Steel Industry Applied to Electric Steelworks. In *Transactions on Engineering Technologies*; Springer: Singapore, 2019; pp. 13–29.
6. Shyamal, S.; Swartz, C.L.E. Real-time energy management for electric arc furnace operation. *J. Process Control* **2019**, *74*, 50–62. [CrossRef]

7. Gajic, D.; Savic-Gajic, I.; Savic, I.; Georgieva, O.; Di Gennaro, S. Modelling of electrical energy consumption in an electric arc furnace using artificial neural networks. *Energy* **2016**, *108*, 132–139. [CrossRef]

8. Zhao, S.; Grossmann, I.E.; Tang, L. Integrated scheduling of rolling sector in steel production with consideration of energy consumption under time-of-use electricity prices. *Comput. Chem. Eng.* **2018**, *111*, 55–65. [CrossRef]

9. Klemen, S.; Kovačič, M.; Peter, K.; Jože, P. Electric-arc-furnace productivity optimization. *Mater. Tehnol.* **2014**, *48*, 3–7.

10. Marchi, B.; Zanoni, S.; Mazzoldi, L.; Reboldi, R. Product-service System for Sustainable EAF Transformers: Real Operation Conditions and Maintenance Impacts on the Life-cycle Cost. *Procedia CIRP* **2016**, *47*, 72–77. [CrossRef]

11. Marchi, B.; Zanoni, S.; Mazzoldi, L.; Reboldi, R. Energy Efficient EAF Transformer—A Holistic Life Cycle Cost Approach. *Procedia CIRP* **2016**, *48*, 319–324. [CrossRef]

12. Belkovskii, A.G.; Kats, Y.L. Effect of the Mass of the Liquid Residue on the Performance Characteristics of an Eaf. *Metallurgist* **2015**, *58*, 950–958. [CrossRef]

13. Wei, G.; Zhu, R.; Dong, K.; Ma, G.; Cheng, T. Research and Analysis on the Physical and Chemical Properties of Molten Bath with Bottom-Blowing in EAF Steelmaking Process. *Metall. Mater. Trans. B* **2016**, *47*, 3066–3079. [CrossRef]

14. Wieczorek, T.; Blachnik, M.; Mączka, K. Building a Model for Time Reduction of Steel Scrap Meltdown in the Electric Arc Furnace (EAF): General Strategy with a Comparison of Feature Selection Methods. In *Artificial Intelligence and Soft Computing—ICAISC 2008*; Springer: Berlin/Heidelberg, Germany, 2008; pp. 1149–1159.

15. Malfa, E.; Nyssen, P.; Filippini, E.; Dettmer, B.; Unamuno, I.; Gustafsson, A.; Sandberg, E.; Kleimt, B. Cost and Energy Effective Management of EAF with Flexible Charge Material Mix. *BHM Berg und Hüttenmännische Monatshefte* **2013**, *158*, 3–12. [CrossRef]

16. Sandberg, E.; Lennox, B.; Undvall, P. Scrap management by statistical evaluation of EAF process data. *Control Eng. Pract.* **2007**, *15*, 1063–1075. [CrossRef]

17. Lee, B.; Sohn, I. Review of Innovative Energy Savings Technology for the Electric Arc Furnace. *JOM* **2014**, *66*, 1581–1594. [CrossRef]

18. Hocine, L.; Yacine, D.; Kamel, B.; Samira, K.M. Improvement of electrical arc furnace operation with an appropriate model. *Energy* **2009**, *34*, 1207–1214. [CrossRef]

19. Feng, L.; Mao, Z.; Yuan, P.; Zhang, B. Multi-objective particle swarm optimization with preference information and its application in electric arc furnace steelmaking process. *Struct. Multidiscip. Optim.* **2015**, *52*, 1013–1022. [CrossRef]

20. Moghadasian, M.; Alenasser, E. Modelling and Artificial Intelligence-Based Control of Electrode System for an Electric Arc Furnace. *J. Electromagn. Anal. Appl.* **2011**, *3*, 47–55. [CrossRef]

21. Mapelli, C.; Baragiola, S. Evaluation of energy and exergy performances in EAF during melting and refining period. *Ironmak. Steelmak.* **2006**, *33*, 379–388. [CrossRef]

22. Kim, D.S.; Jung, H.J.; Kim, Y.H.; Yang, S.H.; You, B.D. Optimisation of oxygen injection in shaft EAF through fluid flow simulation and practical evaluation. *Ironmak. Steelmak.* **2014**, *41*, 321–328. [CrossRef]

23. Cantacuzene, S.; Grant, M.; Boussard, P.; Devaux, M.; Carreno, R.; Laurence, O.; Dworatzek, C. Advanced EAF oxygen usage at Saint-Saulve steelworks. *Ironmak. Steelmak.* **2005**, *32*, 203–207. [CrossRef]

24. Makarov, A.N. Change in Arc Efficiency During Melting in Steel-Melting Arc Furnaces. *Metallurgist* **2017**, *61*, 298–302. [CrossRef]

25. Oda, J.; Akimoto, K.; Tomoda, T.; Nagashima, M.; Wada, K.; Sano, F. International comparisons of energy efficiency in power, steel, and cement industries. *Energy Policy* **2012**, *44*, 118–129. [CrossRef]

26. Balan, R.; Hancu, O.; Lupu, E. Modeling and adaptive control of an electric arc furnace. *IFAC Proc. Vol.* **2007**, *40*, 163–168. [CrossRef]

27. MacRosty, R.D.M.; Swartz, C.L.E. Dynamic Modeling of an Industrial Electric Arc Furnace. *Ind. Eng. Chem. Res.* **2005**, *44*, 8067–8083. [CrossRef]

28. Coetzee, L.C.; Craig, I.K.; Rathaba, L.P. Mpc control of the refining stage of an electric arc furnace. *IFAC Proc. Vol.* **2005**, *38*, 151–156. [CrossRef]

29. Mesa Fernández, J.M.; Cabal, V.Á.; Montequin, V.R.; Balsera, J.V. Online estimation of electric arc furnace tap temperature by using fuzzy neural networks. *Eng. Appl. Artif. Intell.* **2008**, *21*, 1001–1012. [CrossRef]

30. Hanoglu, U.; Šarler, B. Multi-pass hot-rolling simulation using a meshless method. *Comput. Struct.* **2018**, *194*, 1–14. [CrossRef]
31. Koza, J.R. *The Genetic Programming Paradigm: Genetically Breeding Populations of Computer Programs to Solve Problems*; MIT Press: Cambridge, MA, USA, 1992; pp. 203–321.
32. Koza, J.R. *Genetic Programming II: Automatic Discovery of Reusable Programs*; MIT Press: Cambridge, MA, USA, 1994; ISBN 0-262-11189-6.
33. Kovačič, M.; Jager, R. Modeling of occurrence of surface defects of C45 steel with genetic programming. *Mater. Tehnol.* **2015**, *49*, 857–863. [CrossRef]
34. Kovačič, M.; Šarler, B. Genetic programming prediction of the natural gas consumption in a steel plant. *Energy* **2014**, *66*, 273–284. [CrossRef]
35. Kovacic, M.; Brezocnik, M. Reduction of Surface Defects and Optimization of Continuous Casting of 70MnVS4 Steel. *Int. J. Simul. Model.* **2018**, *17*, 667–676. [CrossRef]

energies

MDPI

Article

An Object-Oriented R744 Two-Phase Ejector Reduced-Order Model for Dynamic Simulations

Michal Haida *, Rafal Fingas, Wojciech Szwajnoch, Jacek Smolka, Michal Palacz, Jakub Bodys and Andrzej J. Nowak

Institute of Thermal Technology, Silesian University of Technology, Konarskiego 22, 44-100 Gliwice, Poland; rafal.fingas@gmail.com (R.F.); wojciech.szwajnoch@gmail.com (W.S.); jacek.smolka@polsl.pl (J.S.); michal.palacz@polsl.pl (M.P.); jakub.bodys@polsl.pl (J.B.); andrzej.j.nowak@polsl.pl (A.J.N.)
* Correspondence: michal.haida@polsl.pl; Tel.: +48-237-2810

Received: 1 March 2019; Accepted: 27 March 2019; Published: 3 April 2019

Abstract: The object-oriented two-phase ejector hybrid reduced-order model (ROM) was developed for dynamic simulation of the R744 refrigeration system. OpenModelica software was used to evaluate the system's performance. Moreover, the hybrid ROM results were compared to the results given by the non-dimensional and one-dimensional mathematical approaches of the R744 two-phase ejector. Accuracy of all three ejector models was defined through a validation procedure for the experimental results. Finally, the dynamic simulation of the hybrid ROM ejector model integrated with the R744 refrigeration system was presented based on the summer campaign at three different climate zones: Mediterranean, South American and South Asian. The hybrid ROM obtained the best prediction of ejector mass flow rates as compared with other ejector models under subcritical and transcritical operating conditions. The dynamic simulations of the R744 ejector-based system indicated the ejector efficiency variations and the best efficiency at the investigated climate zones. The coefficient of performance (COP) varied from 2.5 to 4.0 according to different ambient conditions. The pressure ratio of 1.15 allowed a more stabilised system during the test campaign with an ejector efficiency from 20% to over 30%.

Keywords: R744; two-phase ejector; refrigeration; dynamic simulation; low-order model; object-oriented modelling

1. Introduction

In all highly developed societies, energy conversion processes need to be carried out with the highest possible energy efficiency together with a reduction in energy consumption but also with great concern for the environment. This approach is nowadays known as so-called "good industrial practice", and it is utilised in all branches of industry. It is supported and, in many situations, even enforced by the European Commission Regulations. As a typical example, the issue of high energy efficiency of all refrigeration cycles is combined with the reduction of greenhouse gas emissions. Succeeding regulations regarding environmental protection, for instance, as reported in Kigali Amended [1], are very strict regarding the use of synthetic coolants, which are harmful for the environment. One of the solutions is to use natural refrigerants due to their negligible environmental impact and high market availability [2]. This has led many researchers to examine the refrigeration cycles with natural cooling media, e.g., carbon dioxide (denoted as R744), which is featured with a Global Warming Potential (GWP) equal to 1 and an Ozone Depletion Potential (ODP) equal to 0; it is also non-flammable and non-toxic refrigerant.

The first research on R744 as a working fluid in refrigeration cycle, which was conducted by Lorentzen [3], showed great potential for application, although the author pointed out a reduction in cooling capacity at elevated temperatures as a result of the low critical temperature of 31.1 °C and high critical pressure of 73.8 bar. According to Goodarzi et al. [4], the simple vapour compression R744 cycle results in a much smaller coefficient of performance (COP) in comparison to the synthetic equivalents working at similar temperature variations. This effect is connected to high thermodynamic losses during the expansion process due to the transcritical nature of the R744 vapour compression cycle. Therefore, a modification of the R744 system configuration to improve the system's energy performance, especially during transcritical operation, is needed [5]. Moreover, well-designed components installed in the R744 vapour compression cycle influence the COP improvement [6].

COP improvement was attained through implementation of the liquid separator and flash gas bypass valve to the R744 refrigeration system. The R744 booster system is commonly installed in cold and moderate climate zones, especially in the Scandinavian region [7]. Further energy performance improvement can be achieved through application of the parallel compressor, which sucks fluid directly from the liquid separator [8]. This system reduces the electric power consumption by size reduction of the medium temperature (MT) compressor during utilisation in the summer season in hot climates [6]. Sharma et al. [9] stated that the R744 parallel compression system obtains higher COP when compared to the conventional R410A direct expansion system in the northern and central parts of the United States of America. The COP improvement of the R744 parallel compression system was also observed for a heat pump application [10]. Gullo et al. [11] stated that the R744 "all-in-one" transcritical system with mechanical subcooling can reduce the system inefficiencies by up to 59%. However, the high thermodynamic losses during expansion process in the high-pressure expansion valve limit the possible COP improvement. Hence, one of the solutions is to use the ejector technology as the main expansion device and to recover potential expansion work [12].

Gullo and Cortella [13] concluded that the use of a system with a two-phase ejector as the main expansion device is the most efficient choice for upgrading the system's performance. Catalán-Gil et al. [14] confirmed COP improvement of the R744 transcritical system equipped with the gas ejector up to 29.5% when compared to the standard booster system. The main objective of the ejector application is to recover the pressure-related work of the supersonic motive flow and convert it to kinetic energy in order to exchange the momentum with the entrained low pressure suction stream. The consequence is that the pressure at the outlet is higher than the pressure of the suction stream, which results in higher pressure at the compressor inlet and thus less compression work is required as compared with the standard booster type cycle. Complex phenomena appear in the R744 two-phase ejector such as supersonic flow in the converging-diverging nozzle, momentum transfer in the mixing section, and pressure increment of the mixed flow in the diffuser, necessitating the evaluation of the ejector performance using a more advanced mathematical approach [15].

Kornhauser [16] was the first to present a 1D model for a synthetic medium system with the ejector to recover the expansion work. He showed a COP improvement of 21% over the conventional cycle with the expansion valve for different refrigerants. Li and Groll [17] performed a numerical investigation of the R744 ejector-based system and obtained a significant COP improvement compared to the standard direct expansion system or R744 booster system with the flash gas bypass (FGB) concept. The authors stated that the COP improvement was more than 16% when compared to the basic transcritical R744 cycle for air-conditioning operating conditions. A 0D model was also implemented for the dynamic simulation of the R744 ejector-based refrigeration system by Richter [18]. However, that model assumes the efficiency of the ejector given by Elbel and Hrnjak [12], resulting in low accuracy at the wide ranges found in supermarket applications.

Sumeru et al. [19] showed in their paper that the COP improvement of R744 transcritical ejector based system was up to 55% for thermodynamic analyses and up to 20% for experimental investigations. In an experimental study, Elbel and Hrnjak [12] obtained a cooling capacity and COP improvement of up to 8% and 7%, respectively. This discrepancy between theoretical and experimental COP improvements is caused by the idealisation of the refrigeration components, especially the two-phase ejector. Moreover, the ejector models used for such computations do not take the complex ejector geometry and local flow phenomena inside the device into consideration. This suggests a need for conducting more advanced numerical analyses, such as computational fluid dynamics (CFD) models, to allow for the evaluation of the flow behaviour inside the ejector.

The commonly used approach in the numerical investigation is the homogeneous equilibrium model (HEM), assuming mechanical and thermodynamic equilibrium between both phases inside the ejector [20]. This methodology is mostly derived from the concepts of Kornhauser [16]. Smolka et al. [21] used the homogeneous real fluid approach and applied an energy equation expressed in terms of the specific enthalpy instead of the standard temperature basis. The equation was employed in the Ansys Fluent solver utilising the user defined functions (UDF). In this case, the error margin of CFD results validated against real test data was kept below the level of 10% for most of the operating points above the critical point. Similar work was done by the previously mentioned study by Lucas et al. [20], but the authors used an OpenFOAM solver instead of commercial Ansys Fluent. Unfortunately, the HEM featured increasing inaccuracy with a decreasing motive nozzle temperature and a decreasing distance to the saturation line.

To extend a regime in which high quality results are obtained, another approach called the homogeneous relaxation model (HRM) was introduced. The idea was proposed by Bilicki et al. [22], followed later by comparison of the HEM, HRM results and experimental data provided by Downar-Zapolski et al. [23]. The aforementioned model evaluates the metastable effect during the expansion process by an additional vapour mass balance governing equation and the semi-empiric relaxation time (RT) definition [22]. Angielczyk et al. [24] investigated the HRM for the CO_2 supersonic two-phase flow through the ejector motive nozzle and presented a novel correlation for RT by entering additional information, such as temperature and quality profiles as well as the critical mass flow rates (MFR). Colarossi et al. [25] applied the HRM for R744 condensing two-phase ejector simulations for improvement in relation to the HEM in terms of accuracy and a wider operation regime of high quality results. Palacz et al. [26] validated the HRM CFD model of the R744 two-phase ejector with RT defined in [24]. The authors indicated that both nozzles had a mass flow rates accuracy improvement of 5% compared with the HEM. Haida et al. [27] modified the HRM by searching for RT coefficients based on the genetic algorithm optimisation procedure. The authors stated that the modified HRM extended the application range of the CFD model in the subcritical region to motive nozzle pressures above 59 bar.

However, the very time-consuming nature of CFD simulations for single operating points makes these models unfavourable for implementation into system analysis. Under these circumstances, many researchers use the 0D or 1D mathematical model of the two-phase ejector. The implementation of 0D/1D ejector models to system simulations allows system performance evaluation due to several assumptions of the simplified ejector model. Therefore, the dynamic simulations of the R744 ejector-based refrigeration system were performed either based on the assumptions of the two-phase ejector, such as fixed ejector efficiency [28] or accepting constant efficiencies of ejector parts for selected operating regimes [29]. The aforementioned assumptions may cause high discrepancy in theoretical predictions of the motive and suction mass flow rates when compared with the results of experimental tests and more advanced CFD approaches.

As already mentioned, the CFD numerical approach and experimental data enable more accurate evaluation of ejector performance than 0D or 1D models. However, those solutions are not suitable, particularly for dynamic calculations, because of their high computational cost. Hence, there is still a need to develop more versatile computational tools for steady-state and dynamic system analysis of ejector units. One solution is the implementation of a low-order fast and accurate model based on CFD results combined with the experimental data of the system analysis. One of these approaches is the reduced-order model (ROM), proposed by Haida et al. [30], which was developed based on the proper orthogonal decomposition (POD) approximation together with the radial basis interpolation functions (RBF) [31]. Haida et al. [30] showed that the POD-RBF results in mass flow rates with ±10% accuracy compared to the experimental data. It was developed on the basis of a 2D axisymmetric CFD model proposed by Smolka et al. [21], which implements an ejector in the R744 transcritical cycle to give a wide range of operating conditions and a cooling capacity. The single case computational time of ROM was shown to be below 0.05 s, which allows its utilisation in dynamic simulations. The low-order model of the R744 two-phase ejector was also enhanced by hybrid combination of the CFD results together with the experimental data by Haida et al. [32]. The hybrid ROM maintained high accuracy and fast simulations at a very wide operational envelope for the R744 Heat, Ventilation, Air-Conditioning and Refrigeration (HVAC&R)supermarket system. The discussed model was used to generate maps of ejectors performance installed in the multi-ejector module [33] for the R744 supermarket system. Therefore, the implementation of ROM in the system analyses of the R744 ejector-based refrigeration cycles, being a novel approach for non-commercial applications, can be very valuable in terms of the accuracy of computational results in comparison to commonly used methodology, without any negative effect on the computational time.

The main aim of this study was to integrate the two-phase ejector hybrid ROM with the R744 refrigeration system based on object-oriented modelling to produce a dynamic system simulation. First, the three mathematical models of the R744 two-phase ejector, i.e., 0D model, 1D model and the hybrid ROM model, are briefly discussed and confront. Then, the accuracy of all these three ejector models was evaluated by comparing selected numerical results with the experimental measurements. In the next section, these three models are applied into system analysis of the refrigeration cycle to determine its COP. Finally, dynamic simulation of the hybrid ROM ejector model integrated with the R744 refrigeration system was presented based on the summer campaign in three different climate zones. In all system analyses carried out in this work, the OpenModelica (OM) software was utilised.

2. The R744 Two-Phase Mathematical Approaches

In this section, all the mathematical approaches for ejector performance evaluations used in this paper are given. First, the 0D model proposed by Richter [18] is presented with the ejector efficiency definition given by Elbel and Hrnjak [12]. Then, the 1D model based on the Kornhauser approach [16] is described. It comprises a set of equations used to calculate the thermodynamic conditions of stream at the ejector outlet. The third method is the hybrid ROM which uses a POD-RBF approach built on the CFD results and experimental data to find the motive and suction nozzles mass flow rates for a given set of input parameters.

The R744 two-phase ejector geometric assembly is shown in Figure 1. It can be seen that the ejector consists of a converging-diverging motive nozzle, a converging suction nozzle, a pre-mixer with converging cross-section, a mixer with fixed cross-section and a diffuser. The designed fixed ejector consists of part of the the multi-ejector module that was experimentally tested by Banasiak et al. [33]. The multi-ejector module contains four R744 vapour fixed ejectors of different ejector capacity (changed in a binary order 1:2:4:8) to adopt expansion performance for different cooling demands and ambient conditions. Such solution ensures high energy efficiency of the module [34]. The dimensions of the investigated ejector are presented in Table 1.

Table 1. The main geometry parameters of the R744 two-phase ejector installed in the multi-ejector module [33].

Parameter Name	Unit	Dimension
Motive nozzle inlet diameter	mm	3.80
Motive nozzle throat diameter	mm	1.41
Motive nozzle outlet diameter	mm	1.58
Motive nozzle converging angle	°	30.00
Motive nozzle diverging angle	°	2.00
Diffuser outlet diameter	mm	8.40
Diffuser angle	°	5.00

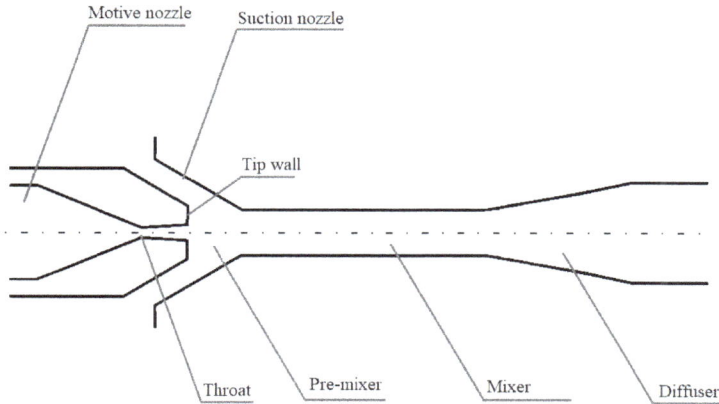

Figure 1. Geometry assembly of the R744 two-phase ejector.

Each of the two-phase ejector mathematical models was compared with the experimental data to define the accuracy range in a wide operating regime. The relative difference was used to define the discrepancy of the MFR and the mass entrainment ratio of the ejector models:

$$\delta_i = \frac{\dot{m}_{model} - \dot{m}_{exp}}{\dot{m}_{exp}} \cdot 100\%, \tag{1}$$

where the subscript i denotes the motive nozzle, the suction nozzle or the mass entrainment ratio.

2.1. 0D Model Using the Bernoulli Equation

The performance of the first theoretical ejector model is a 0D iterative approach that assumes the ejector is adiabatic of constant efficiency and that the mixing of the motive and suction flows is at constant pressure, lower than that of outlet stream. The ejector efficiency was defined by Elbel and Hrnjak [12] as the ratio of the recovered ejector expansion work rate to the maximum possible expansion work rate recovery potential:

$$\eta_{ej} = \frac{\dot{W}_{rec}}{\dot{W}_{rec,max}} = \chi \cdot \frac{h(p_{OUT}, s_{SN}) - h(p_{SN}, s_{SN})}{h(p_{OUT}, s_{MN}) - h(p_{MN}, s_{MN})}, \tag{2}$$

where η_{ej} is the ejector efficiency, $\dot{W}$ is the expansion work rate in W, h is the specific enthalpy in J·kg^{-1}, p is the pressure in Pa and s is the specific entropy in J·kg^{-1}·K^{-1}. The term χ defines the mass entrainment ratio, which is the ratio of the suction nozzle MFR to the motive nozzle MFR:

$$\chi = \frac{\dot{m}_{SN}}{\dot{m}_{MN}}, \tag{3}$$

where $\dot{m}$ is the mass flow rate of the motive nozzle (MN) and the suction nozzle (SN) in kg·s^{-1}.

The main stream flowing through the converging-diverging nozzle of the ejector accelerates due to the Venturi effect and throttles into the two phase region. According to Richter [18], the mass flow rate of the motive nozzle can be calculated from Bernoulli's equation for single phase flow:

$$\dot{m}_{MN} = A_{eff} \cdot \sqrt{2 \cdot \rho_{MN} \cdot (p_{MN} - p_{SN})}, \tag{4}$$

where ρ_{MN} is the fluid density at the motive nozzle in kg·m^{-3} and A_{eff} is the effective cross-section area at the motive nozzle throat in m^2. The effective area was defined as the throat area in the investigated ejector, as shown in Table 1. The suction nozzle MFR is calculated based on Equation (3). According to the mass balance governing equation, the outlet mass flow rate is defined as

$$\dot{m}_{OUT} = \dot{m}_{MN} + \dot{m}_{SN}. \tag{5}$$

Finally, the specific enthalpy of stream at the outlet of the ejector is calculated from the following energy balance:

$$\dot{m}_{MN} \cdot h_{MN} + \dot{m}_{SN} \cdot h_{SN} = (\dot{m}_{MN} + \dot{m}_{SN}) \cdot h_{OUT}. \tag{6}$$

The assumed constant ejector efficiency allows the R744 ejector-based system to be simulated with a defined accuracy level at a specified application and climate zone. Hafner et al. [28] assumed the efficiency of all investigated ejectors to be up to 20% based on the preliminary experimental measurements carried out on the prototype ejector. Hence, a similar efficiency of 20% was used for the investigation presented in this paper.

2.2. 1D Homogeneous Equilibrium Model

The simplified 1D model of the two-phase based on the Kornhauser approach is formulated based on the following assumptions [16]:

- A negligible pressure drop in the gas cooler, evaporator and all connections;
- There is no heat loss to the environment from the system, except via heat rejection in the gas cooler;
- The liquid and vapour streams outflowing from the separator are saturated;
- The flows across expansion devices are isenthalpic;
- The compressor has a given isentropic efficiency;
- The evaporator has a given superheat degree, and the gas cooler has a given outlet temperature;
- The flow in the ejector is considered to be a 1D homogeneous equilibrium flow;
- The motive and suction streams enter the constant area mixing zone with the same pressure, and there is no mixing between them before the inlet of the constant area mixing;
- The expansion efficiencies of the motive and suction streams as well as the efficiency of the ejector diffuser are given constants.

Having specified the aforementioned assumptions, the set of equations describing the ejector performance proposed by Li and Groll [17] is set up. Denoting the pressure at the mixing zone as p_{MN}, the equations describing the ejector section before the inlet to the constant area mixing section (CAMS) are given. The motive stream pressure drops from the motive nozzle inlet to the mixing section inlet as it accelerates in the converging nozzle before entering the mixing section. Since this process is assumed to be isentropic expansion, it is described by

$$s_{MNm,is} = s_{MN}, \tag{7}$$

where $s_{MNm,is}$ is the specific entropy in J·kg^{-1}·K^{-1} at the inlet to the CAMS. The specific enthalpy of the motive stream before entering the mixing zone is determined from the relationship of the properties:

$$h_{MNm,is} = f(s_{MNm,is}, p_{bm}), \tag{8}$$

where the subscript bm determines the beginning of the mixing section. The actual specific enthalpy of the motive stream at the inlet of CAMS is calculated using the definition of expansion efficiency:

$$\eta_m = \frac{h_{MN} - h_{MNm}}{h_{MN} - h_{MNm,is}}. \tag{9}$$

The velocity of the motive stream at the inlet of the CAMS is determined by assuming the conservation of energy across the expansion process:

$$u_{MNm} = \sqrt{2 \cdot (h_{MN} - h_{MNm})}. \tag{10}$$

The specific volume of the motive stream at the inlet of CAMS can be determined as a function of the specific enthalpy and pressure:

$$v_{MNm} = f(h_{MNm}, p_{bm}), \tag{11}$$

where v is the specific volume in $m^3 \cdot kg^{-1}$. The area occupied by the motive stream at the inlet of the CAMS per unit total ejector flow rate is determined using the conservation of mass:

$$a_{MNm} = \frac{v_{MNm}}{u_{MNm} \cdot (1 + \chi)}. \tag{12}$$

The set of equations used to determine the suction stream is similar to the one for the motive stream shown in Equations (7)–(11). The formula for calculating the area occupied by the suction stream at the inlet of CAMS per unit of the total ejector flow rate is given as

$$a_{SNm} = \frac{v_{SNm}}{u_{SNm}} \cdot \frac{\chi}{1 + \chi}. \tag{13}$$

In order to calculate the mixing section outlet conditions, the iteration loop is applied. By assuming that the momentum conservation is satisfied for the mixing process and outlet pressure p_{mix} as a guessed value, the velocity of the stream at the mixing section outlet is calculated as follows:

$$p_{bm} \cdot (a_{MNm} + a_{SNm}) + \frac{1}{1 + \chi} \cdot u_{MNm} + \frac{\chi}{1 + \chi} \cdot u_{SNm} = p_{mix} \cdot (a_{MNm} + a_{SNm}) + u_{mix}. \tag{14}$$

The specific enthalpy of stream at the mixing section outlet is calculated by the following equation:

$$h_{MN} + \chi \cdot h_{SN} = (1 + \chi) \cdot \left(h_{mix} + \frac{1}{2} \cdot u_{mix}^2 \right). \tag{15}$$

The specific volume of mixing stream can be determined as a function of the specific enthalpy and pressure:

$$v_{mix} = f(h_{mix}, p_{bm}). \tag{16}$$

The mixing pressure is then iterated until the following equation has been satisfied:

$$\frac{(a_{MNm} + a_{SNm}) \cdot u_{mix}}{v_{mix}} = 1. \tag{17}$$

The next section describes the calculations of the parameters describing the diffuser section of the ejector. First, the entropy at diffuser outlet is calculated as a function of specific enthalpy and pressure and set equal to the entropy of mixing stream leaving the mixing section:

$$s_{mix} = f(h_{mix}, p_{bm}), \tag{18}$$

$$s_{OUT,is} = s_{mix}. \tag{19}$$

The specific enthalpy at the diffuser outlet can be calculated by applying the rule of conservation of energy across the ejector:

$$(1 + \chi) \cdot h_{OUT} = h_{MN} + \chi \cdot h_{SN}. \tag{20}$$

The isentropic specific enthalpy at the diffuser outlet is calculated from the diffuser efficiency:

$$\eta_{OUT} = \frac{h_{OUT,is} - h_{mix}}{h_{OUT} - h_{mix}}. \tag{21}$$

The diffuser outlet pressure and stream quality are then obtained from the following functions:

$$p_{OUT} = f(h_{OUT,is}, s_{OUT,is}), \tag{22}$$

$$x_{OUT} = f(h_{OUT,is}, p_{OUT}), \tag{23}$$

where x is the vapour quality. Moreover, the stream quality at the diffuser outlet and the mass entrainment ratio must satisfy the following condition:

$$(1 + \chi) \cdot x_{OUT} > 1. \tag{24}$$

The 1D model proposed by Kornhauser [16] requires an iterative solution to reach the condition (24). The motive nozzle MFR is calculated using Equation (4). In addition, information about the thermodynamic efficiency of each ejector part has to be given for the ejector performance calculations. In this paper, an efficiency of 0.8 was assumed for all ejector parts based on the investigation presented Elbel and Hrnjak [12].

2.3. A Hybrid Reduced-Order Model

The numerical model of the ejector used in this paper is a hybrid ROM of the R744 two-phase ejector taken from the scientific work done by Haida et al. [32]. The aforementioned model was developed based on a POD approximation of the CFD results and experimental data. The hybrid ROM evaluates the real ejector efficiency under very wide operating conditions for R744 HVAC&R supermarket application within a short computation time while preserving a high level of accuracy. The operational envelope is presented in Figure 2. The motive nozzle conditions presented in Figure 2a were defined for a pressure range from 50 to 140 bar and for the temperature close to the saturation line, above the critical temperature and for different subcooling degrees. The suction nozzle was defined for different pressures in the range from 26 to 46 bar. Moreover, the different superheat degrees were defined up to 15 K and a two-phase region was also considered at the quality of 0.8. Thus, this model can be used for dynamic simulation of ejector-based refrigeration cycles in refrigeration, air-conditioning and heat pump applications.

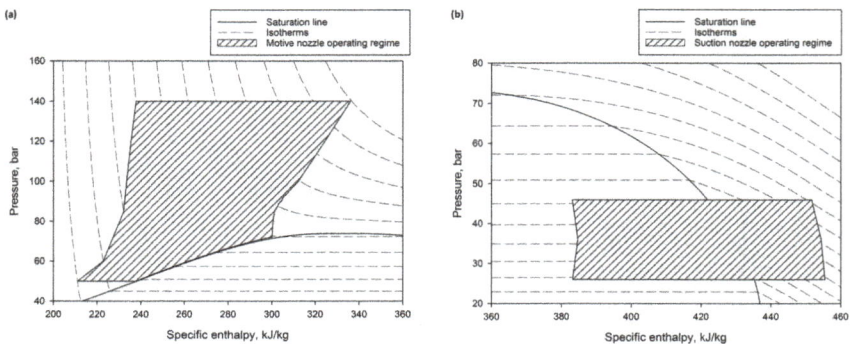

Figure 2. The operational envelope of the hybrid reduced-order model (ROM): (**a**) the motive nozzle conditions; (**b**) the suction nozzle conditions.

The POD approach constructs the optimal approximation base spanning the set of N sampled values of the two-phase flow parameters inside the ejector stored in a single vector called the snapshot [31]. The snapshot vectors are thus related to the input parameters. The POD searches for the orthogonal matrix $\boldsymbol{\Phi}$ reconstructing the rectangular snapshot matrix $\mathbf{U}$ utilising a linear combination of the snapshots. Moreover, the POD model requires an additional interpolation procedure to evaluate the ejector behaviour continuously for different operating conditions. The radial basis interpolation functions were applied for the investigated ROM model. In this study, the thin plate spline radial function with a smoothness factor was employed. The implementation of RBF into the POD model reduces the dimensionality of the ROM to the number of unknown parameters defined as the boundary conditions of the CO_2 two-phase ejector, listed below:

- Motive nozzle pressure,
- Motive nozzle specific enthalpy,
- Suction nozzle pressure,
- Suction nozzle specific enthalpy,
- Outlet pressure.

The snapshot generation for both experimental and CFD values was prepared in a similar way as a set of motive and suction nozzle mass flow rates. More details about the POD-RBF approach can be found in [30].

The CFD results used to build the hybrid ROM were performed using the modified HRM [27] within the operating regime presented in Figure 2 with a defined pressure difference step of 1 bar and a temperature difference step of 5 K. Therefore, the total number of the CFD points was 5380, which were combined with approximately 200 experimental data points to generate the hybrid ROM. Then, the hybrid ROM was validated for all two-phase ejectors installed in the multi-ejector module [32]. The use of experimental data with the CFD results to generate the hybrid ROM allowed the prediction of the MFR of both nozzles with an accuracy level within $\pm 1\%$ at each validated point. The very high accuracy of the hybrid ROM allowed the implementation of the aforementioned model into the R744 supermarket system simulations to evaluate the energy performance of the ejector-based system under different operating conditions and cooling demands.

3. Object-Oriented Modelling of the R744 Transcritical System

The object-oriented model of the R744 transcritical system equipped with the two-phase ejector allows dynamic simulation for performance evaluation and control strategy definition. Moreover, the integration of different ejector models with the R744 cycle allows the sensitivity of the ejector model to be observed with respect to modifications of dynamic system parameters. Hence, the validation procedure of the object-oriented ejector models gives information about the accuracy of each investigated model to select the proper model for system implementation.

The software used for this work was OM, which is the open source environment based on the Modelica 3.3 language for simulating, optimising and analysing complex systems for industrial and academic usage [35]. Moreover, it allows the performance of steady-state as well as dynamic simulations of constructed systems and the analysis of the system as a whole using the transfer behaviour between components as opposed to other well-known equation solvers used for thermodynamic applications. This approach is called object-oriented modelling which is a construction of complex systems based on single objects with well-defined properties or transfer behaviour as a set of interacting and inter-related objects. This kind of modelling allows the visualisation and analysis of complex systems; they can also be modified and their construction can be changed in an easy way. Object-oriented designs are also more maintainable. The CoolProp package [36] containing thermophysical properties of R744 was used to evaluate the flow behaviour inside each system component.

The 0D model and 1D model of the R744 two-phase ejector were directly implemented to the Modelica software due to the simplicity of both approaches. However, the more complex hybrid

ROM was compiled using C-code external functions. Each external function was defined to calculate either the motive nozzle or suction nozzle mass flow rate given from the hybrid ROM under specified operating conditions.

The mathematical approaches of the two-phase ejector were integrated to the R744 transcritical refrigeration system to investigate the system model derivation and ejector work during annual operation. The layout of the R744 ejector-based transcritical system is shown in Figure 3. The R744 loop was designed as a booster type system. The superheated vapour R744 stream from the evaporator outlet is compressed in the semi-hermetic compressor to the discharge pressure level. The high-side pressure is set according to the ambient conditions, either under subcritical or transcritical conditions. The discharged flow outside the compressor rejects the heat in the gas cooler heat exchanger due to the heat transfer between the refrigerant and auxiliary flow, i.e., the glycol–water mixture. Then, the internal heat exchanger (IHX) allows heat transfer between the high-side pressure and intermediate pressure to increase the sub-cooling degree under subcritical conditions and decrease the outlet temperature under transcritical conditions. The sub-cooled flow is expanded either by the high pressure electronic expansion valve (HPV) or by the two-phase ejector. Moreover, the ejector entrains the vapour flow from the evaporator outlet and the mixed flow enters the liquid receiver together with the throttled flow from the HPV. In the liquid receiver, the liquid saturated phase discharges the evaporator, and the vapour phase of R744 flows through IHX to absorb heat from the high-side pressure stream. The vapour outflowing from the IHX is expanded by the flash gas bypass and enters the compression suction side. The R744 liquid outflowing from the liquid receiver is expanded by the metering valve to the MT pressure and absorbs the heat in the evaporator. Finally, the R744 stream outflowing from the evaporator experiences superheat conditions, which prevents any liquid droplets forming in the compressor.

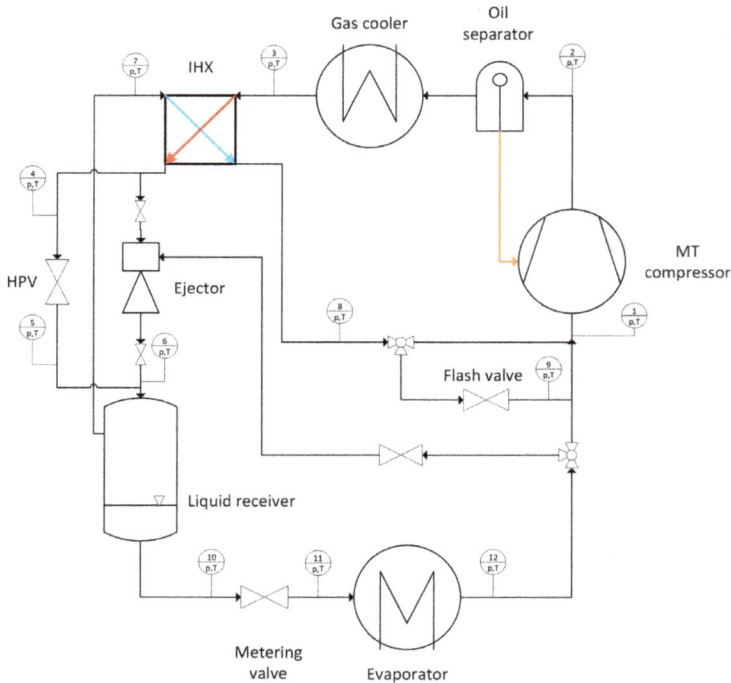

Figure 3. The P & ID diagram of the R744 transcritical cycle equipped with the two-phase ejector.

The system performance analysis was done based on the first-law analysis. Therefore, COP was defined as the ratio of the heat absorbed in the evaporator to the internal work rate of the compressor:

$$COP = \frac{\dot{Q}_{ev}}{\dot{W}_{comp}},$$

(25)

where COP is the coefficient of performance, $\dot{Q}_{ev}$ is the heat rate absorbed in the evaporator in W and $\dot{W}_{comp}$ is the internal work rate of the compressor in W. The heat rate and the internal work rate are defined as follows:

$$\dot{Q}_{ev} = \dot{m}_{11} \cdot [h_{12}(p_{12}, t_{11,sat} + SH) - h_{11}(p_{11}, t_{11,sat})],$$

(26)

$$\dot{W}_{comp} = \dot{m}_1 \cdot [h_2(p_2, t_2) - h_1(p_1, t_1)],$$

(27)

where SH is the superheat degree in K and the indexes are defined according to Figure 3. The specific enthalpy at the compressor outlet was defined based on the isentropic efficiency in the following form:

$$\eta_{is,comp} = \frac{h_{2s}(p_2, s_1) - h_1(p_1, t_1)}{h_2(p_2, t_2) - h_1(p_1, t_1)}.$$

(28)

The rate of heat in the gas cooler was calculated as follows:

$$\dot{Q}_{gc} = \dot{m}_2 \cdot [h_2(p_2, t_2) - h_3(p_3, t_3)].$$

(29)

The outlet conditions are defined either by the set of sub-cooling degrees or by a set of the temperatures and pressures under transcritical conditions. The energy balance equation was defined for IHX in the following form:

$$\dot{m}_3 \cdot (h_3 - h_4) = \dot{m}_7 \cdot (h_8 - h_7).$$

(30)

All valves presented in Figure 3 were defined to control the expansion process and discharge the evaporator as follows:

$$\dot{m}_i = OD \cdot k_v \cdot \rho_{in} \cdot \sqrt{\frac{p_{in} - p_{out}}{\rho_{in}}},$$

(31)

where OD is the opening degree of the valve and k_v is the valve flow coefficient in $m^3 \cdot s^{-1}$. The liquid receiver was evaluated based on the mass balance and energy balance equation:

$$\dot{m}_5 + \dot{m}_6 = \dot{m}_{10} + \dot{m}_7 + V \cdot \frac{d\rho}{d\tau},$$

(32)

$$\dot{m}_5 \cdot h_5 + \dot{m}_6 \cdot h_6 = \dot{m}_{10} \cdot h_{10}(p_{10}, x = 0) + \dot{m}_7 \cdot h_7(p_7, x = 1) + V \cdot h \cdot \frac{d\rho}{d\tau},$$

(33)

where V is the volume of the liquid receiver in m^3, $\frac{d\rho}{d\tau}$ determines the mass accumulation in the liquid receiver in $kg \cdot s^{-1}$ and $V \cdot h \cdot \frac{d\rho}{d\tau}$ is the energy accumulation in W. Finally, the ejector work was calculated using the ejector efficiency defined in Equation (2). Moreover, the ejector performance was also evaluated based on the pressure ratio Π between the outlet and the suction nozzle:

$$\Pi = \frac{p_{OUT}}{p_{SN}}.$$

(34)

Figure 4 presents the flowchart of the computation procedure of the R744 transcritical system at a single time step. The global input data is a weather data to evaluate system work at different test campaigns, i.e., daily or annual. The object-oriented system solved the governing equations for all considered components. The object-oriented ejector model is combined with the C-code hybrid ROM using external functions. Hence, the motive nozzle and the suction nozzle mass flow rates were reached by calling the hybrid ROM at defined input parameters. Based on the input parameters and results given by the hybrid ROM, the ejector model computed continuously the ejector efficiency and the mass entrainment ratio. Finally, the COP value of the investigated system was calculated.

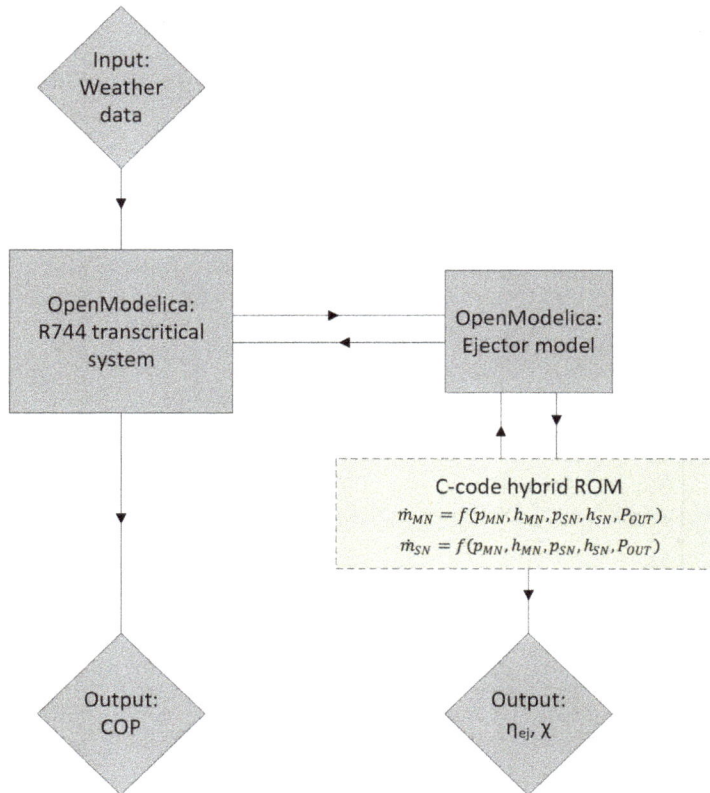

Figure 4. The computational procedure flowchart of the R744 transcritical system at a single time step.

4. Comparison of the Investigated R744 Two-Phase Ejector Numerical Models

In order to compare all three investigated mathematical models of the R744 two-phase ejector, the accuracy of the motive nozzle and the suction nozzle MFRs under different operating conditions (that are typical for refrigeration and air-conditioning applications) are tested. The set of the operating conditions is shown in Table 2. The investigated points were selected from an experimental test campaign carried out at the R744 multi-ejector test rig at the SINTEF/NTNU laboratory in Trondheim [33]. Moreover, the selected points were previously used to validate the HRM CFD model in [26]. The motive nozzle conditions were defined from approximately 54 bar up to approximately 95 bar. Therefore, the accuracy comparison of the ejector models MFRs was defined for ejector utilisation under subcritical conditions, close to the critical point, and transcritical conditions. Moreover, the motive nozzle temperature varied from approximately 279 K to over 308 K. The suction nozzle conditions were defined for refrigeration and air-conditioning applications. Hence, the suction nozzle pressure varied from approximately 27 bar up to approximately 32 bar and the temperature was in the range from approximately 273 K to below 280 K. The outlet pressure was from 32 bar to approximately 39 bar for the evaluation of ejector performance at different ratios of pressure between the outlet and the suction nozzle.

Table 2. The set of operating conditions adapted from [26].

ID	Motive Nozzle Inlet		Suction Nozzle Inlet		Outlet
	p_{MN}	T_{MN}	p_{SN}	T_{SN}	p_{OUT}
	bar	K	bar	K	bar
#1	94.46	308.43	27.21	275.75	32.85
#2	86.04	304.48	27.32	273.61	32.90
#3	91.91	304.13	31.41	278.43	38.24
#4	87.86	301.55	31.55	278.66	38.29
#5	80.62	299.40	31.58	278.49	38.48
#6	78.45	301.71	31.72	278.86	38.28
#7	76.56	301.49	27.33	274.01	32.87
#8	75.79	301.22	28.17	275.73	36.80
#9	66.51	295.56	28.21	275.36	34.85
#10	66.62	295.53	27.87	274.93	32.88
#11	61.79	293.42	29.93	276.73	33.87
#12	59.27	291.58	29.14	277.44	34.83
#13	58.41	283.15	27.82	277.71	34.83
#14	53.93	279.48	27.30	278.85	34.23

All selected operating conditions are also presented in Figure 5 in a pressure-specific enthalpy diagram of R744. In the same figure, the pressure ratio parameter is shown relative to the suction nozzle pressure. The different sub-cooling degrees under subcritical conditions and the different temperatures under transcritical conditions of the motive nozzle were set for validation. The pressure ratio varied from 1.12 to 1.32 due to the pressure difference between the outlet and the suction nozzle. In addition, different pressure ratios were reached at similar suction nozzle conditions, which were strongly related to the value of the mass entrainment ratio.

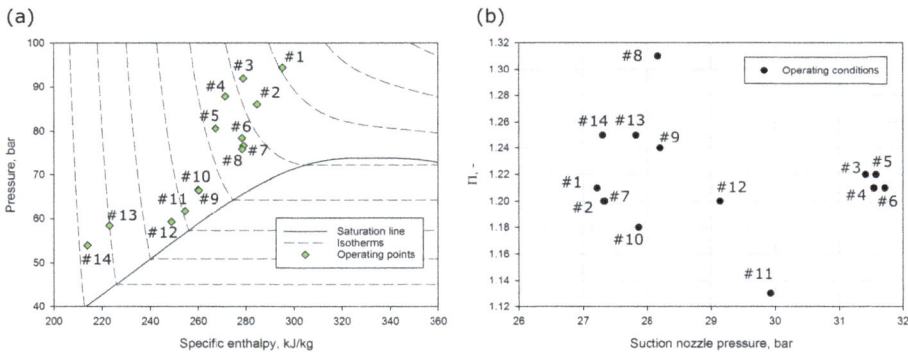

Figure 5. Selected operating conditions: (**a**) the motive nozzle conditions on the R744 pressure-specific diagram; (**b**) the pressure ratio in terms of the suction nozzle pressure.

Table 3 presents the comparison results of each two-phase ejector model together with the experimental data. The relative differences between the motive nozzle and the suction nozzle MFRs together with the mass entrainment ratio were set to evaluate the accuracy of the investigated ejector models. According to the experimental data, the motive nozzle MFR varied from approximately 0.07 kg·s^{-1} to over 0.1 kg·s^{-1} and the suction nozzle MFR was in the range from approximately 0.01 kg·s^{-1} to 0.035 kg·s^{-1}. As the result of both MFRs, the mass entrainment ratio was below 0.42.

The 0D model using a constant value of ejector efficiency obtained a motive nozzle MFR discrepancy from approximately −31% under subcritical conditions to 27% under transcritical conditions. The suction nozzle MFR accuracy of the 0D model was in the range from approximately

−43% to 36.5%. As a result, for the MFR predictions of both nozzles, the relative difference of the mass entertainment ratio varied from −42% to over 47%.

A similar accuracy was reached by the 1D model due to similar calculations of the motive nozzle MFR using Equation (4). The 1D ejector model predicted the suction nozzle MFR with an accuracy in the range from approximately −32% to below 60%. Moreover, the mass entrainment ratio discrepancy of this model was from approximately −21% to over 72%.

The hybrid ROM obtained very high accuracy of the motive nozzle MFR below 1%, which confirmed very good agreement of the results given by the hybrid ROM compared to the experimental data. Very low discrepancy in the hybrid ROM was reached for the suction nozzle MFR and the mass entrainment ratio for all investigated points.

Table 3. The results given by each mathematical model.

ID	Experimental Data			0D Model			1D Model			Hybrid ROM		
	$\dot{m}_{MN}$	$\dot{m}_{SN}$	χ	$\delta_{\dot{m}_{MN}}$	$\delta_{\dot{m}_{SN}}$	δ_χ	$\delta_{\dot{m}_{MN}}$	$\delta_{\dot{m}_{SN}}$	δ_χ	$\delta_{\dot{m}_{MN}}$	$\delta_{\dot{m}_{SN}}$	δ_χ
	$kg \cdot s^{-1}$	$kg \cdot s^{-1}$	-	%	%	%	%	%	%	%	%	%
#1	0.084	0.035	0.417	27.0	5.8	−16.7	27.0	13.1	−11.0	0.9	1.0	1.0
#2	0.079	0.032	0.409	20.3	−12.5	−27.3	20.3	0.7	−16.3	0.8	0.5	0.7
#3	0.095	0.033	0.344	22.6	−15.3	−30.9	22.6	−0.7	−19.0	0.5	0.6	0.6
#4	0.097	0.032	0.326	18.1	−23.0	−34.8	18.1	−6.3	−20.7	0.4	0.2	0.3
#5	0.090	0.025	0.278	11.7	−29.0	−36.4	11.7	−0.6	−11.0	0.1	0.8	0.5
#6	0.073	0.026	0.349	9.9	−35.9	−41.6	9.9	−13.7	−21.5	0.5	0.1	0.3
#7	0.067	0.028	0.411	8.6	−33.1	−38.4	8.6	−13.5	−20.4	0.1	0.5	0.3
#8	0.067	0.011	0.166	6.6	−8.3	−14.0	6.6	44.2	35.3	1.0	0.9	1.0
#9	0.072	0.014	0.192	−11.9	−37.1	−28.6	−11.9	19.7	35.8	0.5	0.8	0.7
#10	0.072	0.022	0.304	−10.9	−42.7	−35.7	−10.9	−21.1	−11.4	0.6	0.5	0.6
#11	0.072	0.019	0.259	−26.8	−40.0	−18.0	−26.8	−32.4	−7.6	0.2	0.4	0.3
#12	0.076	0.009	0.116	−31.3	−26.5	7.1	−31.3	2.6	49.4	0.8	0.1	0.5
#13	0.103	0.007	0.064	−8.9	−11.1	−2.5	−8.9	36.4	49.7	0.1	0.5	0.3
#14	0.100	0.003	0.031	−7.4	36.5	47.4	−7.4	59.4	72.1	0.5	0.1	0.3

Figure 6 presents the relative difference between numerical result produced by each R744 two-phase ejector model and appropriate measurement under all selected operating conditions. The 0D and 1D models reached a motive nozzle MFR relative difference within ±10% for operating conditions close to the critical point (#5÷#8) and for a motive nozzle pressure below 60 bar at #13 and #14, which is shown in Figure 6a. Moreover, both ejector models overestimated the motive nozzle MFR under transcritical conditions and underestimated it under subcritical conditions. Hence, the higher inaccuracy represented by the relative difference for the motive nozzle MFR by the 0D and 1D ejector models was strongly related to the suction nozzle MFR and the mass entrainment ratio predictions. The motive nozzle MFR relative difference shown in Figure 6a confirmed the better agreement of the hybrid ROM. The relative difference for the mass entrainment ratio is shown in Figure 6b together with the ejector efficiency given by the experimental data. The 0D model reached the best agreement close to the ejector efficiency of 0.2 at #12 and #13 due to the assumed constant efficiency based on previous literature [28]. However, the efficiency varied from approximately 0.14 to almost 0.35, which caused underestimation of the mass entrainment ratio for ejector efficiencies above 20% and overestimation for efficiencies below 20% for the 0D model. The 1D model obtained an underestimated mass entrainment ratio under transcritical conditions (#1÷#7), and it also obtained a very low pressure ratio at #10 and #11. The very high pressure ratio above 1.24 caused significant overestimation of the mass entrainment ratio, and the inaccuracy was above 30% for #8 and #9 and for motive nozzle pressure below 60 bar (#12÷#14). The hybrid ROM reached satisfactory agreement at each investigated point due to the hybrid combination of the CFD results and the experimental data.

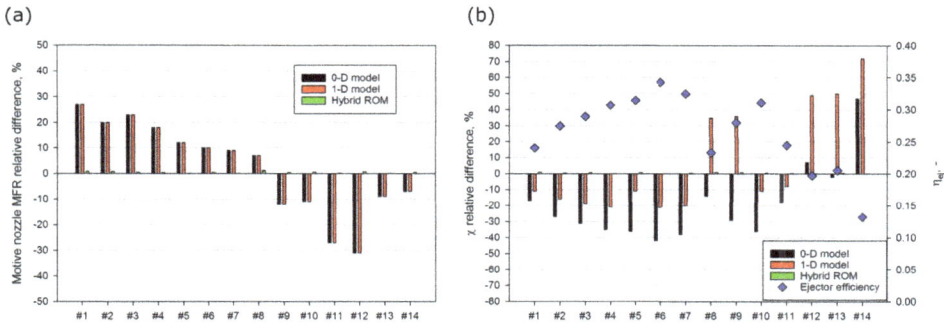

Figure 6. Relative difference between numerical result for each the R744 two-phase ejector mathematical model and measurements: (**a**) the motive nozzle mass flow rate (MFR); (**b**) the mass entrainment ratio together with the ejector efficiency given from the experimental data.

The comparison discussed above proved high accuracy of the hybrid ROM for a very wide operating region. Therefore, the performance evaluation of the R744 two-phase ejector implemented in the refrigeration system using the hybrid ROM allowed ejector efficiency values close to those recorded during experimental tests under different ambient conditions.

5. The System Energy Performance Comparison Using Different Two-Phase Ejector Models

Figure 7 presents the results of the energy performance comparison of the R744 vapour compression system. The comparison was done based on the experimental data given by Banasiak et al. [33]. Two test series were carried out on the R744 system equipped with the two-phase ejector presented in Table 1. The operating conditions are shown in Table 4. The system performance calculations based on the hybrid ROM obtained similar COP values when compared to the experimental data of approximately 2.54 for Test A and 3.3 for Test B. Moreover, the COP accuracy of the hybrid ROM was within ±1% for both cases. The 1D ejector model obtained lower COP values of approximately 2.25 and 2.75 when compared to the experimental data, respectively. Hence, the COP accuracy of 1D model was over −10% for Test A and over −15% for Test B. Similar to the 1D ejector model, the 0D model underestimated COP by up to 2.2. for Test A and approximately 2.73 for Test B. As the results for COP were lower when compared to the experimental data, the COP accuracy of the 0D model was close to −15% and approximately −18% for Tests A and B, respectively. The very high energy performance accuracy of the R744 system based on the hybrid ROM of the two-phase ejector allowed a system performance evaluation under arbitrary ambient conditions and cooling demands due to the wide operating regime of the hybrid ROM.

Table 4. Operating conditions of the R744 vapour compression system adapted from [33].

Parameter	Test A	Test B
Evaporation temperature, °C	−5	8
Evaporator outlet superheat, K	10	10
Gas cooler outlet temperature, °C	25	30
Liquid separator pressure, bar	34	35

(a)

(b)

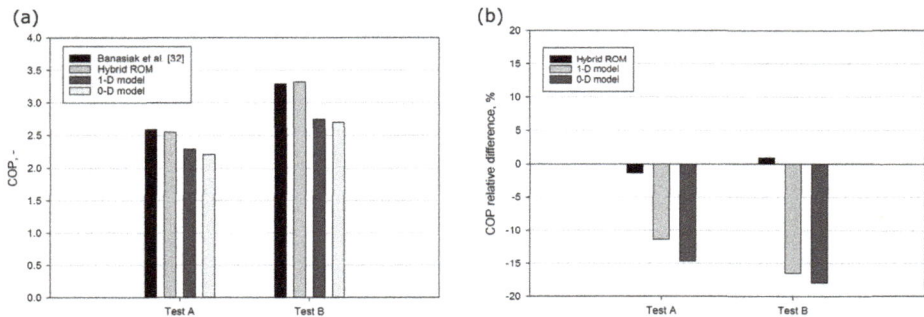

Figure 7. The energy performance of the R744 system based on different two-phase ejector models under the operating conditions given by Banasiak et al. [33]: (**a**) coefficient of performance (COP); (**b**) COP accuracy.

6. Dynamic Simulations of the R744 Two-Phase Ejector Integrated With the Refrigeration System

The implementation of the object-oriented two-phase ejector hybrid ROM to the R744 refrigeration system also allowed dynamic simulations of the ejector-based system for different ambient conditions and cooling demands. Moreover, detailed information about the ejector performance during continuous utilisation was observed. Dynamic investigation of the R744 refrigeration system equipped with the two-phase ejector was completed in three different regions characterised by different ambient temperature ranges: Mediterranean zone (measured in Naples, Italy), South America zone (measured in Rio de Janeiro, Brazil) and South Asia zone (measured in New Delhi, India). The weather data were obtained from the free open-source EnergyPlus platform [37]. The analysis was performed during the summer campaign over a period of two months, which is equivalent to approximately 1488 h. Moreover, the dynamic simulations were done with a time interval of 60 s to evaluate the reaction of the system's operating conditions to the ambient variation.

Figure 8 presents the ambient temperature and the motive nozzle pressure variations during the investigated campaign at each selected climate region. The motive nozzle pressure was strongly related to the gas cooler outlet pressure; therefore, the relationship between the ambient conditions and the motive nozzle pressure controls the ejector and gas cooler utilisation. The ambient temperature in the Mediterranean region shown in Figure 8a was in the range from approximately 15 °C to below 34 °C. Hence, the ejector was utilised under subcritical conditions during the summer campaign. However, the R744 refrigeration system reached a maximum pressure of 85 bar at two temperature peaks halfway through the summer campaign. In Figure 8b, the ambient temperature is shown to vary from 20 °C to approximately 37 °C for the South America region; this variation is smaller than the temperature variation shown in Figure 8a. Hence, transcritical conditions were reached for approximately 50% of time of the investigated campaign. Moreover, the lowest motive nozzle pressure of 60 bar for the South America region was approximately 10 bar higher than the lowest pressure obtained in the Mediterranean region. A highest ambient temperature was reached for a long period of time in the South Asia region, as presented in Figure 8c. The temperature varied from 24 °C to approximately 42 °C. As a result of the high ambient conditions, the two-phase ejector was utilised for transcritical conditions for most of the investigated time. Moreover, the motive nozzle pressure was in the range from 65 bar to over 105 bar, which was strongly related to the system performance as well as the ejector efficiency.

The variations in the motive nozzle conditions in terms of the varied ambient conditions were strongly related to the ejector performance. Moreover, the poor control of the two-phase ejector caused the ejector to have low efficiency, and the system control was far from the optimum COP. One of the most important parameters during ejector operation is the pressure ratio parameter due to the liquid receiver pressure control. Therefore, investigation of the ejector performance during the summer campaign in different regions was done for three different pressure ratios. In addition, the R744 ejector-based system was utilised in a refrigeration application, and the MT evaporator pressure was set to approximately 28 bar with the defined superheat degree of 15 K.

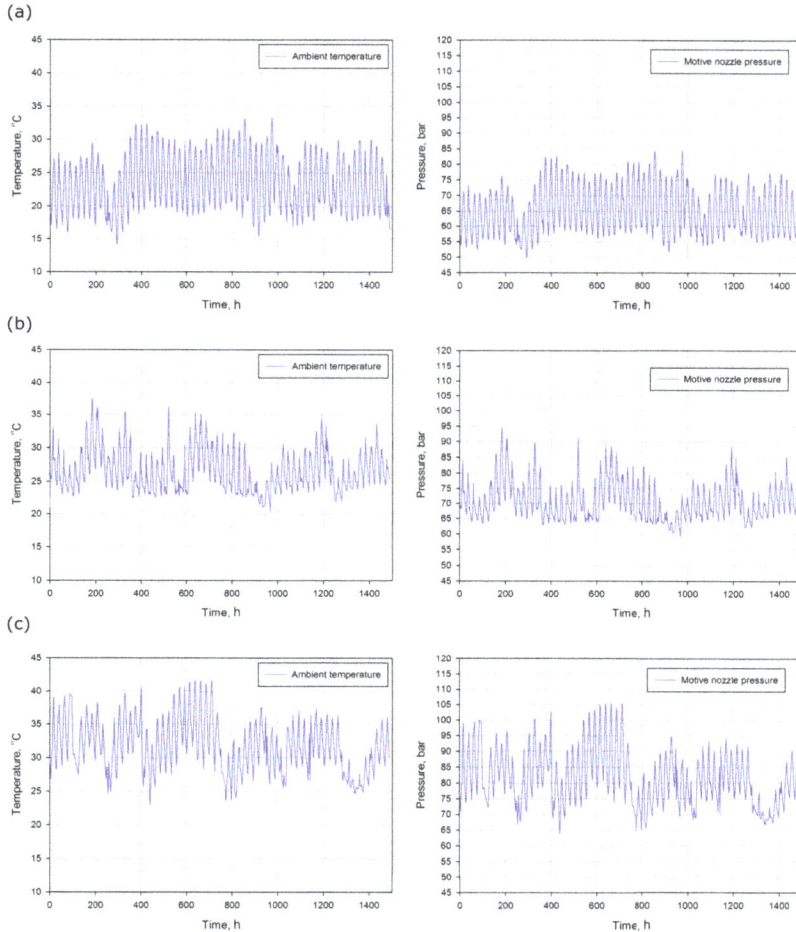

(a)

(b)

(c)

Figure 8. The relationship between the ambient conditions (**left**) and the gas cooler outlet pressure/motive nozzle pressure (**right**) during the summer campaign in different regions: (**a**) Mediterranean zone; (**b**) South America zone; (**c**) South Asia zone.

Figure 9 presents the ejector efficiency for different pressure ratios during the investigation in the Mediterranean region. The lowest pressure ratio of 1.07 was linked to the lowest efficiency of below 0.2 due to the low potential to recover the expansion work. The small value of the pressure ratio caused the ejector to be more stabilised and the efficiency variation was approximately 0.03. An increase in the pressure ratio increased the oscillations of the ejector performance. However, the two-phase ejector with the pressure ratio of 1.15 resulted in an ejector efficiency in the range from approximately 0.22 to below 0.31. The efficiency drop was reached after approximately 300 h as the effect of the significant ambient temperature decreased up to 15 °C. The highest efficiency of the two-phase ejector of approximately 0.33 was obtained for pressure ratio of 1.22. However, the ejector efficiency dropped down to 0.19 following a decrease in the motive nozzle pressure. Hence, the pressure ratio should be related to the ambient conditions and the cooling demand to maintain the high efficiency during temperature decrease.

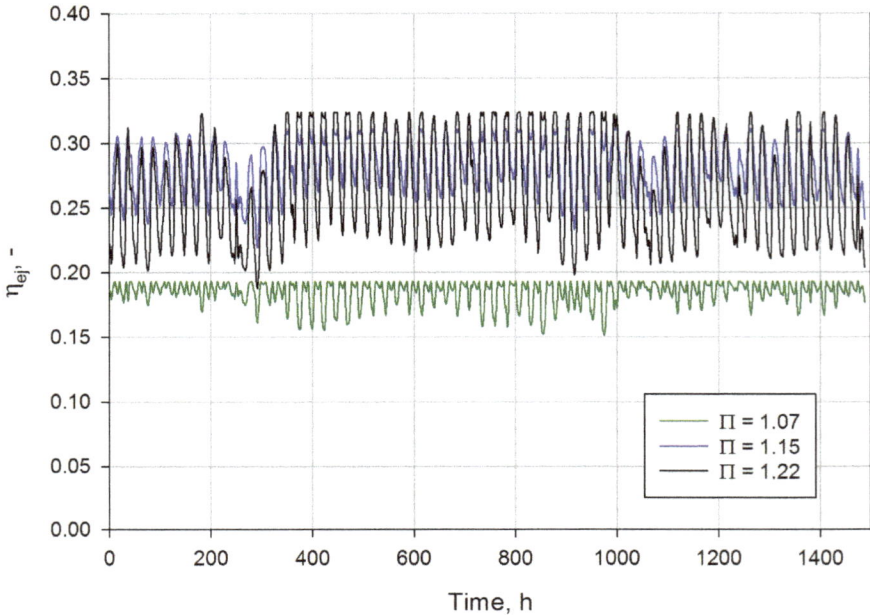

Figure 9. The ejector efficiency at different pressure ratios during the summer campaign in the Mediterranean region.

The performance of the R744 two-phase ejector during the summer campaign in the South America region confirmed the possibility of controlling the ejector in an efficient way, which is shown in Figure 10. Similar to the results presented in Figure 9, the ejector was utilised at the lowest pressure ratio of 1.07 which resulted in the lowest efficiency found during the investigated campaign of below 0.2. Moreover, higher ambient temperatures above 35 °C caused a higher efficiency drop to almost 0.1. The significant efficiency improvement was observed for higher pressure ratios. The two-phase ejector with a pressure ratio of 1.15 reached an efficiency in the range from approximately 0.25 to 0.31. In addition, the ejector at a pressure ratio of 1.15 was more stable compared with the ejector at a pressure ratio of 1.22. The highest ejector efficiency value of approximately 0.33 was obtained at the highest pressure ratio of 1.22, although the ejector performance dropped down to approximately 0.24. Therefore, the possibility to define the ejector operating conditions during continuous work allows stable and high performance of the two-phase ejector to be achieved.

Figure 11 presents the ejector efficiency for different pressure ratios during the investigated campaign in the South Asia region. The worst ejector performance was noticed for the pressure ratio of 1.07, which agrees with the analyses performed for the Mediterranean region and South America region. Moreover, the very high ambient temperature of above 40 °C strongly influenced the efficiency degradation for the pressure ratio of 1.07. As for the results for highly varied ambient conditions, the two-phase ejector with the pressure ratio of 1.07 obtained an efficiency in the range from approximately 0.07 to below 0.2. An increase in the pressure ratio of up to 1.15 significantly improved the ejector efficiency. However, similar performance degradation was observed for very high ambient temperatures for which the efficiency varied from approximately 0.17 to 0.31. The highest value of the pressure ratio of 1.22 caused the best performance for most of the investigated time during the summer campaign. The two-phase ejector reached efficiency in the range from 0.2 up to 0.33. Further improvement of the ejector performance at very high ambient temperatures and a motive nozzle pressure above 100 bar can be achieved by an increase in the pressure ratio. Hence, the R744 two-phase ejector should be controlled to optimise the ejector efficiency by proper setting of the pressure ratio for a specified application.

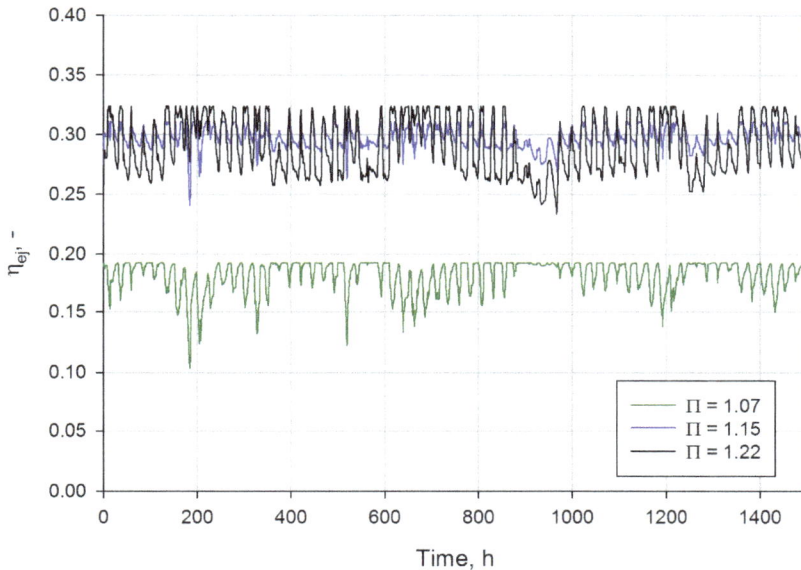

Figure 10. The ejector efficiency at different pressure ratios during the summer campaign in the South America region.

The characteristics of the ejector performance during the summer campaign in different hot climate zones are strongly related to the COP value of the system, especially during work in the daytime. Hence, Figure 12 presents the COP comparison for the R744 ejector-based system at different climate regions for 24 h during the summer campaign. The pressure ratio was set to 1.22 to achieve the highest ejector efficiency. The R744 refrigeration system equipped with the two-phase ejector obtained the highest COP values of above 3.5 in the Mediterranean region due to the subcritical operation for most of the investigated time according to Figure 8a. The lowest COP during the day was observed in the afternoon at approximately 4:00 p.m., whereas the highest peak of COP was reached at 4:00 a.m. The R744 ejector-based system experienced a lower COP in the South America region due to the higher daily ambient temperature, which is shown in Figure 8b. The significant COP drop below 3.0 was observed at approximately 2:00 p.m. The lowest COP of the R744 refrigeration system was obtained for the South Asia climate zone due to the very high ambient temperature above 40 °C. Therefore, the COP dropped from approximately 3.8 to less than 2.5, although the energy performance of the R744 ejector-based system was higher between the second hour and sixth hour as a result of the ejector performance. The significant difference in COP between the selected climate regions forced the control process of the R744 refrigeration system to be improved to operate close to the optimum COP.

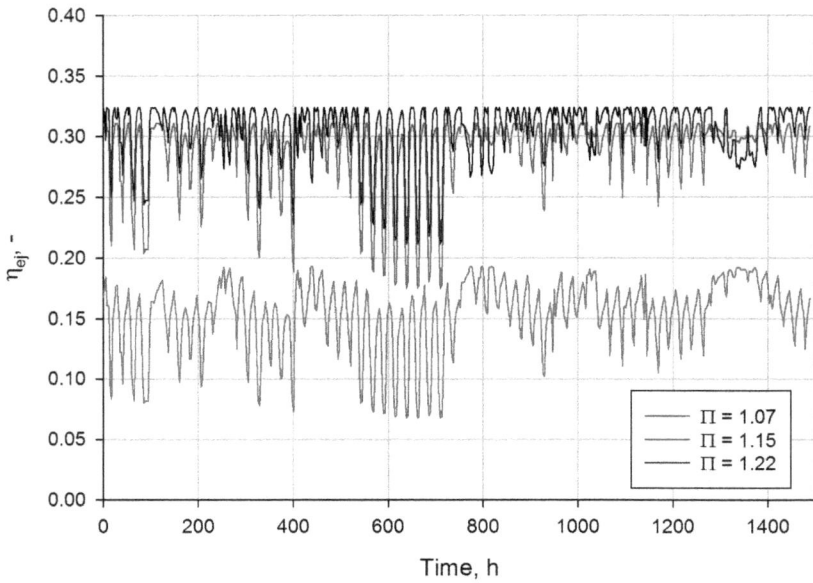

Figure 11. The ejector efficiency at different pressure ratios during the summer campaign in the South Asia region.

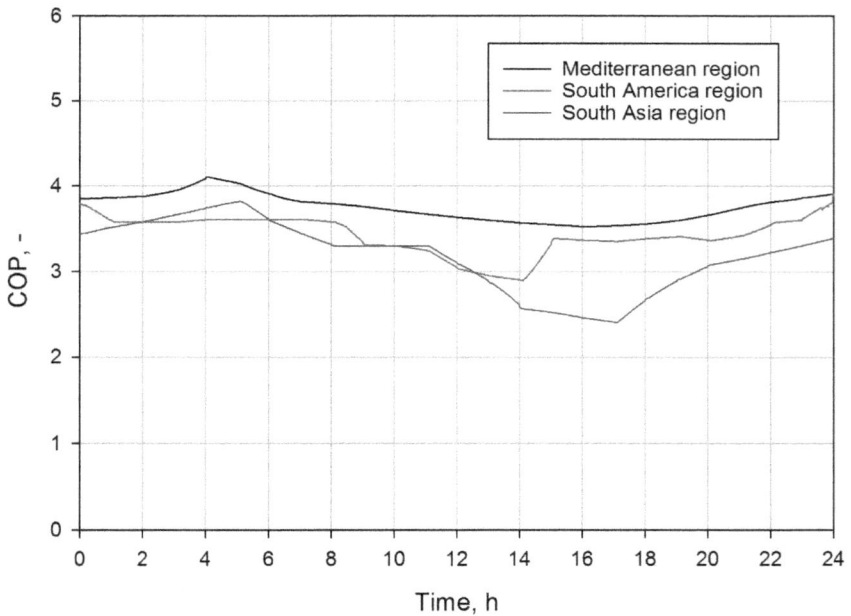

Figure 12. The coefficient of performance for 24 h of the summer campaign in different hot climate zones.

7. Conclusions

The object-oriented hybrid ROM of the R744 two-phase ejector was developed for dynamic simulations. The proposed model was compared with two simpler mathematical approaches of the ejector that have commonly been used in the literature for refrigeration system investigations. Moreover, a dynamic simulation of the R744 refrigeration system equipped with the two-phase ejector for different climate zones was performed to evaluate the ejector and the system's performance.

The object-oriented hybrid ROM was implemented in OpenModelica free software using the external function built in C-code programming. The proposed model allowed the evaluation of ejector performance under arbitrary operating conditions within the operational envelope defined for the HVAC&R supermarket system. Apart from the hybrid ROM, the 0D ejector model assumed a constant efficiency, and the 1D ejector model with the assumption of homogeneous equilibrium two-phase fluid flow was implemented for the comparison of different mathematical approaches to describe ejector performance. In comparison, procedure accuracy, represented by the relative difference between numerical result produced by each two-phase ejector model and appropriate experimental measurement, is determined. The 0D model obtained satisfactory prediction of both nozzles' MFRs only at values close to the assumed ejector efficiency. The more accurate 1D model reached an accuracy level within ±10% for motive nozzle pressures close to the critical point. However, very high discrepancy was observed for high pressure ratios. The hybrid ROM obtained the best accuracy for all investigated points due to the hybrid combination of the experimental data with the CFD results that was used to build the ROM basis. Therefore, the integration of the hybrid ROM allowed the evaluation of the R744 refrigeration system equipped with the designed two-phase ejector.

The investigation of the ejector performance during the summer campaign demonstrates the variation in the work of the ejector. Three different climate zones were selected for investigation: a Mediterranean region (Italy), a South American region (Brazil), and a South Asian region (India). The R744 two-phase ejector was mostly utilised under subcritical conditions in the Mediterranean climate zone. Transcritical conditions were achieved for most of the summer campaign for zones with higher ambient temperatures in South America above 30 °C and in South Asia above 40 °C. The ejector efficiency is strongly related to the pressure ratio, especially during conditions of significant temperature difference. An increase in the pressure ratio allowed a high efficiency of above 0.2 up to 0.35 to be maintained. However, the pressure ratio should be well controlled to achieve an ejector performance close to the optimum during system utilisation. According to the results, the pressure ratio of 1.15 allowed the greatest stabilisation of the system.

The evaluation of energy performance of the R744 refrigeration system equipped with the two-phase ejector confirmed the influence of the high ambient temperature on COP. The highest COP value was obtained in the Mediterranean region as a result of the system operating in subcritical conditions for most of the investigated time. An increase in the ambient temperature generally caused the COP degradation. Hence, the lowest COP value was observed in the South Asia climate zone. Energy performance improvement for the R744 ejector-based system can be achieved by optimisation of the liquid receiver pressure and outlet conditions of the gas cooler as well as the internal heat exchanger.

Author Contributions: Conceptualisation, M.H. and J.S.; methodology, M.H.; resources, M.H., W.S. and R.F; investigation, M.H., W.S. and R.F.; validation, M.H., R.F., W.S., J.S., M.P., J.B. and A.J.N.; writing—original draft preparation, M.H.; writing—review and editing, M.H., W.S., R.F., J.S., M.P., J.B. and A.J.N.; supervision, J.S. and A.J.N.

Funding: The authors gratefully acknowledge the financial support of the National Science Centre through project No. 2017/27/B/ST8/00945. Moreover, the work of M.H. was funded by the Rector's research grant No. 08/060/RGJ18/0157 provided by Silesian University of Technology.

Conflicts of Interest: The authors declare no conflict of interest.

Nomenclature

a	flow rate ratio, $m^2 \cdot kg^{-1} \cdot s^{-1}$
h	specific enthalpy, $kJ \cdot kg^{-1}$
k_v	valve flow coefficient, $m^3 \cdot s^{-1}$
$\dot{m}$	mass flow rate, $kg \cdot s^{-1}$
p	pressure, Pa
s	specific entropy, $kJ \cdot kg^{-1} \cdot K^{-1}$
T	temperature, K
u	stream velocity, $m \cdot s^{-1}$
U	snapshot basis matrix, -
v	specific volume, $m^3 \cdot kg^{-1}$
$\dot{W}$	work rate, W
x	vapour quality, −

Greek Symbols

χ	mass entrainment ratio, −
δ	relative difference, %
η	ejector efficiency, %
Π	pressure ratio, −
ρ	density, $kg \cdot m^{-3}$
τ	time, s

Subscripts

amb	ambient
bm	beginning of the mixing section
comp	compressor

eff	effective cross-section area
ev	evaporator
exp	experimental data
fv	flash valve
lr	liquid receiver
max	maximal value
mix	mixing section
MN	motive nozzle
MNm	motive stream at the inlet of CAMS
OUT	outlet
rec	recovered ejector expansion work rate
SN	suction nozzle
SNm	suction stream at the inlet of CAMS
v	valve

Abbreviations

CAMS	Constant Area Mixing Section
COP	Coefficient of Performance
GWP	Global Warming Potential
HEM	Homogeneous Equilibrium Model
HPV	High Pressure Valve
HRM	Homogeneous Relaxation Model
IHX	Internal Heat Exchanger
MFR	Mass Flow Rate
ODP	Ozone Depletion Potential
ROM	Reduced-Order Model

References

1. Briefing Note on. UN Environment Ozone Secretariat in Nairobi, Kenya: 2017. Ratification of the Kigali Amendment. Available online: http://conf.montreal-protocol.org/meeting/oewg/oewg-39/presession/briefingnotes/ratification_kigali.pdf (accessed on 27 September 2018).
2. ASHRAE. *ANSI/ASHRAE Standard 34, Designation and Safety Classification of Refrigerants*; ASHRAE: Atlanta, GA, USA, 2016.
3. Lorentzen, G. *Throttling, the Internal Haemorrhage of the Refrigeration Process*; Institute of Refrigeration: London, UK, 1983; Volume 80.
4. Goodarzi, M.; Gheibi, A.; Motamedian, M. Comparative analysis of an improved two-stage multi-inter-cooling ejector-expansion trans-critical CO_2 refrigeration cycle. *Appl. Therm. Eng.* **2015**, *81*, 58–65. [CrossRef]
5. Kim, M.H.; Pettersen, J.; Bullard, C.W. Fundamental process and system design issues in CO_2 vapor compression systems. *Prog. Energy Combust. Sci.* **2004**, *30*, 119–174. [CrossRef]
6. Bansal, P. A review—Status of CO_2 as a low temperature refrigerant: Fundamentals and R&D opportunities. *Appl. Therm. Eng.* **2012**, *41*, 18–29. [CrossRef]
7. Sawalha, S.; Karampour, M.; Rogstam, J. Field measurements of supermarket refrigeration systems. Part I: Analysis of CO_2 trans-critical refrigeration systems. *Appl. Therm. Eng.* **2015**, *87*, 633–647. [CrossRef]
8. Chesi, A.; Esposito, F.; Ferrara, G.; Ferrari, L. Experimental analysis of R744 parallel compression cycle. *Appl. Energy* **2014**, *135*, 274–285. [CrossRef]
9. Sharma, V.; Fricke, B.; Bansal, P. Comparative analysis of various CO_2 configurations in supermarket refrigeration systems. *Int. J. Refrig. Revue Int. Du Froid* **2014**, *46*, 86–99. [CrossRef]
10. Rony, R.U.; Yang, H.; Krishnan, S.; Song, J. Recent Advances in Transcritical CO2 (R744) Heat Pump System: A Review. *Energies* **2019**, *12*, 457. [CrossRef]
11. Gullo, P. Advanced Thermodynamic Analysis of a Transcritical R744 Booster Refrigerating Unit with Dedicated Mechanical Subcooling. *Energies* **2018**, *11*, 3058. [CrossRef]
12. Elbel, S.; Hrnjak, P. Experimental validation of a prototype ejector designed to reduce throttling losses encountered in transcritical R744 system operation. *Int. J. Refrig.-Revue Int. Du Froid* **2008**, *31*, 411–422. [CrossRef]

13. Gullo, P.; Cortella, G. Comparative Exergoeconomic Analysis of Various Transcritical R744 Commercial Refrigeration Systems. In Proceedings of the 29th International Conference on Efficiency, Cost, Optimisation, Simulation and Environmental Impact of Energy Systems, Portoroz, Slovenia, 19–23 June 2016; pp. 19–23.

14. Catalán-Gil, J.; Sánchez, D.; Llopis, R.; Nebot-Andrés, L.; Cabello, R. Energy Evaluation of Multiple Stage Commercial Refrigeration Architectures Adapted to F-Gas Regulation. *Energies* **2018**, *11*, 1915. [CrossRef]

15. Elbel, S.; Lawrence, N. Mathematical modeling and thermodynamic investigation of the use of two-phase ejectors for work recovery and liquid recirculation in refrigeration cycles. *Int. J. Refrig.* **2015**, *58*, 41–52. [CrossRef]

16. Kornhauser, A. The use of an ejector as a refrigerant expander. *International Refrigeration and Air Conditioning Conference*; Purdue University, Purdue ePubs: West Lafayette, IN, USA, 1990; pp. 1–11.

17. Li, D.; Groll, E.A. Transcritical CO_2 refrigeration cycle with ejector-expansion device. *Int. J. Refrig.* **2005**, *28*, 766–773. [CrossRef]

18. Richter, C. Proposal of New Object-Oriented Equation-Based Model Libraries For Thermodynamic Systems. Ph.D. Thesis, Braunschweig University of Technology, Braunschweig, Germany, 2008.

19. Sumeru, K.; Nasution, H.; Ani, F.N. A review on two-phase ejector as an expansion device in vapor compression refrigeration cycle. *Renew. Sustain. Energy Rev.* **2012**, *16*, 4927–4937. [CrossRef]

20. Lucas, C.; Rusche, H.; Schroeder, A.; Koehler, J. Numerical investigation of a two-phase CO_2 ejector. *Int. J. Refrig.* **2014**, *43*, 154–166. [CrossRef]

21. Smolka, J.; Bulinski, Z.; Fic, A.; Nowak, A.J.; Banasiak, K.; Hafner, A. A computational model of a transcritical R744 ejector based on a homogeneous real fluid approach. *Appl. Math. Model.* **2013**, *37*, 1208–1224. [CrossRef]

22. Bilicki, Z.; Kestin, J. Physical Aspects of the Relaxation Model in Two-Phase Flow. *Proc. R. Soc. Lond. A Math. Phys. Eng. Sci.* **1990**, *428*, 379–397. [CrossRef]

23. Downar-Zapolski, P.; Bilicki, Z.; Bolle, L.; Franco, J. The non-equilibrium relaxation model for one-dimensional flashing liquid flow. *Int. J. Multiph. Flow* **1996**, *22*, 473–483. [CrossRef]

24. Angielczyk, W.; Bartosiewicz, Y.; Butrymowicz, D.; Seynhaeve, J.M. 1-D modeling of supersonic carbon dioxide two-phase flow through ejector motive nozzle. In *International Refrigeration and Air Conditioning Conference*; Purdue University, Purdue ePubs: Braunschweig, Germany, 2010; Volume 2362, pp. 1–8.

25. Colarossi, M.; Trask, N.; Schmidt, D.P.; Bergander, M.J. Multidimensional modeling of condensing two-phase ejector flow. *Int. J. Refrig.* **2012**, *35*, 290–299. [CrossRef]

26. Palacz, M.; Haida, M.; Smolka, J.; Nowak, A.J.; Banasiak, K.; Hafner, A. HEM and HRM accuracy comparison for the simulation of CO_2 expansion in two-phase ejectors for supermarket refrigeration systems. *Appl. Therm. Eng.* **2017**. [CrossRef]

27. Haida, M.; Smolka, J.; Hafner, A.; Palacz, M.; Banasiak, K.; Nowak, A.J. Modified homogeneous relaxation model for the R744 trans-critical flow in a two-phase ejector. *Int. J. Refrig.* **2018**, *85*, 314 – 333. [CrossRef]

28. Hafner, A.; Forsterling, S.; Banasiak, K. Multi-ejector concept for R-744 supermarket refrigeration. *Int. J. Refrig. Revue Int. Du Froid* **2014**, *43*, 1–13. [CrossRef]

29. Liu, F.; Groll, E.A.; Li, D. Investigation on performance of variable geometry ejectors for CO_2 refrigeration cycles. *Energy* **2012**, *45*, 829–839. [CrossRef]

30. Haida, M.; Smolka, J.; Hafner, A.; Ostrowski, Z.; Palacz, M.; Nowak, A.J.; Banasiak, K. System model derivation of the CO_2 two-phase ejector based on the CFD-based reduced-order model. *Energy* **2018**, *144*, 941–956. [CrossRef]

31. Ostrowski, Z.; Białecki, R.A.; Kassab, A.J. Solving inverse heat conduction problems using trained POD-RBF network inverse method. *Inverse Probl. Sci. Eng.* **2008**, *16*, 39–54. [CrossRef]

32. Haida, M.; Smolka, J.; Hafner, A.; Ostrowski, Z.; Palacz, M.; Madsen, K.B.; Försterling, S.; Nowak, A.J.; Banasiak, K. Performance mapping of the R744 ejectors for refrigeration and air conditioning supermarket application: A hybrid reduced-order model. *Energy* **2018**, *153*, 933–948. [CrossRef]

33. Banasiak, K.; Hafner, A.; Kriezi, E.E.; Madsen, K.B.; Birkelund, M.; Fredslund, K.; Olsson, R. Development and performance mapping of a multi-ejector expansion work recovery pack for R744 vapour compression units. *Int. J. Refrig.* **2015**, *57*, 265–276. [CrossRef]

34. Gullo, P.; Hafner, A.; Banasiak, K.; Minetto, S.; Kriezi, E.E. Multi-Ejector Concept: A Comprehensive Review on its Latest Technological Developments. *Energies* **2019**, *12*, 406. [CrossRef]

35. Fritzson, P. *Principles of Object-Oriented Modeling and Simulation with Modelica 3.3: A Cyber-Physical Approach*, 2nd ed.; Wiley-IEEE Press: Piscataway Township, NJ, USA, 2015.

36. Bell, I.H.; Wronski, J.; Quoilin, S.; Lemort, V. Pure and Pseudo-pure Fluid Thermophysical Property Evaluation and the Open-Source Thermophysical Property Library CoolProp. *Ind. Eng. Chem. Res.* **2014**, *53*, 2498–2508. [CrossRef] [PubMed]

37. EnergyPlus Weather Data. EnergyPlus, 2018. Available online: https://energyplus.net/ (accessed on 1 February 2019).

energies

MDPI

Article

Estimation of the Biot Number Using Genetic Algorithms: Application for the Drying Process

Krzysztof Górnicki *, Radosław Winiczenko and Agnieszka Kaleta

Department of Fundamental Engineering, Warsaw University of Life Sciences, Nowoursynowska 164 St., 02-787 Warsaw, Poland
* Correspondence: krzysztof_gornicki@sggw.pl; Tel.: +48-22-593-46-18

Received: 28 May 2019; Accepted: 17 July 2019; Published: 22 July 2019

Abstract: The Biot number informs researchers about the controlling mechanisms employed for heat or mass transfer during the considered process. The mass transfer coefficients (and heat transfer coefficients) are usually determined experimentally based on direct measurements of mass (heat) fluxes or correlation equations. This paper presents the method of Biot number estimation. For estimation of the Biot number in the drying process, the multi-objective genetic algorithm (MOGA) was developed. The simultaneous minimization of mean absolute error (MAE) and root mean square error (RMSE) and the maximization of the coefficient of determination R^2 between the drying model and experimental data were considered. The Biot number can be calculated from the following equations: $Bi = 0.8193\exp(-6.4951T^{-1})$ (and moisture diffusion coefficient from $D/s^2 = 0.00704\exp(-2.54T^{-1})$) (RMSE = 0.0672, MAE = 0.0535, R^2 = 0.98) or $Bi = 1/0.1746\log(1193847T)$ ($D/s^2 = 0.0075\exp(-6T^{-1})$) (RMSE = 0.0757, MAE = 0.0604, R^2 = 0.98). The conducted validation gave good results.

Keywords: Biot number; genetic algorithms; drying

1. Introduction

The dimensionless Biot number (Bi) is present in partial differential equations in cases when the surface boundary conditions (of the third kind) are written in a dimensionless form. The Biot number informs researchers about the relationship between the internal and external fluxes [1]. As far as heat transfer is concerned, Bi expresses the ratio between the internal and external resistances [2]. Therefore, it can be stated that the heat Biot number is a measure of the temperature drop in the material with respect to the difference of the temperatures between the solid surface and the surrounding medium [3]. Assuming similarity between heat and mass transfer, the Biot number used for mass exchange is gained by equating internal and external mass fluxes at the interface [3,4] and the discussed number is defined as follows:

$$Bi = \frac{h_m L}{D},$$ (1)

The analysis of Biot numbers enables researchers to answer questions regarding the controlling mechanisms employed for heat or mass transfer during the considered process [3]. Dincer [3] stated that the values of Bi for mass transfer can be divided into the following groups:

$Bi < 0.1$ (the surface resistance across the surrounding medium boundary layer is much bigger in comparison with the internal resistance to the mass diffusion within the solid body);

$0.1 < Bi < 100$ (the values of the internal and external resistance can be treated as comparable);

$Bi > 100$ (external (surface) resistance is much lower than the internal resistance).

Ruiz-López et al. [5,6] assumed that for $Bi > 40$, it can be accepted that the internal resistance to mass exchange is the only mechanism controlling the rate of the drying process. In such a case, the moisture content of the solid body surface reaches its equilibrium value at once. Wu and

Irudayaraj [7] experimentally stated that drying can only be treated as an isothermal process for very low Biot numbers.

The mass transfer coefficients (and heat transfer coefficients) are usually determined experimentally based on direct measurements of mass (heat) fluxes or correlation equations. The mass transfer coefficient can also be calculated from the dimensionless Sherwood number (*Sh*). The *Sh* can be expressed as a function of the Reynolds number (*Re*) and the Schmidt number (*Sc*) (forced convection), as a function of the Grashof number (mass) (Gr_m) and the *Sc* (natural convection), and as a function of the Archimedes number (*Ar*) and the *Sc* (vacuum-microwave drying) [8].

The importance of the heat and mass Biot number has been shown in several publications. Rovedo et al. [4] analysed the drying process of shrinking potato slabs. The numerical solving of the drying model using various forms of *Bi* gave a deeper insight into the process. Huang and Yeh [9] considered an inverse problem in simultaneously estimating the heat and mass Biot numbers during the drying of a porous material. Dincer and Hussain [10] developed a new Biot number and lag factor correlation, which gave good agreement between the predicted and measured values of moisture content. Chen and Peng [11] analysed the values of modified Biot numbers during hot air drying of small, moist, and porous objects. Giner et al. [1] considered the variableness of heat and mass *Bi* during the drying of wheat and apple-leather, whereas Xie et al. [12] modelled the pulsed vacuum drying of rhizoma dioscoreae slices using the Dincer model [13], which describes the moisture ratio with a correlation between *Bi* and the lag factor.

An accurate estimation of the Biot number is essential for an efficient heat and mass transfer analysis, leading to optimum operating conditions and an efficient process.

The aim of the multi-objective optimization task is to optimize the several objective functions simultaneously with many criteria. These issues have long been of interest to researchers using traditional optimization and search techniques. In the case of multi-criteria optimization, the concept of the optimal solution is not as obvious as in the case of one criterion. If we do not agree in advance to compare the values of different criteria, then we must propose a definition of optimality that respects the integrity of each of them. This approach is called optimality in the Pareto sense. It is convenient to classify possible solutions of multi-criteria optimization tasks as dominated and non-dominated (*Pareto-optimal*) solutions.

Multi-criteria optimization of drying technology is used for conveyer-belt dryer design [14,15], fluidized bed dryers [16,17], control of a drying process [18], batch drying of rice [19], and drying [20] and rehydrated apple issue [21]. Non-preference multi-criterion optimization methods with the Pareto-optimal set were used in the papers [14–17]. The researchers developed a mathematical model of the fluidized bed dryer and determined the colour deterioration laws for potato slices. A multi-criterion optimization of the thermal processing was conducted by [22] and [23]. The authors developed an intelligent hybrid method for identifying the optimal processing conditions. The complex method applied to different shapes was subjected to the processing boundary conditions to find the best process temperature and to maximize the retention of thiamine. Winiczenko et al. predicted the quality indicators of the drying [20] and rehydrated [21] apple issue using a non-sorting genetic algorithm.

The aim of the present study is to determine the Biot number using the genetics algorithm (GA). The method is applied in the process of drying.

2. Materials and Methods

2.1. Material

The research material was parsley roots of the Berlińska variety. The parsley roots were purchased at a local market in Warsaw. The material was cut into 6 mm thick slices, which were dried in natural convection conditions at the temperatures of 40, 50, 60, and 70 °C.

A detailed description of the equipment used, measurements performed, and their accuracy may be found in [24].

2.2. Moisture Transfer Analysis

It can be assumed that the moisture movement inside the dried solid body is only a diffusion movement in the convection drying of food products. The unsteady-state, one-dimensional mass exchange equation within a slice (treated as an infinite plane) of a thickness of 2 s can be expressed in the following form [25]:

$$\frac{\partial M}{\partial t} = D\frac{\partial^2 M}{\partial x^2} \quad (t > 0; \; -s < x < s).$$ (2)

The following common assumptions were taken in Equation (2):

- Shape and volume of the slice do not change during drying;
- Mass diffusion coefficient is constant;
- Equation (2) is subjected to the following conditions:

Moisture content at any point of the slice is the same at the beginning of drying (initial condition):

$$M(x,0) = M_0 = \text{const},$$ (3)

The mass flux from the surface of the slice is expressed in terms of the moisture content difference between the surface and the equilibrium moisture content (boundary conditions of the third kind):

$$\pm D\frac{\partial M(\pm s,t)}{\partial x} = k[M(\pm s,t) - M_e],$$ (4)

An analytical solution of Equation (2) at the initial condition in the form of Equation (3) and at the boundary conditions given by Equation (4) with respect to the mean moisture content as function of time can be written in the following form [25]:

$$MR = \frac{M - M_e}{M_0 - M_e} = \sum_{i=1}^{\infty} B_i \exp\left[-\mu_i^2\frac{D}{s^2}t\right],$$ (5)

where

$$B_i = \frac{2Bi^2}{\mu_i^2 Bi^2 + Bi + \mu_i^2},$$ (6)

$$ctg\mu_i = \frac{1}{Bi}\mu_i,$$ (7)

The Biot number in Equations (6) and (7) was assumed to be dependent on the temperature using the following formulas:

$$Bi = a_b\exp\left(-\frac{b_b}{T}\right),$$ (8a)

$$Bi = a_b\ln(b_b T),$$ (8b)

$$Bi = a_b\log(b_b T),$$ (8c)

$$Bi = a_b\ln(T) + b_b,$$ (8d)

$$Bi = a_b T^{-b_b},$$ (8e)

$$Bi = \frac{1}{a_b\ln(b_b T)},$$ (8f)

$$Bi = \frac{1}{a_b\log(b_b T)},$$ (8g)

$$Bi = \frac{1}{a_b\exp\left(-\frac{b_b}{T}\right)},$$ (8h)

Dependencies (8a)–(8d) indicate the increase of the Biot number with the increase of the temperature, whereas dependencies (8e)–(8h) indicate the decrease of *Bi* with the increase of the temperature (i.e., [26]).

The moisture diffusion coefficient in Equation (5) was assumed to be dependent on the temperature using the following formula (i.e., [27,28]):

$$\frac{D}{s^2} = a_d \exp\left(-\frac{b_d}{T}\right)$$

(9)

An increase of the temperature according to Equation (9) results in a increase of the moisture diffusion coefficient.

2.3. Optimization

The optimization problem has been divided into two tasks. The first multi-objective optimization (MOO) task was to determine a Biot number and mass diffusion coefficient from the moisture ratio (*MR*) model (5) for each drying temperature: 40, 50, 60, and 70 °C. The second optimization task was to find the constants in equations for calculation of the Biot number (8a)–(8h) and mass diffusion coefficient (9).

The mean absolute error (MAE) and root mean square error (RMSE) were minimized, whereas the coefficient of determination R^2 was maximized for the difference between the data and objective function for both optimized tasks. The algorithm randomly selects a set of models by minimizing the error between the proposed model and experimental data. Then, it evolves them to create the best fit. Therefore, the MOO problem was expressed as follows:

$$\text{Min} = \begin{cases} \text{Min RMSE} = \sqrt{\frac{1}{N}\sum_{i=1}^{N}\left(MR_{\text{pred}} - MR_{\text{exp}}\right)^2} \\ \text{Min MAE} = \frac{1}{N}\sum_{i=1}^{N}\left|MR_{\text{pred}} - MR_{\text{exp}}\right| \\ \text{Max } R^2 = 1 - \frac{\sum_{i=1}^{N}\left(MR_{\text{pred}} - MR_{\text{exp}}\right)^2}{\sum_{i=1}^{N}\left(MR_{\text{pred}} - \overline{MR}_{\text{exp}}\right)^2} \end{cases}$$

(10)

The multi-objective optimization was carried out using an elitist, non-dominated sorting genetic algorithm (NSGA II). The Pareto front for this problem was created using Optimization Toolbox™ realized in MATLAB R2018. The multi-objective genetic algorithm (MOGA) options are shown in Table 1.

Table 1. The multi-objective genetic algorithm (MOGA) settings.

Parameter	Value
Crossover coefficient	0.8
Crossover function	Intermediate
Mutation coefficient	0.2
Mutation function	Uniform
Number generations	500·number of variables
Population size	60·number of variables
Selection function	Turnament size = 2

3. Results

3.1. Case 1. Optimizaton of the Biot Number

The results of optimization described by Equation (10) are presented in Table 2 and Figures 1 and 2. The Pareto set of the Biot number for the T of 40 °C indicates that in the case of searching for the smallest values of RMSE, the best solution is ID 40_1 (0.04441). For ID 40_2, the value of R^2 is the greatest (0.9937). The solution of MAE minimization is ID 40_5 (0.03652). The set of the best solutions for 40 °C (marked in Figure 1) indicates the greatest dispersion of solutions compared with the best solutions for other drying temperatures examined. The Biot number determined for the drying temperature of 50 °C has the smallest RMSE for ID 50_1 (0.03731). The value of R^2 is the greatest for ID 50_2 (0.9955). The solution of MAE minimization is ID 50_3 (0.03228). The dispersion of the best solutions is smaller than for the temperature of 40 °C (Figure 1), and the solutions are characterised by the best statistics. The Biot number determined for the drying temperature of 60 °C has the smallest RMSE for ID 60_1 and 60_2 (0.04072–0.04073, difference 0.00001). For ID 60_4, the value of R^2 is the greatest (0.9948). The solution of MAE minimization is ID 60_4 i 60_3 (0.03640 and 0.03641, difference 0.00001). The set of best solutions for 60 °C (Figure 1) is characterised by small (the smallest) dispersion. The Biot number determined for the drying temperature of 70 °C has the smallest RMSE and the greatest R^2 for ID 70_1–70_3 (0.03888–0.03889 and 0.9943, respectively). The solution of MAE minimization is ID 70_4 (0.03420), with the solutions for ID 70_5, 70_2, 70_3, and 70_1 being slightly worse (0.03424, 0.03427, 0.03428, and 0.03429, difference 0.00009). The set of best solutions for 70 °C (Figure 1) is only characterised by a very small amount of dispersion when the MAE minimization criterion is not considered.

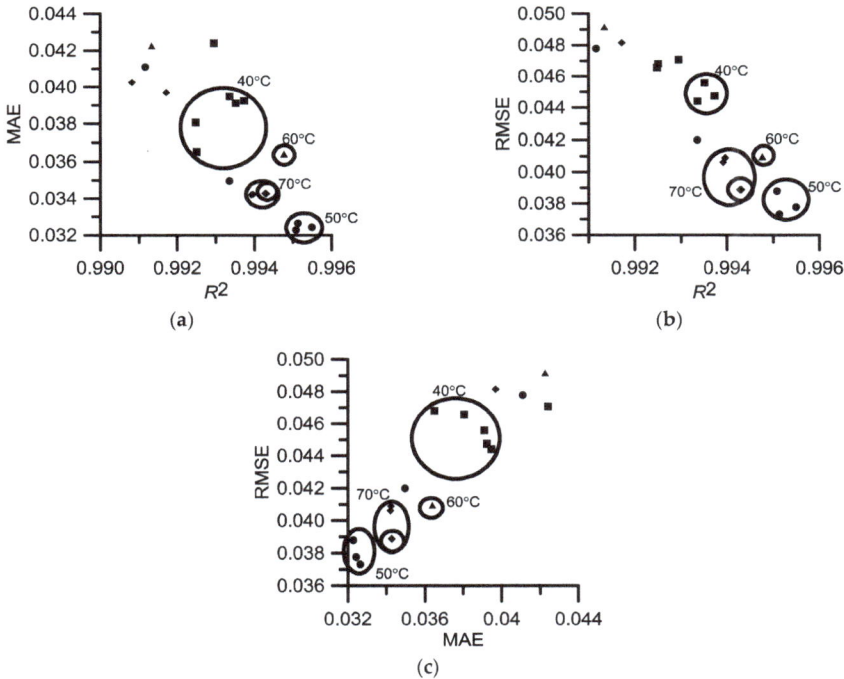

Figure 1. Results of the statistical analysis of Pareto optimal sets for the Biot number: (■) $T = 40°C$, (●) $T = 50°C$, (▲) $T = 60°C$, and (♦) $T = 70°C$.

Table 2. Pareto optimal set for the Biot number (Bi) and D/s^2 and results of the statistical analysis.

Temperature, °C	ID	Bi	D/s^2 (min^{-1})	R^2	RMSE	MAE
40	40_1	0.7233	0.00610	0.9934	0.04441	0.03947
	40_2	0.7243	0.00601	0.9937	0.04474	0.03924
	40_3	0.7016	0.00625	0.9935	0.04558	0.03910
	40_4	0.7548	0.00442	0.9925	0.04656	0.03807
	40_5	0.7450	0.00454	0.9925	0.04680	0.03652
	40_6	0.7555	0.00601	0.9930	0.04707	0.04240
	40_7	0.7749	0.00532	0.9858	0.07744	0.06666
	40_8	0.6745	0.00883	0.9820	0.07770	0.06498
	40_9	0.6901	0.00883	0.9804	0.08622	0.07129
	40_10	0.6555	0.00990	0.9748	0.10327	0.08541
50	50_1	0.7245	0.00615	0.9951	0.03731	0.03264
	50_2	0.7231	0.00605	0.9955	0.03777	0.03243
	50_3	0.7012	0.00635	0.9951	0.03879	0.03228
	50_4	0.6967	0.00698	0.9934	0.04200	0.03496
	50_5	0.6967	0.00747	0.9912	0.04779	0.04110
	50_6	0.7283	0.00734	0.9897	0.06001	0.05115
	50_7	0.6506	0.00884	0.9867	0.06764	0.05661
	50_8	0.7040	0.00824	0.9858	0.07501	0.06318
	50_9	0.6540	0.00912	0.9836	0.07945	0.06742
	50_10	0.7040	0.00873	0.9828	0.08853	0.07409
60	60_1	0.7017	0.00788	0.9944	0.04072	0.03646
	60_2	0.7017	0.00791	0.9943	0.04073	0.03654
	60_3	0.7086	0.00780	0.9946	0.04085	0.03641
	60_4	0.7001	0.00775	0.9948	0.04096	0.03640
	60_5	0.7095	0.00872	0.9913	0.04916	0.04225
	60_6	0.7337	0.00838	0.9917	0.05097	0.04363
	60_7	0.6507	0.01012	0.9889	0.05853	0.04684
	60_8	0.6513	0.01016	0.9888	0.05896	0.04720
	60_9	0.7152	0.00958	0.9875	0.06637	0.05393
	60_10	0.6791	0.01046	0.9845	0.07567	0.06103
70	70_1	0.7003	0.01012	0.9943	0.03888	0.03429
	70_2	0.7005	0.01011	0.9943	0.03888	0.03427
	70_3	0.7014	0.01011	0.9943	0.03889	0.03428
	70_4	0.6789	0.01062	0.9939	0.04061	0.03420
	70_5	0.6898	0.01060	0.9940	0.04089	0.03424
	70_6	0.7043	0.01120	0.9917	0.04815	0.03969
	70_7	0.6525	0.01260	0.9908	0.05378	0.04025
	70_8	0.6705	0.01324	0.9878	0.06406	0.05103
	70_9	0.6914	0.01286	0.9878	0.06602	0.05290
	70_10	0.6636	0.01376	0.9860	0.07202	0.05885

The values of the Biot numbers obtained as a result of optimization for the temperature of 40 °C (ID 40_1–40_5) are within the range of 0.723–0.755; for 50 °C (ID 50_1–50_3), are within the range of 0.701–0.725; for 60 °C (ID 60_1–60_4), are within the range of 0.700–0.709; for 70 °C (ID 70_1–70_3), are within the range of 0.700–0.701; and including the MAE criterion (ID 70_1–70_5), are within the range of 0.679–0.701. The smallest RMSE (0.0373) and MAE (0.0323) were obtained for $T = 50$ °C, whereas the greatest errors for were obtained for $T = 40$ °C. The results of optimization indicate that, initially, the Biot number does not change (or changes slightly) with the increase of temperature (40–50 °C), followed by the decrease in its value (50–70 °C). The obtained results of optimization also indicate that the increase of T results in the increase of the moisture diffusion coefficient (i.e., [26,28]) (Table 2).

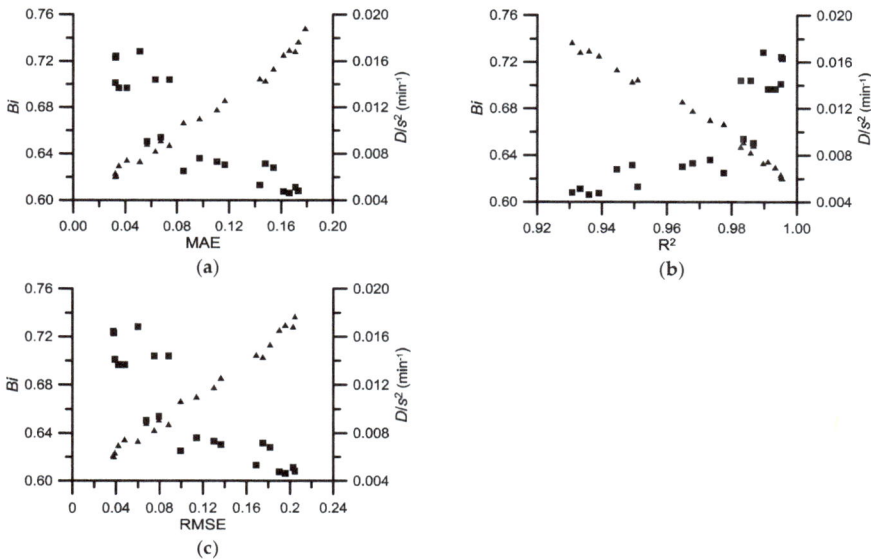

Figure 2. Pareto optimal set for the Biot number (■) and D/s^2 (▲) for $T = 50\,°C$.

In accordance with Equations (5)–(7), the value of the calculated reduced moisture content MR depends not only on the Biot number, but also, assuming the accuracy of the description by the model, on the moisture diffusion coefficient D. Figure 2 shows an example of the relation between Bi and the moisture diffusion coefficient. The increase of the Biot number requires the value of D to be reduced (Figure 2), with better fitting of the model (5) to experimental data being obtained for greater Bi values. For smaller values of Bi, the errors (RMSE and MAE) are greater, and R^2 is smaller, even after the increase of D (Figure 2).

3.2. Case 2a. Optimization of Parameters of the Function for Determining the Biot Number (Equation (8a))

The optimization task described by Equation (10) considers the Biot number from Equation (8a) in MR_{pred} (Equations (5)–(7)). Therefore, the parameters a_b and b_b of the equation (Equation (8a)) were sought in this strategy.

The results of optimization are presented in Table 3, Figure 3, and Figure 4. The best solution is ID 2a_1 (the smallest MAE = 0.05346 and the greatest R^2 = 0.9822) and ID 2a_2 (the smallest RMSE = 0.06695 and slightly worse MAE and R^2 compared with ID 2a_1).

Table 3. Pareto optimal set for constants in Equations (8a) and (9) and results of the statistical analysis.

ID	a_b	b_b	a_d	b_d	R^2	RMSE	MAE
2a_1	0.81932	6.49505	0.00704	2.54000	0.9822	0.06724	0.05346
2a_2	0.81976	6.53818	0.00720	2.54000	0.9817	0.06695	0.05358
2a_3	0.81935	6.48207	0.00717	2.28576	0.9815	0.06717	0.05374
2a_4	0.81959	6.43306	0.00722	2.06814	0.9812	0.06744	0.05405
2a_5	0.82018	6.47067	0.00700	1.57058	0.9814	0.06783	0.05405
2a_6	0.82042	6.47042	0.00705	1.57079	0.9813	0.06779	0.05411
2a_7	0.81959	6.43306	0.00722	1.56814	0.9807	0.06804	0.05455
2a_8	0.81964	6.52297	0.00732	1.91671	0.9807	0.06794	0.05457
2a_9	0.81955	6.46056	0.00735	1.58362	0.9803	0.06864	0.05514
2a_10	0.81958	6.45445	0.00750	1.87225	0.9800	0.06911	0.05557

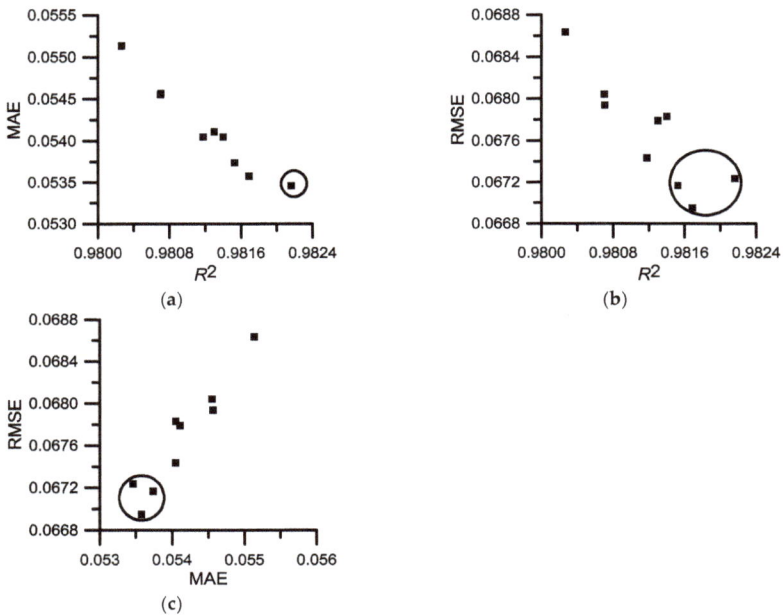

Figure 3. Pareto optimal sets for constants of Equation (8a).

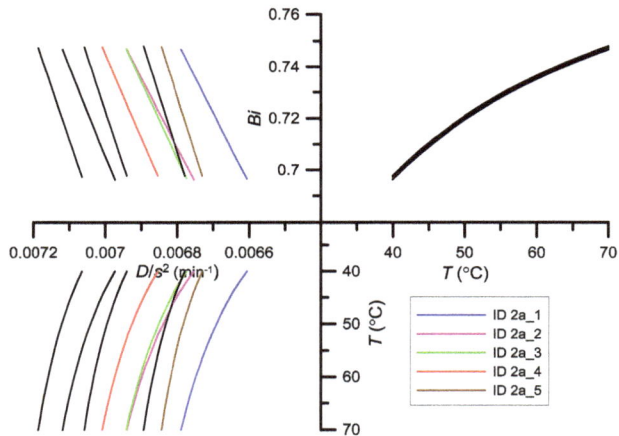

Figure 4. Pareto optimal set for the function of the Biot number and D/s^2: Equations (8a) and (9).

The courses of the function (Equation (8a)) with parameters a_b and b_b obtained as a result of optimization are very similar (Figure 5) (overlap almost entirely–similar values of parameters (Table 3)), and the values of the Biot number obtained from them for the temperatures of 40–70°C are within the range of 0.696523–0.74672 for ID 2a_1 and 0.696144–0.746659 for ID 2a_2. The courses of the function (Equation (9)) with parameters a_d and b_d obtained as a result of optimization differ from each other (Figure 4) and the best solution of the optimization task was obtained for smaller values of the moisture diffusion coefficient. The values of D/s^2 for the temperatures of 40–70°C are within the range of 0.006606–0.006788 for ID 2a_1 and 0.006753–0.00694 for ID 2a_2.

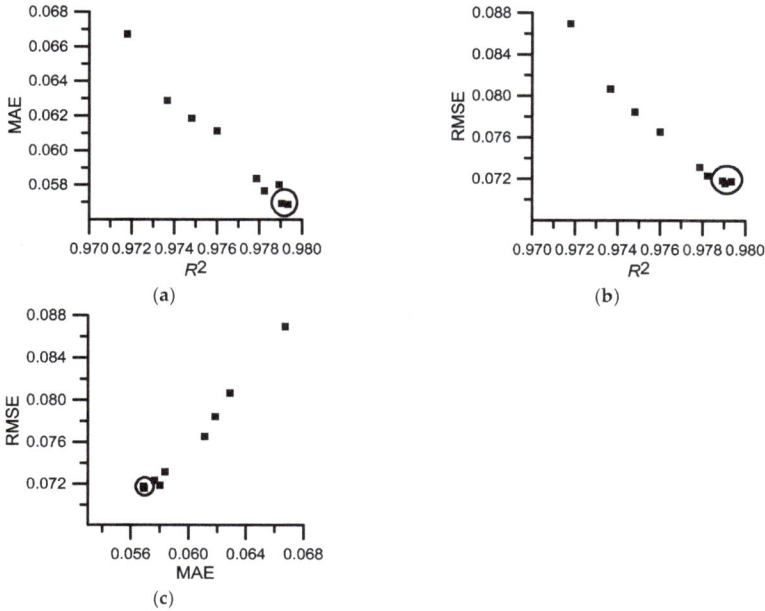

Figure 5. Pareto optimal sets for constants of Equation (8b).

3.3. Case 2b. Optimization of Parameters of the Function for Determining the Biot number (Equation (8b))

The results of optimization are presented in Table 4 and Figures 5 and 6. The best solutions are ID 2b_1 (the smallest MAE = 0.05688 and the greatest R^2 = 0.9794) and ID 2b_2 (the smallest RMSE = 0.07160 and slightly smaller R^2 (difference 0.0004)).

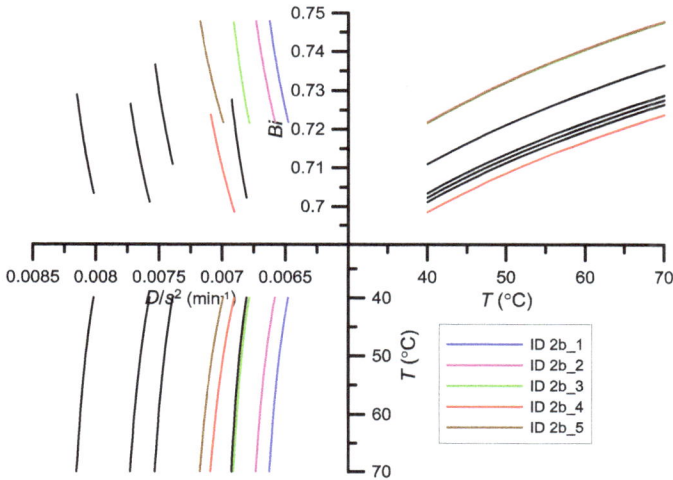

Figure 6. Pareto optimal set for the function of the Biot number and D/s^2: Equations (8b) and (9).

The courses of the function (Equation (8b)) with parameters a_b and b_b obtained as a result of optimization are very similar (Figure 6) (ID 2b_1, 2b_2, and 2b_3 overlap almost entirely - similar values of parameters (Table 4)), and the values of the Biot numbers obtained from the function for the temperatures of 40–70 °C are within the range of 0.721854–0.747842 both for ID 2b_1 and ID 2b_2, and they are the greatest compared with the other solutions. The courses of the function (Equation (9)) with parameters a_d and b_d obtained as a result of optimization differ from each other (Figure 6) and the best solution of the optimization task was obtained for smaller values of the moisture diffusion coefficient. The values of D/s^2 for the temperatures of 40–70°C are within the range of 0.006477–0.006624 for ID 2b_1 and 0.006582–0.006731 for ID 2b_2.

Table 4. Pareto optimal set for constants in Equations (8b) and (9) and results of the statistical analysis.

ID	a_b	b_b	a_d	b_d	R^2	RMSE	MAE
2b_1	0.04644	140795.3	0.006825	2.094928	0.9794	0.07178	0.05688
2b_2	0.04644	140795.5	0.006935	2.091899	0.9790	0.07160	0.05694
2b_3	0.046449	139323	0.007073	1.661161	0.9782	0.07233	0.05766
2b_4	0.045152	130655.1	0.007347	2.498029	0.9789	0.07185	0.05803
2b_5	0.046592	133294.6	0.007432	2.446309	0.9779	0.07316	0.05837
2b_6	0.045209	139490.1	0.007085	1.60367	0.9760	0.07654	0.06113
2b_7	0.045793	138027.1	0.007722	1.747985	0.9748	0.07845	0.06186
2a_8	0.045224	135305.8	0.007933	1.8519	0.9737	0.08067	0.06288
2a_9	0.045296	138802.9	0.008329	1.525632	0.9718	0.08710	0.06673
2b_10	0.046629	139913.5	0.008114	1.605228	0.9718	0.08986	0.06883

3.4. Case 2c. Optimization of Parameters of the Function for Determining the Biot Number (Equation (8c))

The results of optimization are presented in Table 5 and Figures 7 and 8. The best solutions are ID 2c_1 (the smallest MAE = 0.05693 and the greatest R^2 = 0.9793) and ID 2c_2 (the smallest RMSE = 0.07168843 and slightly smaller R^2 (difference 0.0003)).

Table 5. Pareto optimal set for constants in Equations (8c) and (9) and results of the statistical analysis.

ID	a_b	b_b	a_d	b_d	R^2	RMSE	MAE
2c_1	0.105914	161892.9	0.006818	1.999354	0.9793	0.07184	0.05693
2c_2	0.105920	161892.9	0.006939	1.999354	0.9790	0.07168	0.05701
2c_3	0.105956	161896.9	0.007083	1.836184	0.9784	0.07212	0.05751
2c_4	0.105896	162115	0.007115	1.717303	0.9782	0.07242	0.05773
2c_5	0.105920	161892.9	0.007259	1.999354	0.9780	0.07274	0.05803
2c_6	0.102556	162170.2	0.007465	1.623216	0.9777	0.07301	0.05877
2c_7	0.102552	162127.3	0.007589	1.80125	0.9775	0.07331	0.05887
2c_8	0.102543	162188.4	0.007620	1.608689	0.9772	0.07384	0.05919
2c_9	0.102548	162077.8	0.007761	1.823051	0.9770	0.07446	0.05951
2c_10	0.103676	162170.9	0.007290	1.999114	0.9765	0.07493	0.05968

The courses of the function (Equation (8c)) with parameters a_b and b_b obtained as a result of optimization are very similar (Figure 8) (overlap almost entirely–similar values of parameters (Table 5)), and the values of the Biot numbers obtained from the function for the temperatures of 40–70°C are within the range of 0.721413–0.747151 for ID 2c_1 and 0.721450–0.747192 for ID 2c_2, and they are the greatest compared with other solutions. The courses of the function (Equation (9)) with parameters a_d and b_d obtained as a result of optimization differ from each other (Figure 8) and the best solution of the optimization task was obtained for smaller values of the moisture diffusion coefficient. The values of D/s^2 for the temperatures of 40–70°C are within the range of 0.00649–0.00663 for ID 2c_1 and 0.0066–0.00674 for ID 2c_2.

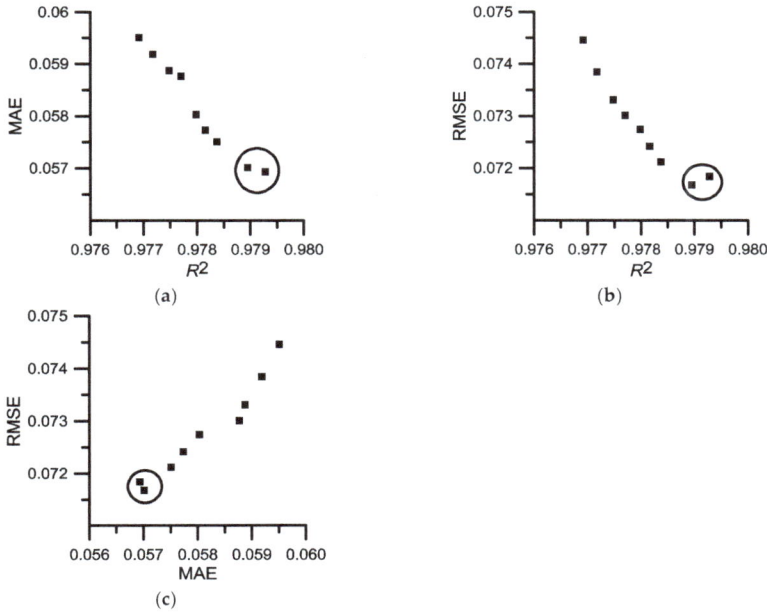

Figure 7. Pareto optimal sets for constants of Equation (8c).

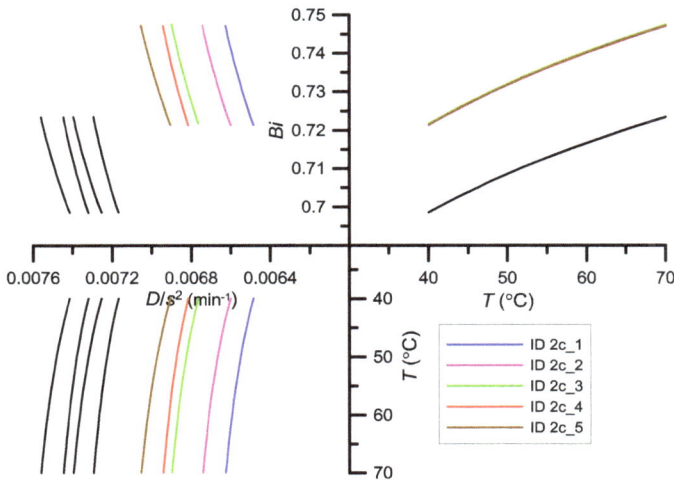

Figure 8. Pareto optimal set for the function of the Biot number and D/s^2: Equations (8c) and (9).

3.5. Case 2d. Optimization of Parameters of the Function for Determining the Biot Number (Equation (8d))

The results of optimization are presented in Table 6 and Figures 9 and 10. The performed optimization also allowed us to obtain constants for Equation (8d) and Equation (9) and then to calculate the Biot number and the moisture diffusion coefficient (Table 6). The best solutions are ID 2d_1 (the smallest MAE = 0.05378 and the greatest R^2 = 0.9818) and ID 2d_2 (the smallest RMSE = 0.06742 and only slightly worse R^2 (difference 0.0004)).

Table 6. Pareto optimal set for constants in Equations (8d) and (9) and results of the statistical analysis.

ID	a_b	b_b	a_d	b_d	R^2	RMSE	MAE
2d_1	0.087671	0.374624	0.006984	1.988010	0.9818	0.06763	0.05378
2d_2	0.087777	0.374868	0.007130	1.988010	0.9814	0.06742	0.05390
2d_3	0.099520	0.329786	0.007546	2.461728	0.9804	0.06837	0.05507
2d_4	0.091440	0.359263	0.007757	2.217054	0.9794	0.07085	0.05688
2d_5	0.100428	0.303184	0.008240	2.052006	0.9783	0.07329	0.05852
2d_6	0.091342	0.356587	0.007770	2.413837	0.9778	0.07300	0.05883
2d_7	0.078979	0.400324	0.007770	1.930612	0.9773	0.07414	0.05956
2d_8	0.072832	0.412350	0.007200	1.973931	0.9772	0.07705	0.06194
2d_9	0.077064	0.381598	0.008456	2.366783	0.9758	0.07764	0.06241
2d_10	0.029274	0.581316	0.008014	1.660092	0.9746	0.07869	0.06313

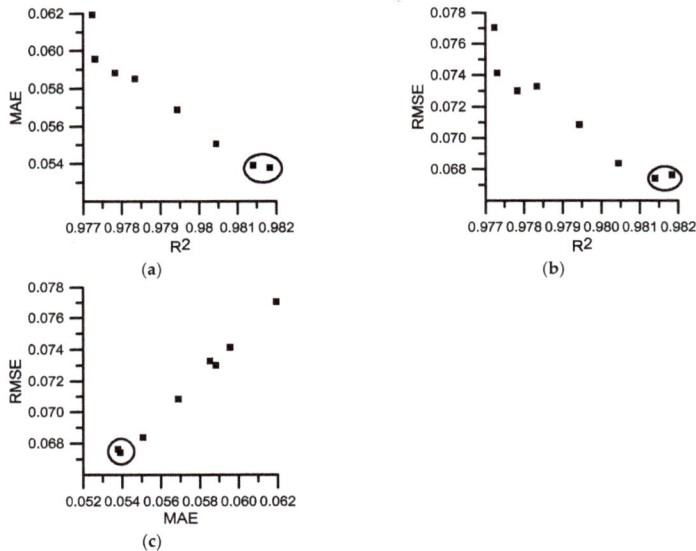

Figure 9. Pareto optimal sets for constants of Equation (8d).

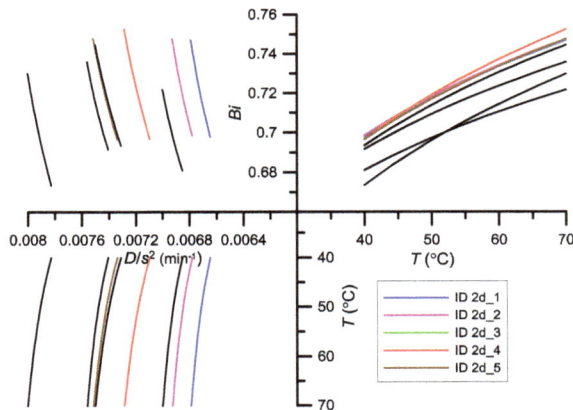

Figure 10. Pareto optimal set for the function of the Biot number and D/s^2: Equations (8d) and (9).

The courses of the function (Equation (8d)) with parameters a_b and b_b obtained as a result of optimization are very similar (Figure 10) (overlap almost entirely–similar values of parameters (Table 6)), and the values of the Biot numbers obtained from the function for the temperatures of 40–70 °C are within the range of 0.698032–0.747095 for ID 2d_1 and 0.698668–0.747789 for ID 2d_2, and are the greatest compared with other solutions. The courses of the function (Equation 9) with parameters a_d and b_d obtained as a result of optimization differ from each other (Figure 10) and the best solution of the optimization task was obtained for smaller values of the moisture diffusion coefficient. The values of D/s^2 for the temperatures of 40–70°C are within the range of 0.00665–0.00679 for ID 2d_1 and 0.00678–0.00693 for ID 2d_2.

3.6. Case 2e. Optimization of Parameters of the Function for Determining the Biot Number (Equation (8e))

The results of optimization are presented in Table 7 and Figures 11 and 12. The optimization also allowed us to obtain constants Equation (8e) and Equation (9) and then to calculate the Biot number and the moisture diffusion coefficient (Table 7). The set of best solutions is made up of ID 2e_1–ID 2e_5, with ID 2e_1 being characterised by the smallest value of MAE = 0.06633, ID 2e_4 by the smallest value of RMSE = 0.08183, and ID 2e_5 by the greatest R^2 = 0.9771. For ID 2e_1, RMSE is greater than for ID 2e_4 by only 0.0002 and R^2 is smaller than for ID 2e_5 by only 0.0003. For ID 2e_4, MAE is greater than for ID 2e_1 by only 0.0002 and R^2 is only 0.0003 smaller than for ID 2e_5. For ID 2e_5, MAE is greater than for ID 2e_1 by only 0.0002 and RMSE is greater than for ID 2e_4 by only 0.00006. For ID 2e_2, ID 2e_3, and ID 2e_6, R^2 is smaller than the greatest one (ID 2e_5) by 0.0002, 0.00007, and 0.0001, respectively; RMSE is greater than the smallest one (ID 2e_4) by 0.00013, 0.00014, and 0.000005, respectively; and MAE is greater than the smallest one (ID 2e_1) by 0.00001, 0.0009, and 0.00025, respectively.

Table 7. Pareto optimal set for constants in Equations (8e) and (9) and results of the statistical analysis.

ID	a_b	b_b	a_d	b_d	R^2	RMSE	MAE
2e_1	1.298971	0.124544	0.006152	4.960183	0.9768	0.08207	0.06633
2e_2	1.449848	0.153382	0.006185	4.996656	0.9769	0.08196	0.06634
2e_3	1.290019	0.131396	0.006165	4.989554	0.9770	0.08197	0.06642
2e_4	1.289663	0.129626	0.006287	4.989283	0.9768	0.08183	0.06651
2e_5	1.456449	0.165094	0.006188	4.984513	0.9771	0.08189	0.06652
2e_6	1.466073	0.166745	0.006246	4.961883	0.9770	0.08184	0.06658
2a_7	1.557768	0.172773	0.006043	3.882209	0.9762	0.08287	0.06694
2e_8	1.562403	0.184398	0.006047	3.886745	0.9765	0.08278	0.06715
2e_9	1.56531	0.183702	0.006119	3.888001	0.9763	0.08269	0.06719
2e_10	1.276498	0.120256	0.005896	2.895483	0.9756	0.08381	0.06743

The courses of the function (Equation (8e)) with parameters a_b and b_b obtained as a result of optimization differ from each other (Figure 12), and the values of Bi obtained from them are the greatest. The values of Bi obtained for the temperatures of 40–70 °C are within the range of 0.823371–0.721929: 0.820488–0.76525 for ID 2e_1, 0.823371–0.755645 for ID 2e_2, 0.794496–0.738172 for ID 2e_3, 0.79948–0.743539 for ID 2e_4, 0.792145–0.722239 for ID 2e_5, and 0.792538–0.721929 for ID 2e_6. The courses of the function (Equation (9)) with parameters a_d and b_d obtained as a result of optimization are similar (Figure 12) and the best solution of the optimization task was obtained for smaller values of the moisture diffusion coefficient (except for ID 2e_4). The values of D/s^2 obtained for the temperatures of 40–70 °C are within the range of 0.00543–0.00573 for ID 2e_1, 0.00546–0.00576 for ID 2e_2, 0.00544–0.00574 for ID 2e_3, 0.00555–0.00585 for ID 2e_4, 0.00546–0.00576 for ID 2e_5, and 0.00552–0.00582 for ID 2e_6.

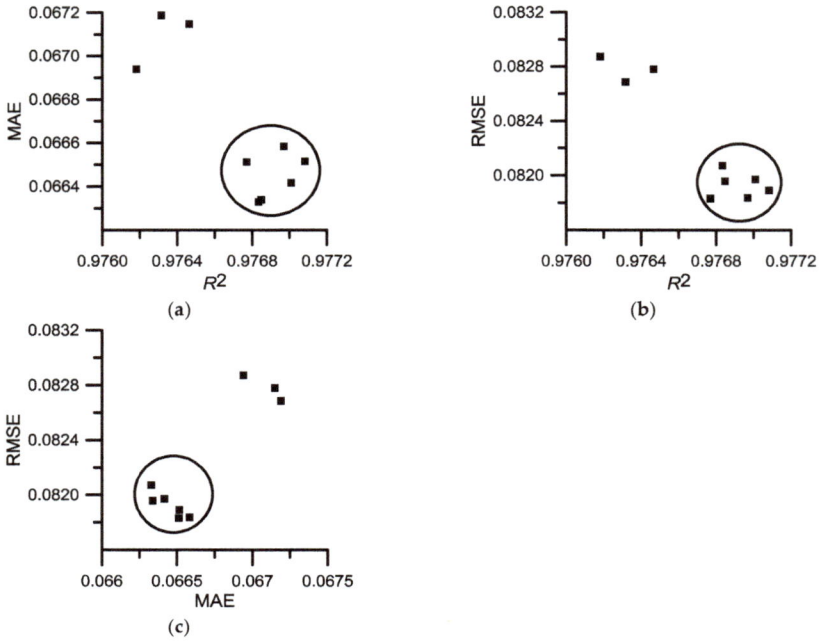

Figure 11. Pareto optimal sets for constants of Equation (8e).

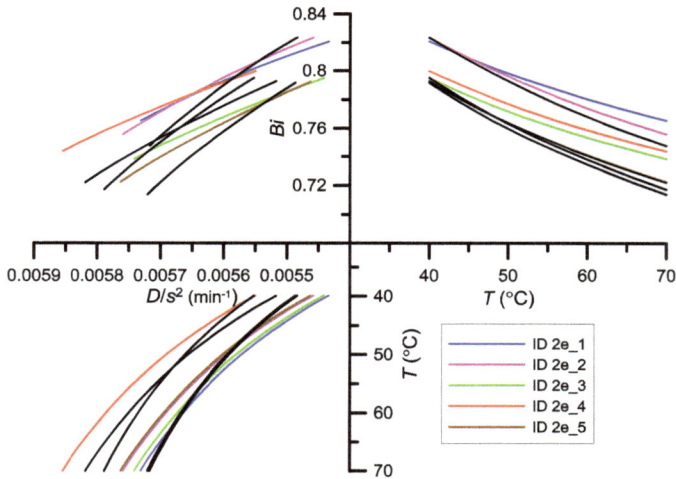

Figure 12. Pareto optimal set for the function of the Biot number and D/s^2: Equations (8e) and (9).

3.7. Case 2f. Optimization of Parameters of the Function for Determining the Biot Number (Equation (8f))

The results of optimization are presented in Table 8 and Figures 13 and 14. The best solutions are ID 2f_1 (the smallest MAE = 0.06056 and the greatest R^2 = 0.9764) and ID 2f_2 (the smallest RMSE = 0.07591, and compared with 2f_1, slightly worse R^2 (smaller by 0.0002) and MAE (greater by 0.00002)).

Table 8. Pareto optimal set for constants in Equations (8f) and (9) and results of the statistical analysis.

ID	a_b	b_b	a_d	b_d	R^2	RMSE	MAE
2f_1	0.077244	862213.6	0.007368	5.739963	0.9764	0.07598	0.06056
2f_2	0.077236	862220.8	0.007449	5.729922	0.9762	0.07591	0.06058
2f_3	0.077207	864646.7	0.007868	5.355261	0.9748	0.07803	0.06199
2f_4	0.073875	1059418	0.006741	2.720529	0.9745	0.07933	0.06235
2a_5	0.077859	862396.3	0.007587	5.708511	0.9730	0.08043	0.06483
2a_6	0.076727	865525.6	0.007648	4.417204	0.9713	0.08331	0.06520
2f_7	0.078282	862213.6	0.007368	5.741489	0.9727	0.08254	0.06684
2f_8	0.076328	960205.5	0.007648	2.969758	0.9698	0.08642	0.06705
2f_9	0.072421	1059416	0.006582	1.131156	0.9690	0.08750	0.06884
2f_10	0.070248	1137211	0.006003	0.042246	0.9696	0.08812	0.06893

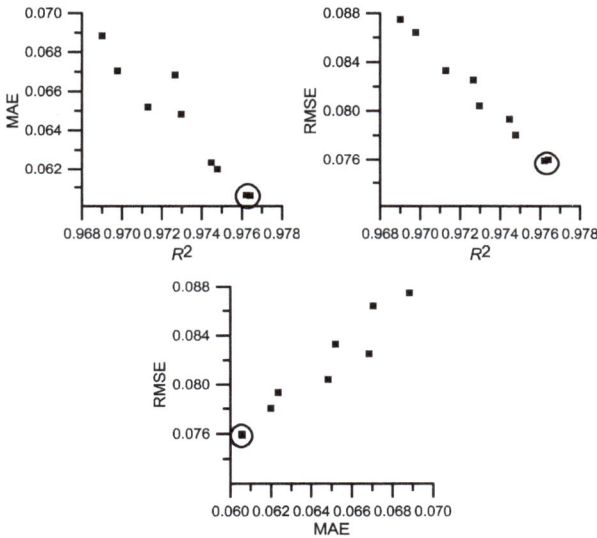

Figure 13. Pareto optimal sets for constants of Equation (8f).

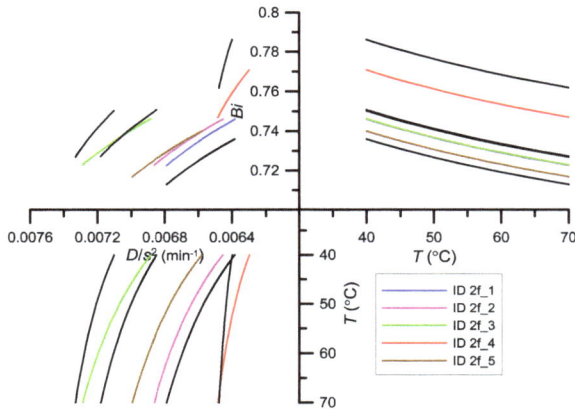

Figure 14. Pareto optimal set for the function of the Biot number and D/s^2: Equations (8f) and (9).

The courses of the function (Equation (8f)) with parameters a_b and b_b obtained as a result of optimization differ from each other (Figure 14). The values of the Biot number obtained for the temperatures of 40–70 °C are within the range of 0.745901–0.722602 for ID 2f_1 and 0.745978–0.722677 for ID 2f_2. The courses of the function (Equation 9) with parameters a_d and b_d obtained as a result of optimization differ from each other (Figure 14). The values of D/s^2 obtained for the temperatures of 40–70 °C are within the range of 0.006383–0.006788 for ID 2f_1 and 0.006454–0.006863 for ID 2f_2.

3.8. Case 2g. Optimization of Parameters of the Function for Determining the Biot Number (Equation (8g))

The results of optimization are presented in Table 9 and Figure 15–16. The best solution is ID 2g_1 (the smallest both MAE = 0.06041, RMSE = 0.07566 and the greatest $R^2 = 0.9764$).

Table 9. Pareto optimal set for constants in Equations (8g) and (9) and results of the statistical analysis.

ID	a_b	b_b	a_d	b_d	R^2	RMSE	MAE
2g_1	0.17462	1193847	0.007492	5.999963	0.9764	0.07566	0.06041
2g_2	0.17462	1193847	0.007492	4.984338	0.9754	0.07677	0.06124
2g_3	0.20101	109835.1	0.007664	7.999996	0.9748	0.07827	0.06191
2g_4	0.20101	109834.6	0.007698	7.999992	0.9747	0.07826	0.06192
2g_5	0.20097	109790.8	0.007727	7.968262	0.9746	0.07830	0.06195
2g_6	0.19994	114067.1	0.007438	6.988878	0.9743	0.07932	0.06265
2g_7	0.18938	116127.4	0.007119	8.999996	0.9760	0.07926	0.06310
2g_8	0.18938	116103.4	0.007247	8.999909	0.9758	0.07911	0.06312
2g_9	0.18944	115792.8	0.007267	8.882749	0.9749	0.07991	0.06341
2g_10	0.19830	114642.6	0.007446	5.755039	0.9731	0.08041	0.06346

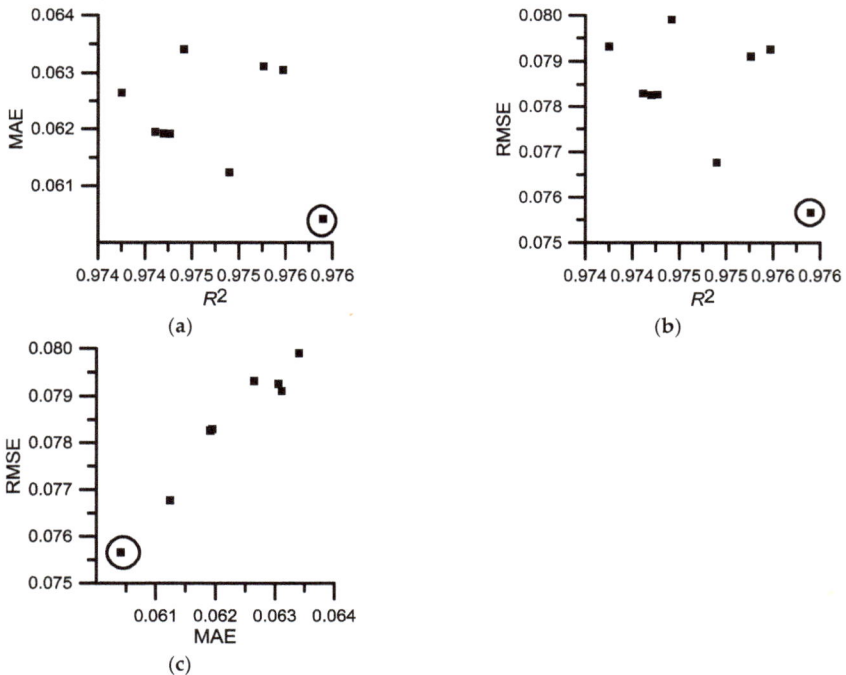

Figure 15. Pareto optimal sets for constants of Equation (8g).

The courses of the function (Equation (8g)) with parameters a_b and b_b and the function (Equation (9)) with parameters a_d and b_d obtained as a result of optimization differ from each other (Figure 16). The values of the Biot number and the values of D/s^2 obtained for the temperatures of 40–70 °C for ID 2g_1 change slightly and are within the range of 0.745762–0.722883 and 0.006448–0.006877, respectively.

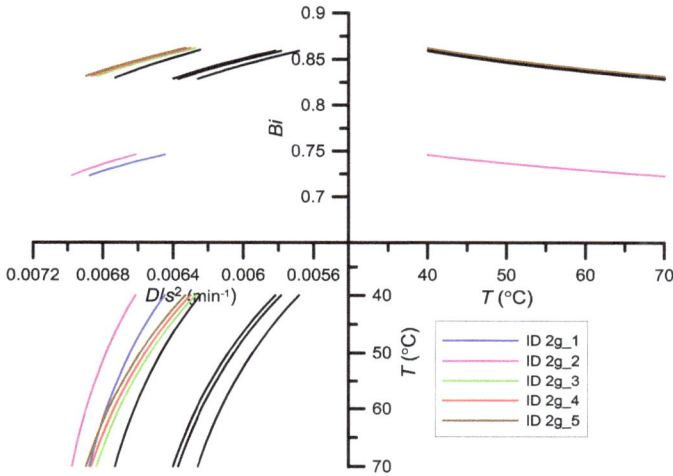

Figure 16. Pareto optimal set for the function of the Biot number and D/s^2: Equations (8g) and (9).

3.9. Case 2h. Optimization of Parameters of the Function for Determining the Biot Number (Equation (8h))

The results of optimization are presented in Table 10 and Figures 17 and 18. The best solutions are ID 2h_1 (the smallest MAE = 0.06293 and the greatest R^2 = 0.9740) and ID 2h_2 (the smallest RMSE = 0.07925, with MAE being slightly worse than for ID 2h_1 (the difference 0.00003)).

Table 10. Pareto optimal set for constants in Equations (8h) and (9) and results of the statistical analysis.

ID	a_b	b_b	a_d	b_d	R^2	RMSE	MAE
2h_1	1.434697	2.507106	0.006923	2.14	0.9740	0.07927	0.06293
2h_2	1.434897	2.509596	0.006967	2.14	0.9739	0.07925	0.06296
2h_3	1.431146	2.528872	0.007108	2.13075	0.9736	0.07946	0.06315
2h_4	1.431491	2.57301	0.007147	2.047283	0.9734	0.07970	0.06330
2h_5	1.429736	2.524792	0.007201	2.118279	0.9733	0.07981	0.06337
2h_6	1.429389	2.530306	0.007283	2.102477	0.9731	0.08025	0.06364
2h_7	1.432067	2.515218	0.007349	2.128414	0.9729	0.08062	0.06385
2h_8	1.430119	2.532143	0.007376	2.103931	0.9728	0.08085	0.06400
2h_9	1.42891	2.540204	0.007397	2.086483	0.9727	0.08104	0.06412
2h_10	1.435259	2.545177	0.007423	2.112372	0.9727	0.08118	0.06418

The courses of the function (Equation (8h)) with parameters a_b and b_b obtained as a result of optimization differ slightly from each other (Figure 18). The values of the Biot number obtained for the temperatures of 40–70 °C are within the range of 0.742097–0.722428 for ID 2h_1 and within the range of 0.742039–0.722353 for ID 2h_2. The courses of the function (Equation (9)) with parameters a_d and b_d obtained as a result of optimization are similar (Figure 18) and the best solution of the optimization task was obtained for smaller values of the moisture diffusion coefficient. The values of D/s^2 obtained for the temperatures of 40–70 °C are within the range of 0.006562–0.006714 for ID 2h_1 and within the range of 0.006604–0.006757 for ID 2h_2.

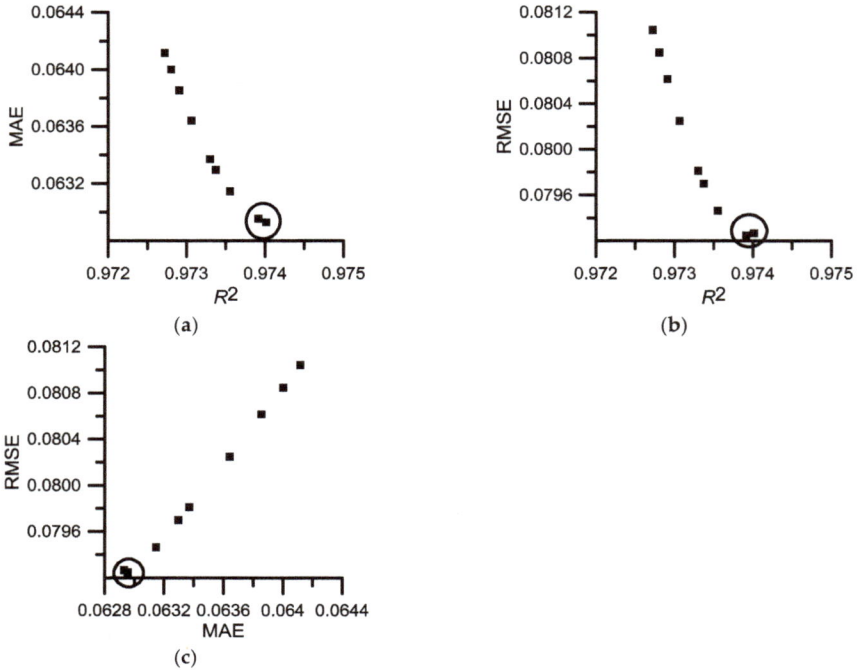

Figure 17. Pareto optimal sets for constants of Equation (8h).

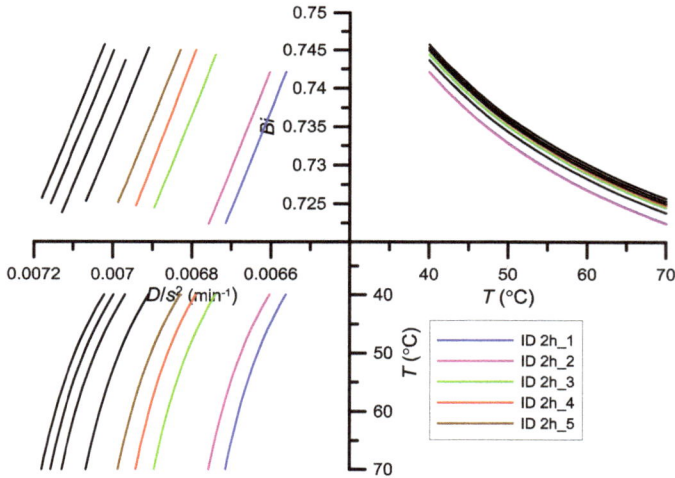

Figure 18. Pareto optimal set for the function of the Biot number and D/s^2: Equations (8h) and (9).

Among the functions (Equation (8a))–(Equation (8d)) indicating the increase of the Biot number with the increase of the temperature, the best one was the function described by the equation (Equation (8a)), and the second best function (Equation (8d)) was only slightly worse. Among the functions (Equation (8e))–(Equation (8h)) indicating the decrease of the Biot number with the increase of the temperature, the best one was the function described by the equation (Equation (8g)), and the second best function (Equation (8h)) was only slightly worse. Figure 19 shows the Biot number and the moisture diffusion coefficient for the best solutions of the optimization task 2a, 2d, 2g, and 2h.

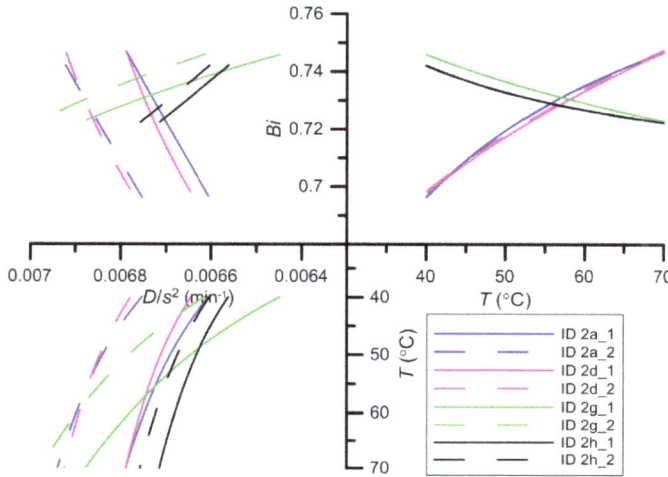

Figure 19. Pareto optimal set for the function of the Biot number and D/s^2: Equations (8a), (8d), (8g), (8h), and (9).

The range of variability of the Biot number for ID 2a_1, 2a_2, and 2d_1 is identical (the courses of the functions are almost identical), and it ranges between 0.70 and 0.75, whereas for ID 2g_1, 2g_2, 2h_1, and 2h_2, the range is smaller, between 0.72 and 0.75. D/s^2 calculated using (Equation (9)) changes most for ID 2g_1 (0.00645–0.00688). For ID 2g_2, the changes are smaller (0.00661–0.00698), and for ID 2a_2 and 2d_2, the changes are the smallest (the courses of the functions are almost identical), ranging between 0.00675 and 0.00685.

According to (Equation (1)), the Biot number depends on the mass transfer coefficient h_m and mass diffusion coefficient D. Therefore, the dependence of Bi on temperature is determined by the temperature influence on both mentioned coefficients. An increase of the temperature (according to the literature (i.e., Equation (9))) results in an increase of the moisture diffusion coefficient. In the definition of the Biot number (Equation (1)), there is a ratio h_m/D and the Biot number depends precisely on this relation. A greater temperature impact on h_m than on D results in the increase of the Biot number with temperature. Many authors have applied a dimensionless number to estimate the value of the mass transfer coefficient [8,24,29,30]. Equation $Sh = f(Gr,Sc)$ is often applied under natural convection conditions [31].

The dependences obtained as a result of optimization indicate that a better fitting of the model (5) to the experimental data is obtained when Bi increases with the increase of the drying temperature according to the (Equation (8a)) with the constants of the equation for ID 2a_1 (Table 3):

$$Bi = 0.8193 \exp\left(-6.4951T^{-1}\right), \tag{11}$$

Simultaneously, D/s^2 is calculated based on (Equation (9)) with the constants of the equation for ID 2a_1 (Table 3):

$$D/s^2 = 0.00704 \exp\left(-2.54T^{-1}\right), \tag{12}$$

However, only slightly worse fitting is obtained when Bi decreases with the increase of the drying temperature according to (Equation (8g)) with the constants of the equation for ID 2g_1 (Table 9):

$$Bi = 1/0.1746 \log(119347T) \tag{13}$$

Simultaneously, D/s^2 is calculated based on (Equation (9)) with the constants of the equation for ID 2g_1 (Table 9):

$$D/s^2 = 0.0075 \exp\left(-6T^{-1}\right), \tag{14}$$

The values of D/s^2 obtained from Equation (12) (for 40–70 °C) are the following: 0.00665–0.00679, and they are in the range of 0.00645–0.00688 obtained from Equation (14). The D values lie within the general range of 10^{13} to 10^6 m^{2}s^{-1} for food materials [32,33]. However the range of the Biot number calculated from Equation (13) (0.7729–0.7458) is wider than that obtained from Equation (11) (0.6965–0.6747).

The acceptance in the drying model (Equation (5)) of the Biot number and D/s^2 determined from Equations (11) and (12) or from Equations (13) and (14) results in various fits of the discussed model at individual considered temperatures. The MAE values for the model (Equation (5)) (Bi and D/s^2 from Equations (11) and (12)) are the following: 0.0690, 0.0340, 0.0426, and 0.0725 (for 40, 50, 60, and 70 °C, respectively), whereas for Bi and D/s^2 determined from Equations (13) and (14), the MAE values are much higher for extreme temperatures, namely 40 and 70 °C, and they are the following: 0.0815, 0.0377, 0.0.0425, and 0.0856 (for the mentioned temperatures, respectively).

The results of the validation of Biot number estimation are the following. The validation was done using the experimental data of drying kinetics of parsley root (Berlińska variety, 6 mm thick slices) dried in natural convection conditions at 55 °C. Biot number values were the following: 0.7281 and 0.7326 (Equations (11) and (13), respectively), whereas D/s^2 equalled 0.0067 (Equations (12) and (14)). Regardless of whether Equations (11) and (12) or (13) and (14) were used, $R^2 = 0.9955$, RMSE = 0.0376, and MAE = 0.0336, so the validation can be treated as satisfactory.

4. Conclusions

The analysis of Biot numbers enables questions about the controlling mechanisms employed for heat or mass transfer during the considered process to be answered.

This paper used a multi-objective optimization method (based on GA and Pareto optimization) for determination of the Biot number. A MOO GA method with a consideration of the simultaneous maximization of R^2 and minimization of RMSE and MAE between experimental data and the drying model was successfully applied.

The optimum values of the Biot number and constants of function used to determine the Biot number, gained by the MOO GA, were found. Eight types of equations for determining the Bi were tested. The Biot number can be calculated from the following equations: $Bi = 0.8193\exp(-6.4951T^{-1})$ (and moisture diffusion coefficient from $D/s^2=0.00704\exp(-2.54T^{-1})$) or $Bi=1/0.1746\log(1193847T)$ ($D/s^2 = 0.0075\exp(-6T^{-1})$). The results of statistical analysis were as follows: RMSE = 0.0672, MAE = 0.0535, and $R^2 = 0.98$ for the first equation for Bi and D/s^2 and RMSE = 0.0757, MAE = 0.0604, and $R^2 = 0.98$ for the second one. The conducted validation gave good results: RMSE = 0.0376, MAE = 0.0336, and $R^2 = 0.9955$.

Author Contributions: K.G.: Proposal of the research topic, experiments, modelling, and writing of manuscript; R.W.: Optimization and writing of manuscript; A.K.: Writing of manuscript and critical revision of manuscript.

Funding: This research received no external funding.

Conflicts of Interest: The authors declare no conflicts of interest.

Nomenclature

a_b, b_b	constants in Equations (8a)–(8h)
a_d, b_d	constants in Equation (9)
$Bi = h_m L/D$	Biot number for mass transfer
c	concentration
D	mass diffusion coefficient ($m^2 s^{-1}$)
D_{AB}	moisture diffusivity in the gaseous phase ($m^2 s^{-1}$)
g	acceleration of gravity (ms^{-2})
$Gr = gL^3 \beta \Delta c / \nu^2$	Grashof number
h_m	mass transfer coefficient (ms^{-1})
L	characteristic dimension (m)
M	moisture content (kg H_2O kg^{-1} d.m.)
M_0	initial moisture content (kg H_2O kg^{-1} d.m.)
M_e	equilibrium moisture content (kg H_2O kg^{-1} d.m.)
MR	moisture ratio
MR_{pred}	predicted moisture ratio
MR_{exp}	moisture ratio from experiment
$\overline{MR}_{exp}$	average value of moisture ratio from experiment
n	number of data
s	half of plane thickness (m)
$Sc = \nu/D_{AB}$	Schmidt number
$Sh = h_m L/D_{AB}$	Sherwood number
T	drying air temperature (°C)
t	time (s)
x	coordinate (m)

Greek symbols

β	coefficient of expansion
υ	kinematic viscosity ($m^2 s^{-1}$)

References

1. Giner, S.A.; Irigoyen, R.M.T.; Cicuttín, S.; Fiorentini, C. The variable nature of Biot numbers in food drying. *J. Food Eng.* **2010**, *101*, 214–222. [CrossRef]
2. Cuesta, F.J.; Lamúa, M.; Alique, R. A new exact numerical series for the determination of the Biot number: Application for the inverse estimation of the surface heat transfer coefficient in food processing. *Int. J. Heat Mass Transf.* **2012**, *55*, 4053–4062. [CrossRef]
3. Dincer, I. Moisture transfer analysis during drying of slab woods. *Heat Mass Transf.* **1998**, *34*, 317–320. [CrossRef]
4. Rovedo, C.O.; Suarez, C.; Viollaz, P. Analysis of moisture profiles, mass Biot number and driving forces during drying of potato slabs. *J. Food Eng.* **1998**, *36*, 211–231. [CrossRef]
5. Ruiz-López, I.I.; Ruiz-Espinosa, H.; Arellanes-Lozada, P.; Bárcenas-Pozos, M.E.; García-Alvarado, M.A. Analytical model for variable moisture diffusivity estimation and drying simulation of shrinkable food products. *J. Food Eng.* **2012**, *108*, 427–435. [CrossRef]
6. Ruiz-López, I.I.; Ruiz-Espinosa, H.; Luna-Guevara, M.L.; García-Alvarado, M.A. Modeling and simulation of heat and mass transfer during drying of solids with hemispherical shell geometry. *Comput. Chem. Eng.* **2011**, *35*, 191–199. [CrossRef]
7. Wu, Y.; Irudayaraj, J. Analysis of heat, mass and pressure transfer in starch based food systems. *J. Food Eng.* **1996**, *29*, 399–414. [CrossRef]
8. Górnicki, K.; Kaleta, A. Some problems related to mathematical modelling of mass transfer exemplified of convection drying of biological materials. In *Heat and Mass Transfer*; Hossain, M., Ed.; IntechOpen: Rijeka, Croatia, 2011.
9. Huang, C.-H.; Yeh, C.-Y. An inverse problem in simultaneous estimating the Biot numbers of heat and moisture transfer for a porous material. *Int. J. Heat Mass Transf.* **2002**, *45*, 4643–4653. [CrossRef]

10. Dincer, I.; Hussain, M.M. Development of a new Biot number and lag factor correlation for drying applications. *Int. J. Heat Mass Transf.* **2004**, *47*, 653–658. [CrossRef]
11. Chen, X.D.; Peng, X. Modified Biot number in the context of air drying of small moist porous objects. *Dry. Technol.* **2005**, *23*, 83–103. [CrossRef]
12. Xie, Y.; Gao, Z.; Liu, Y.; Xiao, H. Pulsed vacuum drying of rhizoma dioscoreae slices. *LWT* **2017**, *80*, 237–249. [CrossRef]
13. Dincer, I.; Dost, S. A modelling study for moisture diffusivities and moisture transfer coefficients in drying of solid objects. *Int. J. Energy Res.* **1996**, *20*, 531–539. [CrossRef]
14. Kiranoudis, C.T.; Markatos, N.C. Pareto design of conveyor-belt dryers. *J. Food Eng.* **2000**, *46*, 145–155. [CrossRef]
15. Kiranoudis, C.T.; Maroulis, Z.B.; Marinos-Kouris, D. Product quality multi-objective dryer design. *Dry. Technol.* **1999**, *17*, 2251–2270. [CrossRef]
16. Krokida, M.K.; Kiranoudis, C.T. Pareto design of fluidized bed dryers. *Chem. Eng. J.* **2000**, *79*, 1–12. [CrossRef]
17. Krokida, M.K.; Kiranoudis, C.T. Product quality multi-objective optimization of fluidized bed dryers. *Dry. Technol.* **2000**, *18*, 143–163. [CrossRef]
18. Quirijns, E.J. Modelling and Dynamic Optimisation of Quality Indicator Profiles during Drying. PhD Thesis, Wageningen University, Wageningen, The Netherland, April 2006.
19. Olmos, A.; Trelea, I.C.; Courtois, F.; Bonazzi, C.; Trystram, G. Dynamic optimal control of batch rice drying process. *Dry. Technol.* **2002**, *20*, 1319–1345. [CrossRef]
20. Winiczenko, R.; Górnicki, K.; Kaleta, A.; Martynenko, A.; Janaszek-Mańkowska, M.; Trajer, J. Multi-objective optimization of convective drying of apple cubes. *Comput. Electron. Agric.* **2018**, *145*, 341–348. [CrossRef]
21. Winiczenko Górnicki, K.; Kaleta, A.; Janaszek-Mankowska, M.; Trajer, J. Multi-objective optimization of the apple drying and rehydration processes parameters. *Emir. J. Food Agric. EJFA* **2018**, *30*, 1–9.
22. Chen, C.R.; Ramaswamy, H.S. Modeling and optimization of variable retort temperature (VRT) thermal processing using coupled neural networks and genetic algorithms. *J. Food Eng.* **2002**, *53*, 209–220. [CrossRef]
23. Erdoğdu, F.; Balaban, M.O. Complex method for nonlinear constrained multi-criteria (multi-objective function) optimization of thermal processing. *J. Food Process Eng.* **2003**, *26*, 357–375. [CrossRef]
24. Górnicki, K.; Kaleta, A. Modelling convection drying of blanched parsley root slices. *Biosyst. Eng.* **2007**, *97*, 51–59. [CrossRef]
25. Luikov, A.V. *Analytical Heat Diffusion Theory*; Academic Press Inc.: New York, NY, USA, 1970.
26. Zielinska, M.; Markowski, M. Drying behavior of carrots dried in a spout–fluidized bed dryer. *Dry. Technol.* **2007**, *25*, 261–270. [CrossRef]
27. Celma, A.R.; Cuadros, F.; López-Rodríguez, F. Convective drying characteristics of sludge from treatment plants in tomato processing industries. *Food Bioprod. Process.* **2012**, *90*, 224–234. [CrossRef]
28. Nguyen, T.H.; Lanoisellé, J.L.; Allaf, T.; Allaf, K. Experimental and fundamental critical analysis of diffusion model of airflow drying. *Dry. Technol.* **2016**, *34*, 1884–1899. [CrossRef]
29. Górnicki, K.; Kaleta, A. Drying curve modelling of blanched carrot cubes under natural convection condition. *J. Food Eng.* **2007**, *82*, 160–170. [CrossRef]
30. Chen, X.D.; Lin, S.X.Q.; Chen, G. On the ratio of heat to mass transfer coefficient for water evaporation and its impact upon drying modeling. *Int. J. Heat Mass Transf.* **2002**, *45*, 4369–4372. [CrossRef]
31. Bird, R.B.; Stewart, W.E.; Lightfoot, E.N. *Transport Phenomena*; Rev. 2nd ed.; Wiley: New York, NY, USA, 2007; ISBN 978-0-470-11539-8.
32. Zogzas, N.P.; Maroulis, Z.B.; Marinos-Kouris, D. Moisture diffusivity data compilation in foodstuffs. *Dry. Technol.* **1996**, *14*, 2225–2253. [CrossRef]
33. Mujumdar, A.S. *Handbook of Industrial Drying*, 4th ed.; CRC Press, Taylor & Francis Group: Boca Raton, FL, USA, 2015; ISBN 978-1-4665-9665-8.

Article

Experimental Study of a Bubble Mode Absorption with an Inner Vapor Distributor in a Plate Heat Exchanger-Type Absorber with NH$_3$-LiNO$_3$

Jorge J. Chan [1,*], Roberto Best [2], Jesús Cerezo [3], Mario A. Barrera [2] and Francisco R. Lezama [1]

[1] Facultad de Ingeniería, Universidad Autónoma de Campeche, Av. Agustín Melgar s/n Col, Buenavista 24030, Campeche, México; frlezama@uacam.mx

[2] Instituto de Energías Renovables, Universidad Nacional Autónoma de México, Temixco 62580, Morelos, México; rbb@ier.unam.mx (R.B.); mabch@ier.unam.mx (M.A.B.)

[3] Centro de Investigación en Ingeniería y Ciencias Aplicadas, Universidad Autónoma del Estado de Morelos, Av. Universidad 1001, Cuernavaca 62209, México; jesus.cerezo@uaem.mx

* Correspondence: jorjchan@uacam.mx; Tel.: +52-981-811-9800 (ext. 3030100)

Received: 16 July 2018; Accepted: 10 August 2018; Published: 16 August 2018

Abstract: Absorption systems are a sustainable solution as solar driven air conditioning devices in places with warm climatic conditions, however, the reliability of these systems must be improved. The absorbing component has a significant effect on the cycle performance, as this process is complex and needs efficient heat exchangers. This paper presents an experimental study of a bubble mode absorption in a plate heat exchanger (PHE)-type absorber with NH$_3$-LiNO$_3$ using a vapor distributor in order to increase the mass transfer at solar cooling operating conditions. The vapor distributor had a diameter of 0.005 m with five perforations distributed uniformly along the tube. Experiments were carried out using a corrugated plate heat exchanger model NB51, with three channels, where the ammonia vapor was injected in a bubble mode into the solution in the central channel. The range of solution concentrations and mass flow rates of the dilute solution were from 35 to 50% weight and 11.69 to 35.46 $\times$ 10^{-3} kg·s^{-1}, respectively. The mass flow rate of ammonia vapor was from 0.79 to 4.92 $\times$ 10^{-3} kg·s^{-1} and the mass flow rate of cooling water was fixed at 0.31 kg·s^{-1}. The results achieved for the absorbed flux was 0.015 to 0.024 kg m^{-2}·s^{-1} and the values obtained for the mass transfer coefficient were in the order of 0.036 to 0.059 m·s^{-1}. The solution heat transfer coefficient values were obtained from 0.9 to 1.8 kW·m^{-2}·K^{-1} under transition conditions and from 0.96 to 3.16 kW·m^{-2}·K^{-1} at turbulent conditions. Nusselt number correlations were obtained based on experimental data during the absorption process with the NH$_3$-LiNO$_3$ working pair.

Keywords: bubble absorber; absorption cooling; ammonia-lithium nitrate; plate heat exchanger

1. Introduction

The increasing awareness of global warming is inspiring the population to find better solutions to the problems related to the use of clean energy. For cooling needs, industries and residential sectors use mostly mechanical vapor compression refrigeration systems driven by electricity as well as conventional refrigerants. This field generates many opportunities for research on the use of alternatives energy sources friendly to the environment [1]. The opportunity arises to develop refrigeration systems that fit into the model of the three Es (energy, economy, and ecology) [2,3].

Recently, the demand for absorption refrigeration systems has increased mainly for small capacity (5 to 20 kW) units that can be driven by low and medium temperature heat sources such as waste heat, solar, geothermal or biomass. These thermal cooling systems are an attractive option in order to reduce electric power consumption.

The design of efficient heat exchangers is vital for absorption cooling systems to compete against compression systems, since they constitute the main and most expensive components [4]. The commercial absorption systems use working pairs such as NH_3-H_2O for cooling and freezing applications; however, this system has the need for an ammonia vapor rectification component that consequently produces a rise in cost and a reduction in the coefficient of performance. However, NH_3-$LiNO_3$ could be an alternative working mixture, since it has a better thermodynamic properties than NH_3-H_2O and a rectifier is not necessary [5–7]. The absorber is a critical component of the absorption system [8] due to the complex heat and mass transfer processes during the absorption of ammonia vapor in the dilute solution. The bubbling absorption mode has been proposed as a viable option with NH_3-H_2O [9–13], rather than the traditional falling film absorption systems because it not only provides high heat transfer coefficients but also provides good mixing between the vapor and the liquid. However, it requires of good vapor distribution, whilst falling film transfer requires an excellent liquid distribution [11].

Plate heat exchangers (PHEs) have been used as an absorber in bubble mode because the corrugation of the channels produces high turbulence values. Lee et al. [9] analyzed an absorber using a flat plate heat exchanger with ammonia-water; they concluded that the increase of solution flow rate slightly affected mass transfer, but improved heat transfer; besides the heat transfer was improved when the vapor flow rate was increased. The authors proposed experimental correlations for the Nusselt and Sherwood numbers.

Cerezo et al. [8] presented a mathematical model to analyze the bubble absorption process using a PHE. They carried out a comparative study using different mixtures as working fluids: NH_3-H_2O, NH_3-$NaSCN$ and NH_3-$LiNO_3$; the system was simulated at typical refrigeration operating conditions. The results showed that the mixtures NH_3-H_2O and NH_3-$NaSCN$ had greater absorption heat rates and ammonia vapor mass absorption compared to the mixture NH_3-$LiNO_3$. The low values obtained by the last mixture were mainly caused by the high viscosity of the solution, which decreased the absorption process. On the other hand, the NH_3-$LiNO_3$ mixture resulted with the highest COP values. Ayub [14] presented a study of available correlations for PHE´s in only one phase in a format that is easily used by engineers to design and analyze systems of this kind. Oronel et al. [15] presented in 2012 a study of NH_3-$LiNO_3$ absorption in the bubbling mode with a PHE with Chevron-L type corrugation (30° from the plate vertical axis) and using a simple tube as a gas distributor with an internal diameter of 1.7 mm. The results showed empirical relationships for the Nusselt and Sherwood numbers for the NH_3-$LiNO_3$ solutions. Best and Rivera [16] carried out an extensive review of the theoretical and experimental thermal cooling systems. In particular, systems at small and medium size capacities operating with water/lithium bromide, and ammonia/lithium nitrate.

This paper carries out an experimental evaluation of an absorber with a type L plate heat exchanger (PHE) with NH_3-$LiNO_3$ using a vapor distributor with five perforations located on the bottom side of the absorber in order to increase the mass transfer. According to the literature review performed for plate heat exchangers that operate as an absorber with the NH_3-$LiNO_3$ working pair, the mass transfer coefficients calculated in this work reached higher values than other studies. The modified plate exchanger with an inner gas distributor will be installed as an absorber within an absorption cooling system driven by solar energy. The results of this work are presented as a good engineering option: to incorporate an efficient absorber into a compact refrigeration system by means of NH_3-$LiNO_3$ absorption. These results are important for the absorption refrigeration research and development.

2. Description of the Experimental System

2.1. Test Rig Description

A test rig was designed to study the absorption process in a PHE. It consisted basically of three circuits: the ammonia vapor (yellow), ammonia-lithium nitrate solution (orange) and the cooling water as can be seen in Figure 1. In the ammonia vapor circuit, liquid ammonia was stored in a tank and pumped through a micrometric expansion valve, which reduced the pressure and changed to a vapor phase, it was injected into the bottom side of the absorber using a vapor distributor as it is shown in Figure 2. The absorption process was carried out in the solution circuit. The poor in ammonia solution was stored in a container and it was pumped to the bottom side of the absorber, where ammonia vapor and solution were mixed. The concentrated solution left the absorber and was sent to a storage container. The heat generated by the absorption process was removed by the cooling water circuit, which was pumped from a cold water reservoir (15,000 L) to the side plates of the PHE, extracting the absorption heat. A needle valve regulated the cooling water mass flow. The ammonia vapor and solution circuits were made of stainless steel, whereas the cooling circuit was made of PVC. In Figure 1, we can also observe the control and measurement instruments that were placed in the experimental system (pressure, temperature, mass flows, needle valves, diaphragm pumps).

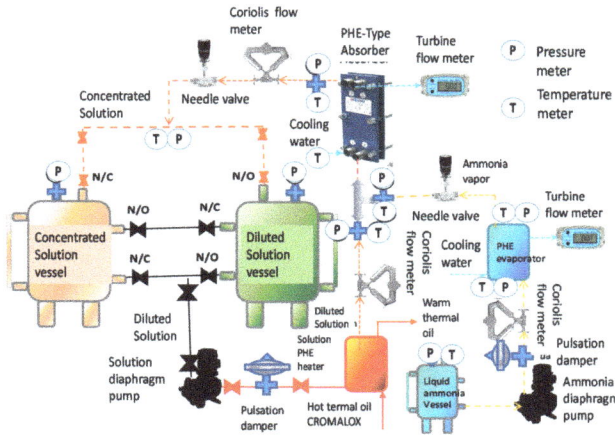

Figure 1. Schematic diagram of the device to study the bubble absorption process in a plate heat exchanger.

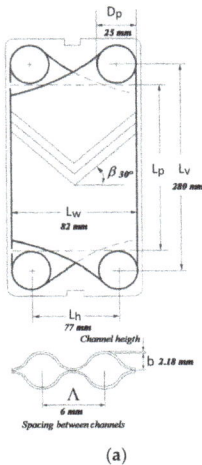

Figure 2. Diagram and photograph of a chevron plate heat exchanger. (**a**) Detailed inner main parameters; (**b**) the assembly of the modified PHE, with the measuring and control instruments.

2.2. Plate Heat Exchanger as Absorber

A stainless steel plate heat exchanger model T2-BFG from Alfa Laval was used as an absorber. Table 1 shows the dimension details of a plate. The PHE consisted of three channels, ammonia vapor and NH_3-$LiNO_3$ flowed in the central channel, and cooling water flowed into the side channels. Details of the absorber are shown in Figure 2.

Table 1. Main dimensions of a plate heat exchanger.

Parameter	Dimension
Spacing between channels, Λ [mm]	6
Inter-plate channel height, b [mm]	2.18
Effective width length, L_h [mm]	77
Pattern length, L_p [mm]	275
Effective height length, L_v [mm]	280
Inside width between gaskets edges L_w [mm]	82
Port diameter, D_p [mm]	25
Number of passes, N_p	1
Chevron angle, β [degree]	30

The vapor distributor, shown in Figure 3, consisted of a 50 mm long and 5 mm inside diameter steel tube with five equidistant perforations: two of 1 mm, two of 2 mm and one of 3 mm [10,11]. Temperature, pressure, and mass flow rate sensors were placed at the inlet and outlet of the cooling water and solution flows in the plate heat exchanger as it can be seen in Figure 1. Information on the accuracy of the sensors can be seen in Table 2, and the operating parameters and their respective accuracies summary can be seen in Appendix A.

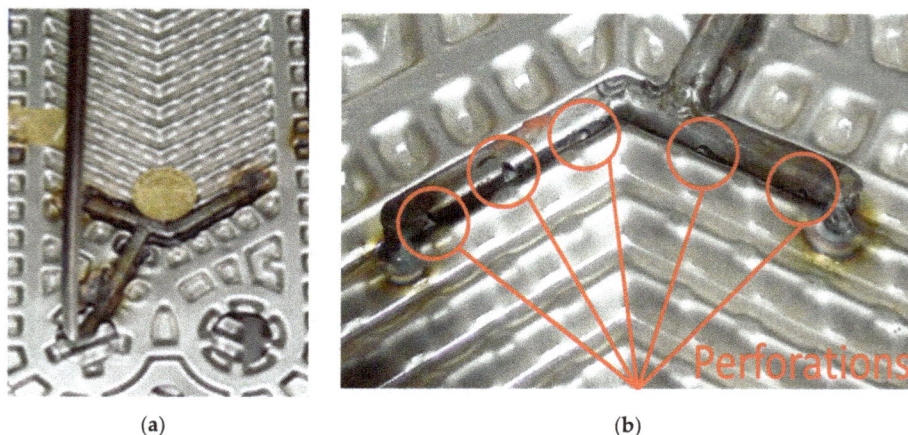

(a)   (b)

Figure 3. Inner distributor of ammonia vapor placed in the central channel of the PHE. (**a**) general view of the vapor distributor; (**b**) close up of the tube with perforations.

Table 2. Parameters of the sensors used in the experimental study and associated uncertainty values.

Sensor	Device	Operating Range	Accuracy
Temperature	RTD	-180 a $520\ ^\circ\text{C}$	$\pm 0.2\ ^\circ\text{C}$
Mass flow	Coriolis	0 a 5 kg·min^{-1}	$\pm 0.1\%$
Density	Coriolis	700 a 1200 kg·m^{-3}	$\pm 0.1\%$
Pressure	piezoelectric	0 a 10 bar	$\pm 0.15\%$
Mass flow	Turbine	0 a 30 kg·min^{-1}	$\pm 0.2\%$

3. Data Reduction

The thermodynamic properties of NH_3-$LiNO_3$ solutions were obtained from Libotean [17], Infante Ferreira [18] and Conde [19]. The experimental data correspond to steady state conditions during at least 15 min for each experimental run. The heat transfer area (A_{exch}) was calculated using the surface enlargement factor (ϕ) defined as the proportion of increased length with respect to a plane plate length or projected length [20]. A_{eff} is the projected area of the plate. The heat transfer area was 0.0504 m^2:

$$A_{exch} = \varnothing * A_{eff}\ [\text{m}] \tag{1}$$

$$\varnothing = \frac{increased\_length}{projected\_length}\ [dimensionless] \tag{2}$$

The equivalent diameter, D_e, is defined as:

$$D_e = \frac{4bL_w}{2(b + L_w\varnothing)} \approx \frac{2b}{\varnothing}\ [m];\ \text{Considering that}:\ \text{b} << \text{Lw} \tag{3}$$

The Reynolds number (Re), channel mass velocity (Gc_{sol}) and the number of channels per pass (N_{cp}), are calculated as [21–25]:

$$Re = \frac{G_{Csol}D_e}{\mu}\ [dimensionless] \tag{4}$$

$$Gc_{sol} = \frac{\dot{m}_{sol}}{N_{cp}bL_w}\ \left[\text{kg m}^{-2}\text{s}^{-1}\right] \tag{5}$$

$$N_{cp} = \frac{N_t - 1}{2N_p}\ [dimensionless] \tag{6}$$

G_{Csol} is a relationship of the solution mass flow rate with respect to the physical characteristics of the heat exchanger. D_e is equivalent diameter, b is the inter-plate channel height, μ is the dynamic viscosity, $\dot{m}_{sol}$ is the mass flow of solution, N_{cp} is the number of channels per pass, N_t is the total number of plates, N_p is the number of passes and, L_w is the effective height length.

The logarithmic mean temperature difference ($LMTD$) was used considering the inlet and outlet temperatures of the absorber [24]:

$$LMTD = \frac{[T_{sol,in} - T_{c,out}] - [T_{sol,out} - T_{c,in}]}{ln\left[\frac{[T_{sol,in} - T_{c,out}]}{[T_{sol,out} - T_{c,in}]}\right]}\ [dimensionless] \tag{7}$$

The Nusselt number is stated by Equations (8) and (9), where μ_{SOL} is dynamic viscosity of solution, μ_W is the dynamic viscosity of water, L_{cha} is the channel length:

$$Nu_{sol} = aRe^b Pr^c \left(\frac{\mu_{sol}}{\mu_w}\right)\ [dimensionless] \tag{8}$$

$$Nu_{sol} = \frac{h_{sol}L_{cha}}{k_{sol}} \quad [dimensionless] \tag{9}$$

$$Pr_{sol} = \frac{\mu_{sol}Cp_{sol}}{k_{sol}} \quad [dimensionless] \tag{10}$$

The overall heat transfer coefficient, U, was calculated by Equation (11):

$$U = \frac{1}{\frac{1}{h_w} + \frac{x_{ss}}{k_{ss}} + \frac{1}{h_{sol}}} \quad \left[kWm^{-2}K^{-1}\right] \tag{11}$$

Ammonia mass absorption flux, F_{ABS}, indicates the capacity of the solution to absorb the ammonia vapor per unit of heat transfer area. $\dot{m}_{NH3}$ is absorbed ammonia mass flow rate, A_{exch} is the heat exchange area:

$$F_{ABS} = \frac{\dot{m}_{NH3abs}}{A_{exch}} \quad \left[kgm^{-2}s^{-1}\right] \tag{12}$$

The overall mass transfer coefficient, K_m, defined by Equation (13):

$$K_m = \frac{\dot{m}_{NH3abs}}{\rho_{NH3}A_{exch}LMCD} \quad \left[m\,s^{-1}\right] \tag{13}$$

$$LMCD = \frac{\left[X_{sol,in}^{equi} - X_{sol,in}\right] - \left[X_{sol,out}^{equi} - X_{sol,out}\right]}{In\left[\frac{\left[X_{sol,in}^{equi} - X_{sol,in}\right]}{\left[X_{sol,out}^{equi} - X_{sol,out}\right]}\right]} \quad [dimensionless] \tag{14}$$

where, $\dot{m}_{NH3abs}$ is the absorbed ammonia mass flow, A_{exch} is the exchange area, ρ_{NH3} is ammonia density, LMCD is logarithmic mean concentration difference, X, is solution concentration [26–29].

4. Discussion

The absorber was characterized using water on both sides of the PHE in order to calculate a Nusselt Number correlation for cooling water. Previously, experiments were carried out introducing water in both cold and hot sides of the channels in order to calculate the cooling water heat transfer coefficient. A correlation of the Nusselt number was obtained as a function of the Reynolds and Prandtl numbers:

$$Nu_{water} = 0.3417Re_{water}^{0.5891} Pr_{water}^{0.1726} \quad [dimensionless]. \tag{15}$$

The range of operation conditions for the experimental test started with a solution of NH_3-$LiNO_3$ at 35% ammonia weight and it was progressively increased in each experiment until reaching a maximum concentration of 50%. The range of the ammonia mass flow rates injected was from 0.0794×10^{-3} to 1.217×10^{-3} kg s^{-1}. The flow rates of the dilute ammonia solution flow rate were from 0.0117 to 0.035 kg s^{-1}.

In Figure 4 the x-axis represents in ascending order the experimental tests. Each test was considered valid after being kept in steady state for at least fifteen minutes; data acquisition was performed at ten-second intervals. The values of the concentrated (X_{sc}) and diluted (X_{sd}) solutions, for each test are shown on the left-y-axis. In this work, the concentration gradually increased. The experiments started from an initial diluted concentration of 35% up to a concentration of 50%. The concentrated solution also increased step by step. The difference in concentrations (ΔX) for each valid test is shown in the lower part of the graph; the average difference in concentrations was $0.011 \pm 0.005\%$. On the other hand, the behavior presented by the Reynolds number of the solution, Re_{sol}, in each of the test points is also shown; it is observed that it was increasing as the concentration increased. The latter was to be expected taking into account the decrease in the viscosity of the ammonia solution as a result of the increase in the ammonia concentration. Two behaviors can be seen for the Re_{sol}, the transition zone that goes from $20 \leq Re_{sol} \leq 90$ and the turbulent zone $90 < Re_{sol} <$

400. The transition zone showed a linear behavior with a slightly steep slope. On the other hand, the turbulent zone had a polynomial behavior.

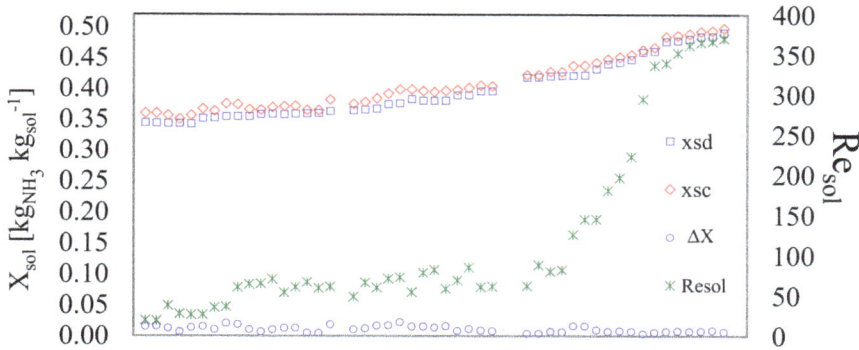

Figure 4. Experimental valid tests, their respective concentrations, concentration difference and Reynolds number of solution.

Empirical correlations for the solution Nusselt numbers were proposed on the basis of the experimental data presented here for the absorption of ammonia vapor by ammonia-lithium nitrate solution in a PHE-type absorber. The correlations were made using the experimental data. Equation (16) is valid for the transition zone and Equation (17) for the turbulent zone. These empirical relationships in conjunction with Equation (9) allowed finding the h_{sol}:

$$Re_{sol} < 90 \; Nu_{sol} = 0.1301 Re_{sol}^{0.3523} Pr^{0.333} \left(\frac{\mu_{sol}}{\mu_w} \right)^{0.17} \; [dimensionless]. \tag{16}$$

$$Re_{sol} \geq 90 \; Nu_{sol} = 0.0089 Re_{sol}^{0.8213} Pr^{0.333} \left(\frac{\mu_{sol}}{\mu_w} \right)^{0.17} \; [dimensionless]. \tag{17}$$

Figure 5 shows the behavior of the solution heat transfer coefficient, h_{sol}, and the channel mass velocity of the solution, G_{Csol}, against the Reynolds solution number, Re_{sol}. The solution heat transfer coefficient, h_{sol}, increased linearly from 0.96 to 2.46 kW m^{-2}K^{-1} at Re_{sol} range from 20 to 90 and G_{Csol} shows a linear increase for the same range of Re from 40 to 130 kg m^{-2} s^{-1}. Subsequently, h_{sol} showed a slight increase and the channel mass velocity of the solution, G_{Csol} remained almost constant. This behavior was caused by the increment of the ammonia concentration in the solution, which reduced the viscosity and influenced the thermodynamic behavior; these variations influence the Prandtl number as shown in Figure 5, which shows that the thermal boundary layer will increase as a result of the decrease in the viscosity of the solution. The h_{sol}, increased as the result of the combination of an increase in Re_{sol} and the decrease in G_{Csol}. It was observed in the range of Re_{sol} from 50 to 80 a compact grouping of measured points came were obtained, as a result of the variation of solution mass flow rate which were chosen for the study (which translate into mass velocities), similar results have been reported in previous studies [22–25]. Figure 6 shows the behavior of the convective heat transfer coefficient of the solution with respect to the Re_{sol} number. Likewise, it is shown how the Prandtl number of the solution was modified. As the concentration of ammonia increased during the experimental tests, the viscosity of the solution decreased and consequently the Reynolds number increased.

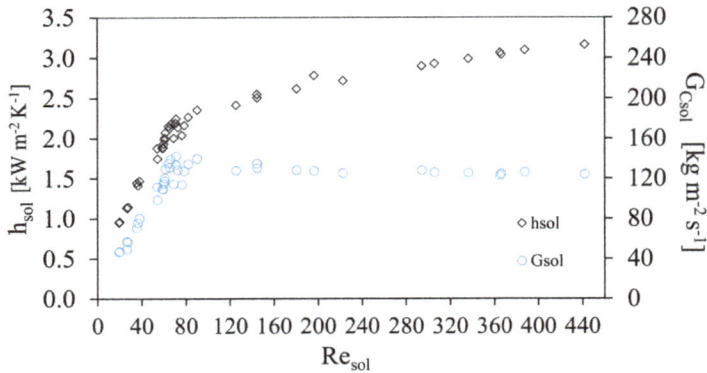

Figure 5. Solution heat transfer coefficient and the channel mass velocity of the solution, G_{Csol}, as a function of Reynolds Number.

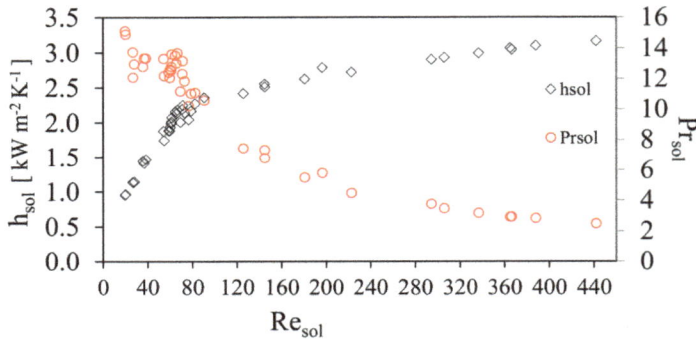

Figure 6. Prandtl number as a function of the solution Reynold number.

It can be seen that as well as the change in viscosity in the solution, μ_{sol}, other solution properties such as the thermal conductivity, K_{sol}, and specific heat, Cp_{sol}, were also modified. It can be seen that there was a clear decrease of the Pr_{sol} due to the thermodynamic and thermal properties of the solution reaching lower values and with the effect that the heat begins to diffuse faster with respect to convection. This behavior results in the increase of the convective heat transfer coefficient in the transition zone with a very steep slope and later in the turbulent zone, this slope decreases.

Figure 7 shows the effect on the mass absorption flux, F_{ABS}, and mass velocities, G_{Csol}, as a function of Re_{sol}. It can be observed that the average value of $1.93 \pm 0.23 \times 10^{-2}$ kg m^{-2}s^{-1} was obtained, for different mass velocities for all interval of Re_{sol}. It can be seen that, although the ammonia solution concentration increased during the experiments, the amount of absorbed ammonia was similar during the transition region [15] as well as in the turbulence zone. The ammonia flow rate was controlled during all experimental tests at different operating conditions.

The results showed that there was a larger absorption potential using higher mass velocities as there exists a larger concentration gradient. At the beginning of the experiments, there was a low concentration of ammonia; therefore, the concentration gradient was high. In addition, during the experiments it was observed that if a larger residence time was used for the ammonia mass flow inside the absorber, a better absorption was obtained although the mass velocity in the channel, G_{Csol} decreased. One option to maintain the higher residence time of the ammonia flow without affecting the channel mass velocity is to employ an absorber with a larger effective length (L_{eff}) than the one used in this work. Figure 8 shows the mass transfer coefficient (K_m), and the ammonia vapor flow rate ($\dot{m}_{NH3}$), with respect to the Re_{sol}. The mass transfer coefficient reached values that ranged from 0.009 to 0.015 m·s^{-1} with an average of 0.011 ± 0.002 m·s^{-1}.

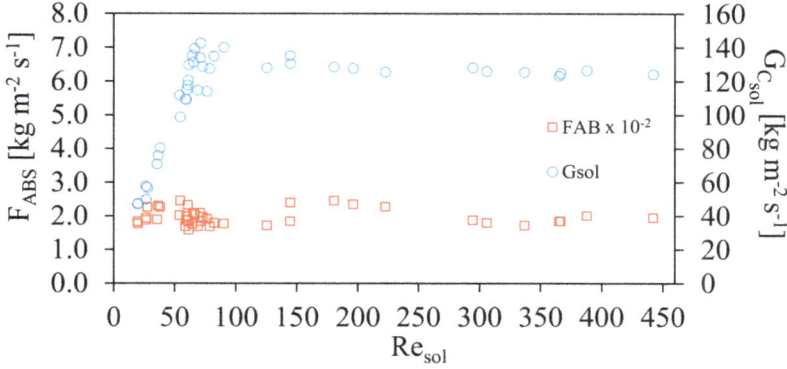

Figure 7. Mass absorption flux, F_{ABS} to different conditions of G_{Csol} and both with respect to Re_{sol}.

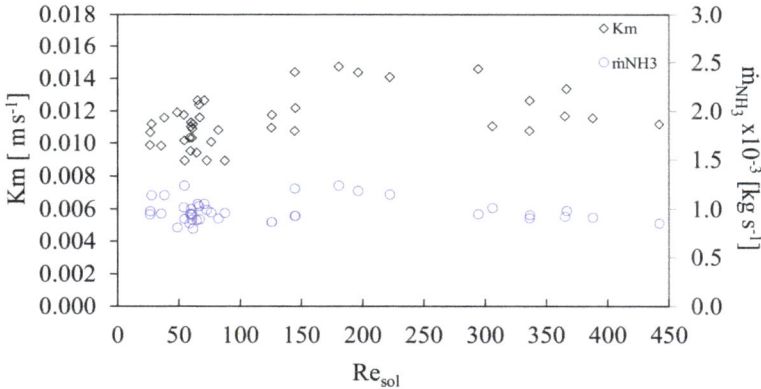

Figure 8. Mass transfer coefficient and vapor flow rate as a function of solution Reynolds number.

Additionally, Figure 9 shows K_m and LMTD as a function of Re_{sol}. The LMTD values obtained were from 6.9 to 21.6 °C. The effect on the mass transfer coefficient, K_m, with respect the logarithmic mean temperature differential, LMTD was subsequently analyzed. Oronel et al. [15] presented a comparison of K_m with respect to sub cooling temperatures of the solution. This work proposes that it is more suitable to compare the behavior of K_m, including all temperatures in the heat exchanger, which is given by the LMTD.

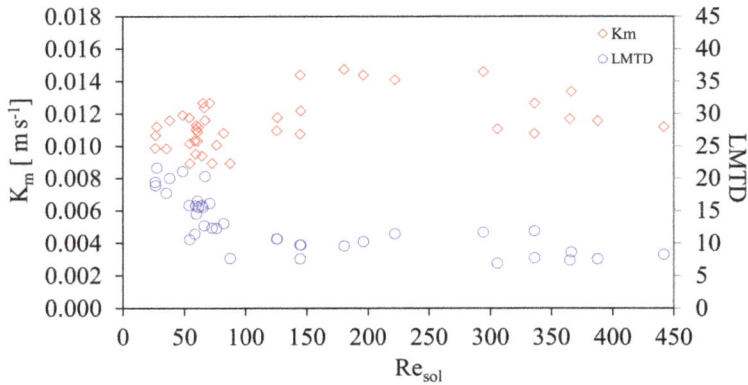

Figure 9. Mass transfer coefficient, K_m and LMTD as a function of Re_{sol}.

Figure 9 shows the K_m and LMTD variation as a function of Re_{sol}. It can be seen that when the LMTD increases K_m is affected positively, both, in the transition and turbulent zones. It is clear that the values of LMTD are lower in the transition than turbulent zone, $90 < Re_{sol} \leq 450$, K_m is increased as a result of the dominance of shear forces over the viscous forces, improving the mixing of vapor with the solution, even though it already has a higher ammonia concentration. It can be deduced that for the transition zone, $20 \leq Re_{sol} \leq 90$, higher values of LMTD are required in order to achieve values of K_m similar to those of the turbulence zone. High viscosity and the temperatures that were registered in the transition area (more exothermic heat generation) caused higher LMTD values.

Table 3 shows h_{sol} and K_m values obtained from Oronel et al. [15] and the present work at similar Reynolds number. The values of h_{sol} were lower than those reported by Oronel et al. [15]. At higher Reynolds numbers, h_{sol} values reached 3.0 kW m^{-2}K^{-1}. However, F_{ABS} values were higher than reported by Oronel et al. due to the improved mixing of vapor and solution, besides it obtained higher values of K_m than those reported in the literature related with absorption in cooling systems.

Table 3. Comparative chart of Some Mass and Convective Heat transfer coefficients K_m reported in the literature with NH$_3$-LiNO$_3$, NH$_3$-H$_2$O and H$_2$O-LiBr.

Parameter	Re_{sol}	h_{sol} [kW m^{-2}K^{-1}]	$K_m \cdot 10^{-5}$ [m s^{-1}]	Method	Solution
Oronel et al. [15]	10–70	2.5–8.0	16.7–38.9	Experimental bubble mode absorption	NH$_3$-LiNO$_3$
Cerezo et al. [22]	170–370	2.7–5.4	100–200	Experimental bubble mode absorption	NH$_3$-H$_2$O
Infante Ferreira [30]		0.115		Experimental falling film	NH$_3$-LiNO$_3$
Venegas et al. [31]	10–250	5.8	18.6	Numerical spray absorption	NH$_3$-LiNO$_3$
Zacarias et al. [32]	6–17		34–101	Experimental spray flat fan	NH$_3$-LiNO$_3$
Jiang J.A. et al. [33]	300–950	0.97–1.95	–	Experimental flow boiling in smooth horizontal tubes	NH$_3$-LiNO$_3$
Jiang J.A. et al. [34]	300–950	0.95–1.95	–	Experimental flow boiling in horizontal tubes	NH$_3$-LiNO$_3$
Palacios E. et al. [35]			30	Experimental spray flat fan	H$_2$O-LiBr
This work	20–450	0.96–3.00	820–1500	Experimental bubble mode absorption	NH$_3$-LiNO$_3$

In this work, a review of the experimental works was carried out with the NH$_3$-LiNO$_3$, H$_2$O-LiBr and NH$_3$-H$_2$O mixtures and with the absorption preferably in the bubbling absorption mode. The review did not find many PHE-type absorbers and even less with the NH$_3$-LiNO$_3$ mixture. Table 3 shows a summary of results, which are interesting as a reference frame, not all of them are experimental studies of a bubble mode absorption in a PHE-type absorber with NH$_3$-LiNO$_3$. Infante Ferreira [30] reported experimental results of NH$_3$-LiNO$_3$ absorption in a falling film absorber in which upward ammonia vapor is absorbed in countercurrent by a laminar film of diluted ammonia solution; it had many technical limitations to achieve it; hence, the small value of h_{sol}. It does not provide more data such as k_m. Venegas et al. [31] performed a numerical study of the spray absorption mode in a very short concentration range of 36.7 to 39% in the concentrated solution. They studied a range of the Re_{sol} also relatively small, of 10 to 70. The cooling water temperature was 34 °C. They considered it as an improvement to the falling film system. It considered an adiabatic chamber where the diluted ammonia solution was contacted, sprayed in small drops with the ammonia vapor. Zacarias et al. [32] carried out experimentally what was stated by Venegas et al., of the spray absorption mode and managed to obtain k_m values an order of magnitude higher than what was predicted by the numerical analysis. Cerezo [22] presented an experimental work with the NH_3-H_2O mixture with cooling water temperatures of 30 °C and mass flow rate of 0.038 kg s^{-1}. He used an ammonia distributor in a PHE-type absorber with an area of 0.1 m^2 of a single central hole and with mass flow rate of 0.0139 kg s^{-1} solution, varying its concentrations from 29 to 33.4% of concentration by weight. He reported h_{sol} and K_m values higher than other studies that used NH_3-$LiNO_3$. The works of Jiang et al. [33,34] served us as a reference for the values of h_{sol}. They used the mixture NH_3-$LiNO_3$ for a boiling process. They did not report values for K_m.

5. Conclusions

An experimental study was carried out of a bubble mode absorption with an internal ammonia vapor distributor in a PHE-type absorber with the NH_3-$LiNO_3$ solution. The following conclusions can be drawn from the present study:

- The results obtained show that h_{sol} increased from a value of 0.9 to 1.8 kW m^{-2}K^{-1} for Re_{sol} values of 20 up to 80. For values of Re_{sol} greater than 120 and up to 440, due to limitations in the experimental rig, h_{sol} increased slowly from 1.8 to 3.0 kW m^{-2}K^{-1}.
- The ammonia absorption flow F_{ABS} in the study had an almost constant value of 0.038 ± 0.004 kg m^{-2}s^{-1}, for all the mass flow velocities, G_{Csol} throughout the experiments. The values are in the same range as the work by Oronel et al. [15] although for low values of Re_{sol} the values in this work are higher. This could be due to the vapor distributor and other operating conditions such as the higher pressure.
- It was found that the value of the mass transfer coefficient, K_m, had a relatively large interval from 0.009 to 0.015 m s^{-1} with an average of 0.011 ± 0.002 m s^{-1}. Although mass flows of ammonia and mass velocities remained almost constant, there were fluctuations in K_m, which demonstrated that an additional factor affected its behavior. The additional factor was the logarithmic mean temperature difference (LMTD). It was observed that K_m behaved inversely proportional to the LMTD.
- It was concluded that by increasing the effective height, L_p, of the modified heat exchanger, the absorption of ammonia vapor could be increased, as the residence time will be increased. In addition, the registered thermodynamic conditions of the dilute ammonia solution were suitable for a better absorption. In order to sustain the previous hypothesis, that, even by increasing the turbulence – that is increasing the mass flow of dilute ammonia solution—the vapor absorption of ammonia remained almost constant, showing a slight decrease at higher Re_{sol} values; and this was mainly due to the fact that there was no greater contact area between the ammonia solution and the ammonia vapor.
- A correlation for the Nusselt number governing the absorption of ammonia vapor by the NH_3-$LiNO_3$ solution in a PHE-type absorber was proposed at two ranges of Reynolds number.

Author Contributions: J.C.G. contributed to the data reduction, the design, the bubble absorber mathematical model, analysis of results, and the operation of the rig test. J.C. contributed to the data reduction, the design, the bubble absorber mathematical model, analysis of results, and the operation of the rig test. J.C. and R.B. contributed to the data reduction, the design and the operation of the rig test. M.A.C. carried out the experimental, construction and the installation. F.L. contributed to the bubble absorber mathematical model and analysis of results.

Funding: The authors would like to thank CONACyT scholarship 84196, Secretaría de Educación Pública for the PROMEP/103.5/09/4287 scholarship and the Universidad Autónoma de Campeche.

Acknowledgments: The authors would like to thank CONACyT scholarship 84196, Secretaría de Educación Pública for the PROMEP/103.5/09/4287 scholarship and the Universidad Autónoma de Campeche. Likewise thanks to Víctor Gómez, Jorge Hernández and Carmen Huerta for their technical support.

Conflicts of Interest: The authors declare no conflict of interest.

Nomenclature

Notation		Subscripts	
A	Area [m^2]	*abs*	Absorption
b	Channel height [m]	*AW*	Air-Water
Cp	Specific heat [J kg^{-1} K^{-1}]	*chann*	Channel
De	Equivalent diameter [m^2]	*CW*	Cooling Water
Dp	Port diameter. [m]	*CS*	Concentrated solution
G_C	Mass speed [kg m^{-2}s^{-1}]	*DS*	Diluted solution
H	Enthalpy [kJ kg^{-1}]	*eff*	Effective
h	Heat transfer coefficient [kW m^{-2}K^{-1}]	*equi*	Equilibrium
k	Thermal conductivity [kWm^{-1} K^{-1}]	*exch*	Exchange
k_m	Overall mass transfer [kg m^{-2}s^{-1}]	*f*	Final
L	Plate length [m]	*G*	Gas
Lh	Effective width length [m]	*Gen*	Generation
Lv	Effective height length [m]	*h*	Effective height
$\dot{m}$	Mass flow rate [kg s^{-1}]	*i*	Initial
Ncp	Passage channel number [dimensionless]	*in*	Input
Nu	Nusselt number [dimensionless]	*l*	Liquid
Pr	Prandtl number [dimensionless]	*out*	Out
Q	Heat transfer rate [kW]	*sol*	Solution
Q_C	Heat transfer rate transferred to cooling water [kW]	*ss*	Stainless Steel
Re_l	Reynolds number [dimensionless]	*v*	Vapor
Sh	Sherwood number [dimensionless]	*w*	Water
T	Temperature [°C]	*equi*	Equilibrium
U	Overall heat transfer coefficient [W m^{-2}K^{-1}]	**Greek letters**	
W	Absorber width [m]	Λ	Spacing between channels [m]
x	Ammonia concentration [% by weight]	β	Chevron angle [degree]
Δt	Time increment [s]	v	Kinematic viscosity [m^2 s^{-1}]
$LMCD$	Logarithmic mean concentration difference	μ	Dynamic viscosity [kg m^{-1}s^{-1}]
$LMTD$	Logarithmic mean temperature difference	ρ	Density [kg m^{-3}]
		$\varnothing$	Superficial enlargement factor [dimensionless]

Appendix

Table A1. Operating parameters and their respective accuracies summary.

Parameter	Nomenclature	Operating Range	Accuracy
Exchange area	A_{exc}	0.0504 m^2	±1.2%
Convective Coeficient of solution heat transfer	h_{SOL}	0.8 a 2.0 kW·m^{-2}·K^{-1}	±2.6%
Absorption flow	F_{ABS}	0.073 a 0.033 kg·m^{-2}·s^{-1}	±1.5%
Convective coeficient of mass transfer	K_m	48.9 a 17.5 kg·m^{-2}·s^{-1}	±1.8%
Equivalent diameter	De	-	±0.5%
Mass velocity in the channel	G_C	40 a 120 kg·m^{-2}·s^{-1}	±1.3%
Solution viscosity	μ_{sol}	$(9\ a\ 1)\cdot10^{-3}$ kg·m^{-1}·s^{-1}	±5.6%
logarithmic mean temperature difference	LMTD	5 to 15	±0.7%
logarithmic mean concentration difference	LMCD	2 to 5	±0.6%
Solution Nussel number	Nu_{SOL}	1.4 to 1.9	±2.7%
Solution Prandtl number	Pr_{SOL}	15.1 to 2.5	±3.1%
Solution Reynolds number	Re_{SOL}	447 to 38	±7.5%

Some views of the experimental system

Figure A1. Calibration work of Coriolis type flow meters.

Two views of the experimental system are shown. We observed PHE-type absorber, with its measuring and control instruments. The photos of the system are of the absorber without thermal insulation. Visual and digital measuring instruments were included for better operation, control and supervision during the experiments.

Figure A2. Two views of the experimental system.

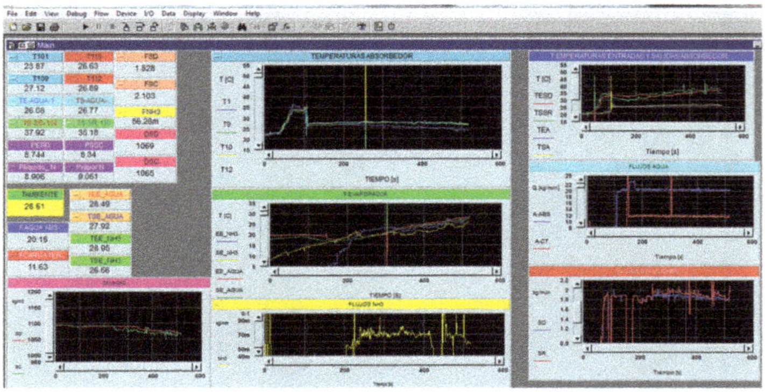

Figure A3. Screenshot of HP VEE real-time graphs.

The screenshots presented in this appendix were generated with the HP VEE data acquisition program during the absorption experiments. They correspond only to representative examples of all tests performed. The HP VEE data acquisition program used for the acquisition of all data for the statistical analysis is shown below:

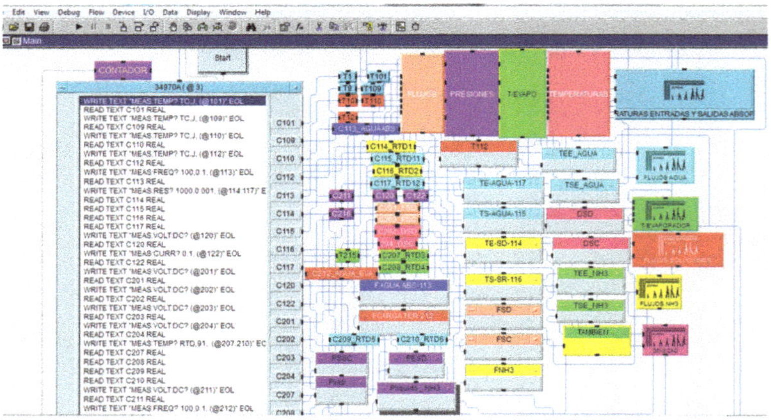

Figure A4. Screenshot of the HP VEE program used to acquire all necessary data.

References

1. García-Valladares, O.; González, J.C.; Hernández, J.I.; y Brown, R.B. The Evaluation of a Small Capacity Shell and Tube Ammonia Evaporator. *Appl. Therm. Eng.* **2003**, *23*, 2151–2167. [CrossRef]
2. McMullan, J.T. Refrigeration and environment-issues and strategies for the future. *Int. J. Refrig.* **2002**, *25*, 89–99. [CrossRef]
3. Rivero, R. Programas integrales de ahorro de energía (exergía) en la industria petrolera. *Revista del IMIQ año XXXVII* **1996**, *3–4*, 26–30.
4. Summerer, F. Evaluation of absorption cycles with respect to COP and economics. *Int. J. Refrig.* **1996**, *19*, 19–24. [CrossRef]
5. Ayala, R. An Experimental Study of Heat Driven Absorption Cooling System. Ph.D. Thesis, University of Salford, Salford, UK, 1995.
6. Antonopoulos, K.A.; Rogdakis, E.D. Performance of solar-driven ammonia-lithium nitrate and ammonia—Sodium thiocyanate absorption systems operating as coolers or heat pumps in Athens. *Appl. Therm. Eng.* **1996**, *16*, 127–147. [CrossRef]

7. Sun, D.-W. Comparison of the performances of nh3-h2o, nh3-lino3 and nh3-nascn absorption refrigeration systems. *Energy Convers. Manag.* **1998**, *39*, 357–368. [CrossRef]
8. Cerezo, J.R.; Best, R. *Estudio del Proceso de Absorción con Diferentes Fluidos de Trabajo en Intercambiadores de Placas para Equipos de Refrigeración por Absorción*; Internal Report, Post PhD; Centro de Investigación en Energía de la UNAM: Temixco, Mexico, 2010.
9. Lee, K.B.; Chun, B.H.; Lee, J.C.; Lee, C.H.; Kim, S.H. Experimental analysis of bubble mode in a plate-type absorber. *Chem. Eng. Sci.* **2002**, *57*, 1923–1929. [CrossRef]
10. Kang, Y.T.; Nagano, T.; Kashiwagi, T. Visualization of bubble behavior and bubble diameter correlation for nh3–h2o bubble absorption. *Int. J. Refrig.* **2002**, *25*, 127–135. [CrossRef]
11. Kang, Y.T.; Nagano, T.; Kashiwagi, T. Mass transfer correlation of NH_3-H_2O bubble absorption. *Int. J. Refrig.* **2002**, *25*, 878–886. [CrossRef]
12. Kang, Y.T.; Christensen, R.N. Ammonia-water (NH_3-H_2O) absorber with a plate heat exchanger. *ASHRAE Trans.* **1998**, *11*, 1565–1575.
13. Cerezo, J.R. Estudio del Proceso de Absorción con Amoniaco Agua en Intercambiadores de Placa para Equipos de Refrigeración por Absorción. Ph.D. Thesis, Universitat Rivira i Virgili, Tarragona, Spain, 2006.
14. Ayub, Z.H. Plate heat exchanger literature survey and new heat transfer and pressure drop correlations for refrigerant evaporators. *Heat Transf. Eng.* **2003**, *24*, 3–16. [CrossRef]
15. Oronel, C.; Amaris, C.; Bourouis, M.; Vallès, M. Heat and mass transfer in a bubble plate absorber with NH_3/$LiNO_3$ and NH_3/($LiNO_3$+H_2O) mixtures. *Int. J. Therm. Sci.* **2013**, *63*, 105–114. [CrossRef]
16. Best, R.; Rivera, W. A review of thermal cooling systems. *Appl. Therm. Eng.* **2014**, *75*, 1162–1175. [CrossRef]
17. Libotean, S.; Martín, A.; Salavera, D.; Valles, M.; Esteve, X.; Coronas, A. Densities, viscosities, and heat capacities of ammonia + lithium nitrate and ammonia + lithium nitrate + water solutions between (293.15 and 353.15) k. *J. Chem. Eng. Data* **2008**, *53*, 2383–2388. [CrossRef]
18. Infante Ferreira, C.A. Thermodynamic and physical property data equations for ammonia-lithium nitrate and ammonia-sodium thiocyanate solutions. *Sol. Energy* **1984**, *32*, 231–236. [CrossRef]
19. Conde, M. Thermodynamic Properties of {NH3+ H2O} Mixtures for the Industrial Design of Absorption Refrigeration Equipment. Available online: http://www.mie.uth.gr/ekp_yliko/NH3_H2OProperties_1.pdf (accessed on 16 July 2018).
20. Kakaç, S.; Liu, H. *Heat Exchangers: Selection, Rating, and Thermal Design*, 2nd ed.; CRC Press: Boca Raton, FL, USA, 2002; pp. 373–404. ISBN 0-8493-0902-6.
21. Wang, L.; Sunden, B.; Manglik, R.M. *Plate Heat Exchangers: Design, Applications and Performance*, 2nd ed.; WIT Press: Southampton, UK, 2007; ISBN 978-1-85312-737-3.
22. Cerezo, J.; Bourouis, M.; Vallès, M.; Coronas, A.; Best, R. Experimental study of an ammonia–water bubble absorber using a plate heat exchanger for absorption refrigeration machines. *Appl. Therm. Eng.* **2009**, *29*, 1005–1011. [CrossRef]
23. Kang, Y.T.; Akisawa, A.; Kashiwagi, T. Analytical investigation of two different absorption modes: Falling film and bubble types. *Int. J. Refrig.* **2000**, *23*, 430–443. [CrossRef]
24. Kang, Y.; Kunugi, Y.; Kashiwagi, T. Review of advanced absorption cycles: Performance improvement and temperature lift enhancement. *Int. J. Refrig.* **2000**, *23*, 388–401. [CrossRef]
25. Lee, J.C. A study on numerical simulations and experiments for mass transfer in bubble mode absorber of ammonia and water. *Int. J. Refrig.* **2003**, *26*, 551–558. [CrossRef]
26. Treybal, R.E. *Operaciones de Transferencia de Masa*, 2nd ed.; McGraw-Hill: Mexico City, Mexico, 1987.
27. Mejbri, K.; Bellagi, A. Modelling of the thermodynamic properties of water-ammonia mixture by three different approaches. *Int. J. Refrig.* **2006**, *29*, 211–218. [CrossRef]
28. Çengel, Y.A. *Heat Transfer a Practical Approach*, 2nd ed.; McGraw-Hill: Mexico City, Mexico, 2003.
29. Incropera, F.P.; DeWitt, D.P. *Fundamentals of Heat and Mass Transfer*, 5th ed.; John Wiley & Son: Hoboken, NJ, USA, 2002.
30. Infante Ferreira, C.A. Operating Characteristics of NH_3-$LiNO_3$ and NH_3-NaSCN Absorption Refrigeration Machines. In Proceedings of the Nineteenth International Congress of Refrigeration, Hague, The Netherlands, 20–25 August 1995; pp. 321–328.
31. Venegas, M.; Arzoz, D.; Rodriguez, P.; Izquierdo, M. Heat and mass transfer in $LiNO_3$-NH_3 spray absorption system. *Int. Commun. Heat Mass Transf.* **2003**, *30*, 805–815. [CrossRef]

32. Zacarías, A.; Venegas, M.; Ventas, R.; Lecuona, A. Experimental assessment of ammonia adiabatic absorption into ammonia–lithium nitrate solution using a flat fan nozzle. *Appl. Therm. Eng.* **2011**, *31*, 3569–3579. [CrossRef]

33. Jiang, J.; He, G.; Liu, Y.; Liu, Y.; Cai, D. Flow boiling heat transfer characteristics and pressure drop of ammonia-lithium nitrate solution in a smooth horizontal tube. *Int. J. Heat Mass Transf.* **2017**, *108*, 220–231. [CrossRef]

34. Jiang, J.; Liu, Y.; He, G.; Liu, Y.L.; Cai, D. Experimental investigations and an updated correlation of flow boiling heat transfer coefficients for ammonia/lithium nitrate mixture in horizontal tubes. *Int. J. Heat Mass Transf.* **2017**, *112*, 224–235. [CrossRef]

35. Palacios, E.; Izquierdo, M.; Marcos, J.D.; Lizarte, R. Evaluation of mass absorption in LiBr flat-fan sheets. *Appl. Energy* **2009**, *86*, 2574–2582. [CrossRef]

![energies logo]

MDPI

Technical Note

Thermal Conductivity of Korean Compacted Bentonite Buffer Materials for a Nuclear Waste Repository

Seok Yoon *, WanHyoung Cho, Changsoo Lee and Geon-Young Kim

Division of Radioactive Waste Disposal Research, Korea Atomic Energy Research Institute (KAERI), 989-111, Daedeok-daero, Yuseong-gu, Daejeon 34057, Republic of Korea; cho0714@kaeri.re.kr (W.C.); leecs@kaeri.re.kr (C.L.); kimgy@kaeri.re.kr (G.-Y.K.)
* Correspondence: busybeeyoon@kaist.ac.kr; Tel.: +82-42-868-2946

Received: 27 June 2018; Accepted: 24 August 2018; Published: 29 August 2018

Abstract: Engineered barrier system (EBS) has been proposed for the disposal of high-level waste (HLW). An EBS is composed of a disposal canister with spent fuel, a buffer material, backfill material, and a near field rock mass. The buffer material is especially essential to guarantee the safe disposal of HLW, and plays the very important role of protecting the waste and canister against any external mechanical impact. The buffer material should also possess high thermal conductivity, to release as much decay heat as possible from the spent fuel. Its thermal conductivity is a crucial property since it determines the temperature retained from the decay heat of the spent fuel. Many studies have investigated the thermal conductivity of bentonite buffer materials and many types of soils. However, there has been little research or overall evaluation of the thermal conductivity of Korean Ca-type bentonite buffer materials. This paper investigated and analyzed the thermal conductivity of Korean Ca-type bentonite buffer materials produced in Gyeongju, and compared the results with various characteristics of Na-type bentonites, such as MX80 and Kunigel. Additionally, this paper suggests various predictive models to predict the thermal conductivity of Korean bentonite buffer materials considering various influential independent variables, and compared these with results for MX80 and Kunigel.

Keywords: bentonite buffer material; Ca-type bentonite; thermal conductivity; predictive models

---

1. Introduction

Spent fuels from nuclear energy sources release decay heat and harmful radiation for extended periods, creating longstanding issues with high-level waste (HLW). Among the various types of disposal systems, deep geological repositories based on the concept of engineered barrier system (Figure 1), which safely isolates HLWs from human society using a surrounding buffer, backfill, and near-field rock, are preferred in most countries, owing to their safety and reliability [1,2]. In EBS system, canisters packed with spent fuel are sealed with buffer and gap materials. The buffer is an important component of the repository. By filling the void between the near-field rock and the canister it minimizes groundwater inflow from intact rock while protecting the disposed HLW from any mechanical impact. For this reason, buffers must possess low hydraulic conductivity, to minimize the inflow of water from surrounding rocks saturated with groundwater. Furthermore, the buffer plays an important role in dissipating decay heat, and for this reason buffers must have high thermal conductivity to release as much as decay heat as possible from the spent fuel [3,4].

In order to satisfy these buffer material criteria, researches have been conducted to determine the most adequate candidate materials. The studies found that bentonite is the most suitable material [5,6]. Bentonite belongs to the smectite group, which contains large amounts of montmorillonite.

Bentonite forms 2:1 layer platy structures, and consists of two silica tetrahedral layers and an aluminum hydroxide octahedral layer. Anions generated from the isomorphous substitution in the montmorillonite readily absorb cations (Na^+, Mg^{2+}, Ca^{2+}, etc.) between the layers to become electrostatically neutral. Bentonite can be largely classified as Na-type bentonite or Ca-type bentonite depending on the exchangeable cations absorbed to become neutral. In Korea, Ca-type bentonite has been produced in the Gyeongju region by CLARIANT KOREA, and since Ca-type bentonite is known to satisfy the appropriate performance criteria, it can be considered one of the candidate buffer materials for HLW repository facilities in Korea [7]. Ca-type bentonite produced before 2015 in Korea is called KJ-I, and after 2015, it is known as KJ-II.

Many studies have been conducted to evaluate the complex thermal-hydro-mechanical (THM) behaviors of such buffers. The thermal conductivity of bentonite buffer materials considering the temperature limit is one of the most important design parameters to guarantee the entire safety performance in a disposal system [8]. However, even though many studies have investigated Na-type bentonite thermal conductivity [4,9–12], relatively few have investigated the thermal conductivity behavior models of the Ca-type bentonite produced in Korea. Therefore, this study measured the thermal conductivity of the Ca-type Korean bentonite and compared results with Na-type bentonite. Furthermore, this study suggested various predictive models for thermal conductivity considering dry density and degree of saturation, which are the main governing factors used to describe the thermal conductivity behavior of the buffer.

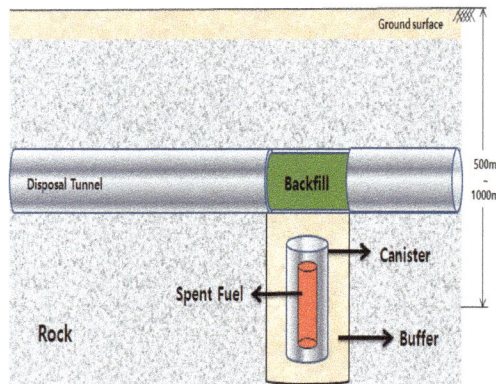

Figure 1. Engineered barrier system.

2. Laboratory Experiment

2.1. Mineral Composition of Bentonite

Ca-type bentonites (KJ-I and KJ-II) produced from Gyeongju contains montmorillonite (70%), feldspar (29%), and small amounts of quartz (~1%). Figure 2 shows XRD analysis results for KJ-I and KJ-II, and Table 1 presents a quantitative analysis of the thermal conductivity of the constitutive minerals. The quantitative analyses were conducted three times. The amount of montmorillonite was a little higher in KJ-I, but there was not a big difference in mineral composition between KJ-I and KJ-II.

Unlike the Ca-type bentonite in this study, for Na-type bentonite the percent of montmorillonite varies. According to the previous researches conducted by Villar [12] and Tang et al. [11], MX80, which is a very well-known commercial Na-type bentonite, contains montmorillonite (92%) and quartz (3%). Another Na-type bentonite considered a potential buffer material in Japan, called Kunigel, has a relatively smaller portion of montmorillonite. It mainly contains montmorillonite (46~49%) and quartz (29~38%) [3,11]. Table 2 shows the clay properties considering the Atterberg limit [13,14] Every bentonite is classified as CH with very high plasticity based on the unified soil classification system (USCS) [15]. On the whole, the Atterberg limit values of KJ-I and KJ-II were much higher than Na-type bentonites, such as MX-80 and Kunigel. It is thought that Na-type bentonite has higher compressibility and a little less tendency to be mechanically stable than Ca-type bentonite. Furthermore, MX80 showed the highest Atterberg limit, which means that MX80 has much more expansive characteristics.

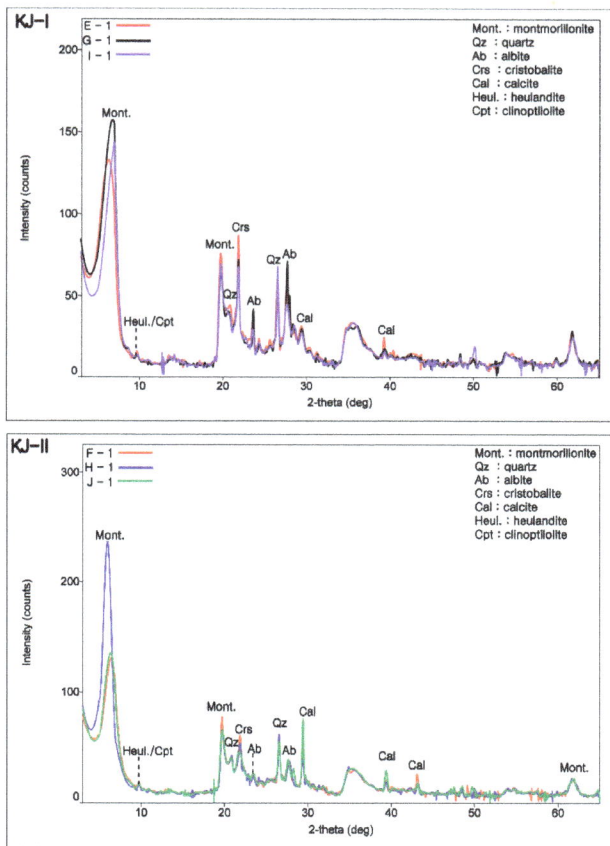

Figure 2. X-ray diffraction pattern of the KJ-I and KJ-II powders.

Table 1. Quantitative XRD analysis for mineral constituents of KJ-I and KJ-II powders.

Bentonite Type	KJ-I				KJ-II			
Sample No.	1	2	3	Avg.	4	5	6	Avg.
Montmorillonite	60.0	67.4	62.1	63.2	63.4	61.7	60.5	61.9
Albite (λ = 1.96 W/mK)	27.2	22.2	27.5	25.6	19.4	22.8	20.4	20.9
Quartz (λ = 7.69 W/mK)	5.0	4.8	5.0	4.9	5.8	4.9	5.3	5.3
Cristobalite (λ = 6.15 W/mK)	3.6	1.8	3.5	3.0	4.0	4.5	3.7	4.1
Calcite (λ = 3.59 W/mK)	2.4	2.0	2.0	2.1	4.3	3.3	6.8	4.8
Heulandite (λ = 1.09 W/mK)	1.8	1.7	Miner	1.8	3.0	2.7	3.3	3.0

Table 2. Geotechnical properties of bentonites.

	Specific Gravity	Liquid Limit (%)	Plastic Limit (%)	Plastic Index (%)	USCS
KJ-I	2.74	244.5	46.1	198.4	CH
KJ-II	2.71	146.7	28.4	118.3	CH
MX-80 [11]	2.76	520	42	478	CH
Kunigel [11]	2.79	415	32	383	CH

2.2. Equipment for Measuring Thermal Conductivity

The thermal conductivity of the compacted bentonite buffer materials were measured using QTM-500 (Kyoto Electronics Manufacturing Company, Kyoto, Japan), based on the transient line source method [16]. In this approach, an impulse of thermal flow is supplied by hot wire into the specimen, and temperature rise is measured within a certain time. As the temperature rises, the thermal conductivity is measured using Equation (1):

$$\lambda = \frac{Q}{4\pi\Delta T} \ln\left(\frac{t_2}{t_1}\right) \tag{1}$$

where λ is the thermal conductivity (W·m·K^{-1}), Q is the heat capacity per unit length (W·m^{-1}), T is the temperature (K), and t is the time (s). The bentonite powders were compressed using the floating die method, and the sample size was 100 mm × 50 mm × 20 mm.

3. Experimental Results

Thermal Conductivity

It is known that the thermal conductivity of the compacted bentonite buffer materials mainly depends on the degree of saturation and dry density [3,5,17,18]. Thus, thermal conductivity of the KJ-II bentonite was measured with various water contents and dry densities. This paper collected 142 datasets for KJ-I [5,18], and 34 datasets for KJ-II was obtained using the QTM-500 equipment. Table 3 provides a summary of the statistical quantities used for the analysis, and Figure 3 depicts the thermal conductivity in proportion to dry density and degree of saturation. On the whole, the thermal conductivity of KJ-II was slightly higher than that of KJ-I because KJ-II has more minerals with high thermal conductivity, including quartz, cristobalite, and calcite, than KJ-I. TANG et al. [11] also explained that the thermal conductivity of the constitutive minerals can affect the total thermal conductivity of the bentonite buffer materials. In addition, the thermal conductivity of KJ-II was measured by drying path, while that of KJ-I was measured by wetting path. It is known that thermal conductivity is higher when measured by drying than by wetting [11,19]. However, there was no great difference in thermal conductivity values between KJ-I and KJ-II except for the low saturation. In comparison, the thermal conductivities of KJ-I and KJ-II were slightly higher than that of MX80 because of mineral composition and the high degree of saturation.

Table 3. Summary of descriptive statistical quantities.

		N	Minimum	Maximum	Average	Standard Deviation	Skewness	Kurtosis
	Dry density (Mg/m^3)		1.200	1.800	1.510	0.154	−0.223	0.095
KJ-I	Degree of saturation (%)	142	0.000	1.000	0.469	0.244	0.159	−0.022
	Thermal conductivity (W/(m·K))		0.301	1.445	0.722	0.248	0.686	−0.203
	Dry density (Mg/m^3)		1.572	1.803	1.693	0.068	−0.024	−0.963
KJ-II	Degree of saturation (%)	34	0.000	0.678	0.177	0.233	1.185	−0.184
	Thermal conductivity (W/(m·K))		0.627	1.046	0.805	0.116	0.761	−0.360

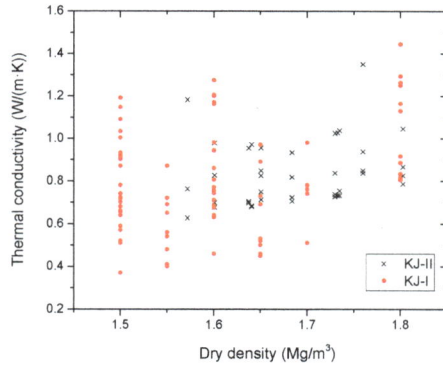

(**a**) Thermal conductivity vs. dry density

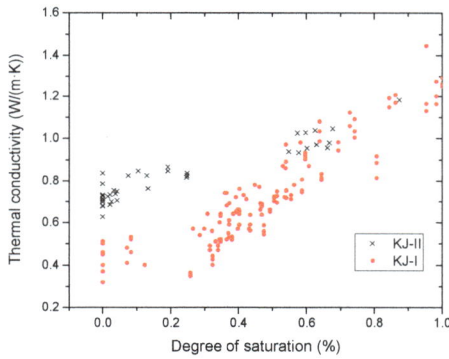

(**b**) Thermal conductivity vs. degree of saturation

Figure 3. Thermal conductivity variation KJ bentonites.

4. Thermal Conductivity Models for KJ-I and KJ-II

4.1. Models with 1.6 Mg/m³ of Dry Density

To satisfy the required functional criteria, the suggested dry density of compacted bentonite buffer for Korea disposal systems is more than 1.6 Mg/m³ [20]. Accordingly, the thermal conductivity model was derived for a bentonite buffer dry density of 1.6 g/cm³. Since there were only three datasets for KJ-II, as shown in Table 4, both the KJ-I and the KJ-II datasets were used, making a total of 23. Many predictive models have been proposed to predict the thermal conductivity of compacted bentonite buffer materials, and this paper mainly used three models. To begin with, a linear regression analysis was applied. Equation (2) represents the linear regression model:

$$\lambda = 0.6683S + 0.4977 \tag{2}$$

Here, λ is thermal conductivity (W/(m·K)), and S means the degree of saturation. The R^2 value was around 0.77. Wilson et al. [21] and Lee et al. [22] also used the following empirical formula, which is well known to be adequate for predicting the thermal conductivity of compacted bentonite buffer materials:

$$\lambda = \frac{A_1 - A_2}{1 + \exp((S - S_{av})/B)} + A_2 \tag{3}$$

A_1 represents the value of λ when $S = 0$, A_2 means λ when $S = 1$. S_{av} is the degree of saturation when the thermal conductivity is the average of the two extreme values, and B is a fitting parameter. Furthermore, with the A_1 and A_2 values, thermal conductivity can also be predicted, as in Equation (4) [2]:

$$\lambda = A_1{}^{1-S}A_2{}^S \qquad (4)$$

Figure 4 shows fitting curves, and Table 5 represents the summary of the predictive models and the fitting parameters, especially for Equations (3) and (4). The parameters for Equations (3) and (4) were derived using the curve fitting tool from the MATLAB program. This tool has a function for deriving the optimum equation that yields the highest R^2 value.

Table 4. Summary of descriptive statistical quantities for KJ-I and KJ-II with 1.6 Mg/m^3 of dry density.

		N	Minimum	Maximum	Average	Standard Deviation	Skewness	Kurtosis
KJ-I	Degree of saturation (%)	20	0	0.982	0.625	0.252	−0.4555	0.480
	Thermal conductivity (W/(m·K))		0.460	1.274	0.889	0.238	0.177	−1.153
KJ-I + KJ-II	Degree of saturation (%)	23	0	0.982	0.556	0.301	−0.393	−0.470
	Thermal conductivity (W/(m·K))		0.460	1.274	0.869	0.229	0.391	−0.971

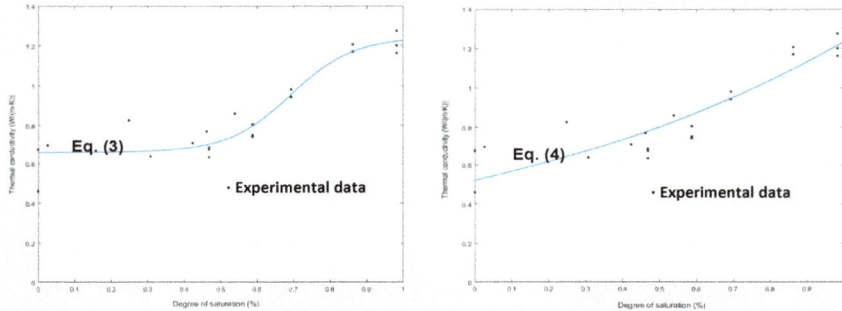

Figure 4. Fitting curves for Equations (3) and (4).

Table 5. Summary of predictive models.

	Equation (3)				Equation (4)	
	A_1	A_2	B	S_{av}	A_1	A_2
Parameters	0.6608	1.2410	0.0878	0.6906	0.5205	1.234
R^2	0.9082				0.8329	

4.2. Models Considering Various Dry Density and Degree of Saturation

Since Equations (2)–(4) only consider the degree of saturation as an independent variable, a multiple regression analysis to predict thermal conductivity was conducted considering dry density and degree of saturation as independent variables. 176 datasets of KJ-I and KJ-II bentonites were used in the regression analysis, and Equation (5) was suggested:

$$\lambda = 0.641\gamma_d + 0.624S - 0.510 \qquad (5)$$

Figure 5 shows the plot of predictive values from Equation (5) and measured values. Table 6 presents the results of the regression analyses, and the R^2 value was 0.739. Based on the t analyses from Table 6 the P-value of the independent variable coefficients were lower than 0.05, which means that the two independent variables can be used statistically to predict the dependent variables [23,24].

The variance inflation factor (VIF) was lower than 10, which means there was no multicollinearity among the independent variables [24]. Table 7 shows the ANOVA analysis. Since the P-values were less than 0.01, there is a statistical significance between the independent and dependent variables[30]. A residual analysis was also conducted. The skewness was 0.210, and kurtosis was = −0.441. Since the absolute value of skewness and kurtosis was less than 2, it can be assumed that the residuals are normally distributed [24,25]. Figure 6 draws the homoscedasticity plot of the residuals, and it can be assumed that the residuals followed the homoscedasticity condition since they do not exhibit a specific pattern [25].

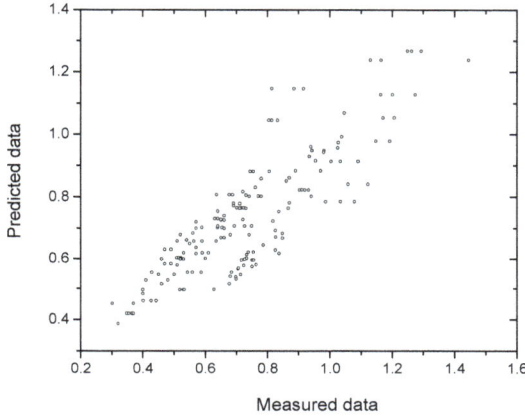

Figure 5. Thermal conductivity of predictive and measured values.

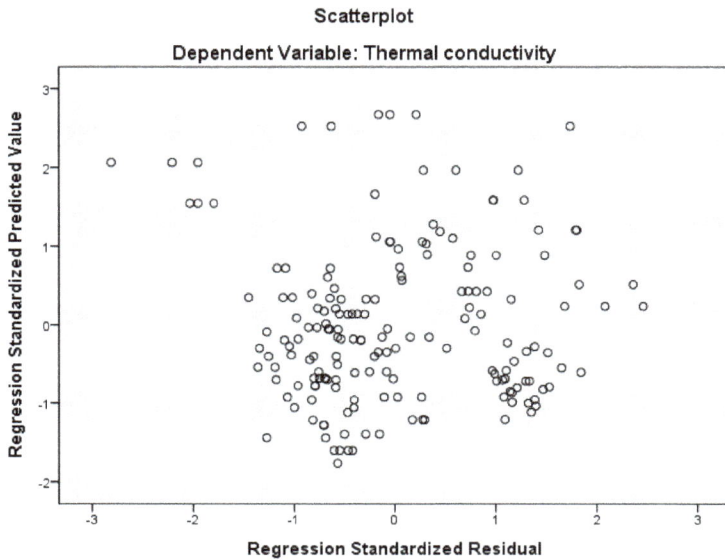

Figure 6. Homoscedasticity plot of residuals.

Table 6. Results of multiple regression analysis model for KJ-I.

	B	Standard Error	t	p-Value	VIF
Constant	−0.510	0.088	−5.761	<0.01	
X1 (dry density)	0.641	0.057	11.334	<0.01	1.001
X2 (degree of saturation)	0.624	0.034	18.611	<0.01	1.001
R^2	0.739				
adjR^2	0.736				

B: non-standardized coefficient, t: B/standard error, VIF: variance inflation factor.

Table 7. Results of ANOVA analysis.

	DF	SS	MS	F	p-Value
Regression	2	6.892	3.446	245.033	<0.01
Residuals	173	2.433	0.014		
Total	175	9.325			

4.3. Model Comparison with Na-Type Bentonite

Many studies have also investigated the thermal conductivity of Na-type bentonites as mentioned in the introduction. Tang et al. [11] also measured the thermal conductivity of MX-80 and suggested various predictive models. Among them, the volume fraction of air and thermal conductivity showed good correlations. The empirical formula can be derived as Equation (6):

$$\lambda = \alpha(V_a/V) + K_{sat} \tag{6}$$

Here, α is the slope of the K-V_a/V plot. K_{sat} is the thermal conductivity at the saturated condition. Equation (6) was also applied to predict thermal conductivity using the 176 datasets of the KJ-I and KJ-II bentonites. Once the dry density and the degree of saturation are obtained, the volume fraction of air can be easily obtained according to the basic geotechnical relation [26].

Table 8 summarizes the parameters from Equation (6) for the KJ, MX80 and Kunigel bentonites, and α and K_{sat} were = −2.05/1.22 for KJ-I and = −1.20/1.18 for KJ-II. Thermal conductivity is inversely proportional to the air fraction because the thermal conductivity of air is much smaller than that of water and soil particles [5,22]. On average, the thermal conductivity of the Kunigel bentonite showed the highest thermal conductivity. Since Kunigel contains about ten times more quartz mineral than MX80, and 6–8 times more than KJ, it is thought that this is the reason Kunigel had a higher thermal conductivity than KJ and MX80. Therefore, it can be inferred that the thermal conductivity of the compacted bentonite buffer materials does not depend on the type of exchangeable cation, such as Ca-type or Na-type.

Table 8. Parameters from Equation (6).

	KJ (Present Work)		MX80 [11]		Kunigel [11]	
	α	K_{sat}	α	K_{sat}	α	K_{sat}
Parameters	−1.76	1.19	−1.79	1.10	−2.36	1.39

5. Conclusions

This paper analyzed thermal conductivity results for Ca-type Korean bentonite buffer materials produced in the Gyeongju region, and suggested various predictive models with three influential independent variables: dry density, degree of saturation, and volume fraction of air. The main conclusions can be summarized as follows:

First, 142 thermal conductivity datasets for the KJ-I bentonite were collected according to dry density and degree of saturation, 34 datasets for KJ-II bentonite were measured using the QTM-500 equipment which is based on the transient hot-wire method. The KJ-I and KJ-II bentonites are composed of more than 60% montmorillonite, and 5% quartz. The thermal conductivity of KJ-I and KJ-II was proportional to dry density and degree of saturation. On average, the thermal conductivity of KJ-II was higher than that of KJ-I because KJ-II has a higher content of minerals with higher thermal conductivity, such as quartz, cristobalite, and calcite. Additionally, the thermal conductivity of KJ-II was measured by the drying path while KJ-I was measured by the wetting path. Compared to MX80, KJ-I and KJ-II had slightly higher thermal conductivities, and it is thought that this was mainly caused by the different mineral compositions of the KJ and MX80 bentonites.

The thermal conductivity estimation models for KJ-I and KJ-II were derived from this research. To satisfy the required functional criteria, the dry density of the buffer materials in Korean disposal systems was at least 1.6 Mg/m^3. This paper used three main equations which are known to be adequate to predict the thermal conductivity of the compacted bentonite buffer materials when the dry density is 1.6 Mg/m^3. In addition, this paper conducted a multiple regression analysis, and suggested a regression model for KJ-I and KJ-II which considered various dry densities and degrees of saturation as independent variables. The R^2 value was 0.739, and satisfied the statistical significance of the regression analyses. It is thought that the regression model proposed in this research can be used as an effective method to estimate the thermal conductivity of KJ-I and KJ-II. In order to compare with Na-type bentonites, such as MX80 and Kunigel, the empirical equation was applied with the volume fraction of air as the independent variable, since the volume fraction of air can be easily calculated using dry density and degree of saturation. Thermal conductivity is inversely proportional to the air fraction. The thermal conductivity of the Korean Ca-type bentonite was 10~15% higher than MX80, but lower than Kunigel, which had different thermal conductivity values because of its mineral composition. It is thought that Kunigel contains about ten times more quartz than MX80, and 6–8 times more than KJ.

The purpose of this paper was to investigate the thermal conductivity of Korean Ca-type bentonite, which can be considered one of the candidate buffer materials for safe HLW disposal in Korea. The predictive models of the Korean Ca-type bentonite suggested by this research can be applied in the design of such disposal systems, since they consider changes in saturation from the inflow of groundwater, and dry density from decay heat. The Na-type bentonite is known to have more swelling, but buffer materials are also required to have high thermal conductivity in order to release as much decay heat as possible from the spent fuel. It is thought that bentonite, which contains minerals with high thermal conductivity, will be adequate as a buffer material in terms of thermal properties. Furthermore, in the future, it will be necessary to determine optimum buffer materials which contain minerals with high thermal conductivity and swelling capacity.

Author Contributions: S.Y. and W.C. conducted experiment; C.L. and G.-Y.K. analyzed the data; and the authors made equal contribution and efforts on writing the manuscript.

Funding: This research was fundeded by the Nuclear Research and Development Program of the National Research Foundation of Korea (NRF-2017M2A8A5014857).

Conflicts of Interest: The authors declare no conflict of interest.

References

1. Cho, W.J.; Kwon, S. Effects of Variable Saturation on the Thermal Analysis of the Engineered Barrier System for a Nuclear Waste Repository. *Nucl. Technol.* **2012**, *2*, 245–256. [CrossRef]
2. Gens, A.; Sánchez, M.; Do, L.; Guimarães, N.; Aloson, E.E.; Lloret, A.; Olivella, S.; Villar, M.V.; Huertas, F. A full-scale in situ heating test for high-level nuclear waste disposal: Observation, analysis and interpretation. *Geotechnique* **2009**, *59*, 377–399. [CrossRef]

3. Japan Nuclear Cycle Development Institute (JNC). *H12 Project to Establish Technical Basis for HLW Disposal in Japan (Supporting Report 2)*; JNC TN1400-99-020; Japan Nuclear Cycle Development Institute: Ibaraki, Japan, 1999.

4. Ye, W.M.; Chen, Y.G.; Chen, B.; Wang, Q.; Wang, J. Advances on the knowledge of the buffer/backfill properties of heavily-compacted GMZ bentonite. *Eng. Geol.* **2010**, *116*, 12–20. [CrossRef]

5. Cho, W.J.; Kwon, S. An empirical model for the thermal conductivity of compacted bentonite and a bentonite-sand mixture. *Heat Mass Transf.* **2011**, *47*, 1385–1393. [CrossRef]

6. Hoffmann, C.; Alonso, E.E.; Romero, E. Hydro-mechanical behavior of bentonite pellet mixtures. *Phys. Chem. Earth* **2007**, *32*, 832–849. [CrossRef]

7. Lee, J.O.; Choi, H.J.; Kim, G.Y. Numerical simulation studies on predicting the peak temperature in the buffer of an HLW repository/Numerical simulation studies on predicting the peak temperature in the buffer of an HLW repository. *Int. J. Heat Mass Trasnf.* **2017**, *115*, 192–204. [CrossRef]

8. Zheng, L.; Rutqvist, J.; Birkholzer, J.T.; Liu, H.H. On the impact of temperature up to 200 °C in clay repositories with bentonite engineer barrier systems: A study with coupled thermal, hydrological, chemical, and mechanical modeling. *Eng. Geol.* **2015**, *97*, 278–295. [CrossRef]

9. Börgesson, L.; Chijimatsu, M.; Fujita, T.; Nguyen, T.S.; Rutqvist, J.; Jing, L. Thermo-hydro-mechanical characterization of a bentonite-based buffer material by laboratory tests and numerical back analyses. *Int. J. Rock Mech. Min. Sci.* **2001**, *38*, 95–104. [CrossRef]

10. Tang, A.M.; Cui, Y.J. Effects of mineralogy on thermo-hydro-mechanical parameters of MX 80 bentonite. *Int. J. Rock Mech. Geotech. Eng.* **2010**, *2*, 91–96.

11. Tang, A.M.; Cui, Y.J.; Le, T.T. A study on the thermal conductivity of compacted bentonites. *Appl. Clay Sci.* **2008**, *41*, 181–189. [CrossRef]

12. Villar, M.V. *MX-80 Bentonite. Thermo-Hydro-Mechanical Characterization Performed at CIEMAT in the Context of the Prototype Project*; Centro de Investigaciones Energeticas, Medioambientalesy Tecnologicas: Madrid, Spain, 2005.

13. Hrubesova, E.; Lunackova, B.; Brodzki, O. Comparison of liquid limit of soils resulted from Casagrande test and modified cone penetrometor methodology. *Procedia Eng.* **2016**, *142*, 364–370. [CrossRef]

14. Andrade, F.A.; Al-Qureshi, H.A.; Hotza, D. Measuring the plasticity of clays: A review. *Appl. Clay Sci.* **2011**, *51*, 1–7. [CrossRef]

15. *ASTM D2487/17—Standard Practice for Classification of Soils for Engineering Purpose (Unified Soil Classification System)*; ASTM International: West Conshohocken, PA, USA, 2017.

16. *ASTM C1113/C1113M-09—Standard Test Method for Thermal Conductivity of Refractories by Hot Wire (Platinum Resistance Thermometer Technique)*; ASTM International: West Conshohocken, PA, USA, 2013.

17. Ballarini, E.; Graupner, B.; Bauer, S. Thermal-hydraulic-mechanical behavior of bentonite and sand-bentonite materials as seal for a nuclear waste repository: Numerical simulation of column experiments. *Appl. Clay Sci.* **2017**, *135*, 289–299. [CrossRef]

18. Lee, J.P.; Choi, J.W.; Choi, H.J.; Lee, M.S. Increasing of Thermal Conductivity from Mixing of Additive on a Domestic Compacted Bentonite Buffer. *J. Nucl. Fuel Cycle Waste Technol.* **2013**, *11*, 11–21. [CrossRef]

19. Farouki, O.T. *Thermal Properties of Soils*; Series on Rock and Soil Mechanics; Trans Tech Publications: Zürich, Switzerland, 1986.

20. Cho, W.J.; Kim, G.Y. Reconsideration of thermal criteria for Korean spent fuel repository. *Ann. Nucl. Energy* **2016**, *88*, 73–82. [CrossRef]

21. Wilson, J.; Savage, D.; Bond, A.; Watson, S.; Pusch, R.; Bennet, D. *Bentonite: A Review of Key Properties, Process and Issues for Consideration in the UK Context*; QRS-1378zG-1.1; Quintessa Limited: Oxfordshire, UK, 2011.

22. Lee, J.O.; Choi, H.; Lee, J.Y. Thermal conductivity of compacted bentonite as a buffer material for a high-level radioactive waste repository. *Ann. Nucl. Energy* **2016**, *94*, 848–855. [CrossRef]

23. Hair, J.F., Jr.; Black, W.C.; Babin, B.J.; Anderson, R.E. *Multivariate Data Analysis*, 7th ed.; Prentice-Hall: Upper Saddle River, NJ, USA, 2009.

24. Yoon, S.; Lee, S.R.; Kim, Y.T.; Go, G.H. Estimation of saturated hydraulic conductivity of Korean weathered granite soils using a regression analysis. *Geomech. Eng.* **2015**, *9*, 101–113. [CrossRef]

25. Park, J.Y. A statistical Entrainment Growth Rate Estimation Model for Debris-Flow Runout Prediction. Master's Thesis, Korea Advanced Institute of Science and Technology, Daejeon, Korea, 2015; 84p.
26. Das, B.M. *Principle of Geotechnical Engineering*, 6th ed.; Nelson: Nelson, New Zealand, 2006.

Article

BFC-POD-ROM Aided Fast Thermal Scheme Determination for China's Secondary Dong-Lin Crude Pipeline with Oils Batching Transportation

Dongxu Han [1], Qing Yuan [2], Bo Yu [1,*], Danfu Cao [3] and Gaoping Zhang [3]

[1] School of Mechanical Engineering, Beijing Key Laboratory of Pipeline Critical Technology and Equipment for Deepwater Oil & Gas Development, Beijing Institute of Petrochemical Technology, Beijing 102617, China; handongxubox@bipt.edu.cn

[2] National Engineering Laboratory for Pipeline Safety, Beijing Key Laboratory of Urban Oil and Gas Distribution Technology, China University of Petroleum, Beijing 102249, China; 2015314026@student.cup.edu.cn

[3] Storage and Transportation Company, Sinopec Group, Xuzhou 221000, China; caodf.gdcy@sinopec.com (D.C.); zhanggp.gdcy@sinopec.com (G.Z.)

* Correspondence: yubobox@bipt.edu.cn; Tel.: +86-10-8129-2805

Received: 14 August 2018; Accepted: 3 October 2018; Published: 7 October 2018

Abstract: Since the transportation task of China's Secondary Dong-Lin crude pipeline has been changed from Shengli oil to both Shengli and Oman oils, its transportation scheme had to be changed to "batch transportation". To determine the details of batch transportation, large amounts of simulations should be performed, but massive simulation times could be costly (they can take hundreds of days with 10 computers) using the finite volume method (FVM). To reduce the intolerable time consumption, the present paper adopts a "body-fitted coordinate-based proper orthogonal decomposition reduced-order model" (BFC-POD-ROM) to obtain faster simulations. Compared with the FVM, the adopted method reduces the time cost of thermal simulations to 2.2 days from 264 days. Subsequently, the details of batch transportation are determined based on these simulations. The Dong-Lin crude oil pipeline has been safely operating for more than two years using the determined scheme. It is found that the field data are well predicted by the POD reduced-order model with an acceptable error in crude oil engineering.

Keywords: fast thermal simulation; crude oil pipeline; batch transportation; body-fitted coordinate-based proper orthogonal decomposition reduced-order model (BFC-POD-ROM); transport scheme determination

---

1. Introduction

This paper focuses on the fast thermal scheme determination for oils batching transportation in China's Secondary Dong-Lin crude pipeline, which is obtained by the proper orthogonal decomposition based reduced-order model (POD-ROM) method. Thus, in this section, the Secondary Dong-Lin crude oil pipeline and the thermal simulations for batch transportation are briefly reviewed first. The POD reduced-order model and applications on the crude pipeline's thermal simulation are reviewed subsequently.

The Secondary Dong-Lin pipeline (also called Dongying-Linyi parallel pipeline), owned by the Sinopec Company, is an important crude oil transportation pipeline across the Shandong province in China. The pipeline is designed to transport the Shengli (SL) crude oil to Linyi, which is produced in the Shengli oilfields in Dongying. Since the fluidity of SL oil is poor, the heated transportation process was adopted before October 2015 [1].

The imported Oman oil (OM), however, with a low condensation point (0 °C) and good fluidity, was planned to be transported in the Secondary Dong-Lin pipeline by October 2015. Thus, its task turned to the transportation of both the imported OM oil and produced SL oil, which totally deviated from the original design. The "batch transportation with different oil temperatures" must be adopted and tremendous corresponding thermal simulations were required.

Different from the batch transportation of petroleum products [2], the thermal characteristics of crude oils' batch transportation are very complex because the different crude oils must be transported in different temperatures due to their large fluidity differences. Batch pipelining with different oil temperatures was first applied to the Pacific Pipeline System, situated in California USA. and commissioned in 1999 [3,4]. Unfortunately, few technique reports are available for its thermal characteristics. Recently, some studies have been performed to discover the thermal behaviors of batch pipelining of crude oils. Cui et al. [5] studied the thermal periodic characteristics for crude oils' batch transportation. Wang et al. [6] gave a report on the thermal and hydraulic behaviors. Yuan [7] studied the thermal characteristics of crude oil batch pipelining with inconstant flow rates.

All the thermal simulations in the above references, however, use the finite volume method (FVM), which is not very suitable for thermal scheme determinations of real engineering pipelines, such as the Secondary Dong-Lin pipeline in the present paper. Moreover, to obtain a proper operational scheme, thousands of thermal simulations should be done using FVM, consuming hundreds of days of simulation even with 10 computers' parallel computing. Thus, to overcome this problem, this paper adopts the POD reduced-order model to significantly improve the simulation speed.

To describe a physical problem with a reduced-order model, the first step is to obtain a series of basis functions, which can express accurately the problem with a small degree of freedom. Normally, the basis function is extracted from a large amount of data by mathematical methods, such as POD [8,9], empirical mode decomposition (EMD) [10], or dynamic mode decomposition (DMD) [11]. POD is adopted in this paper for model reduction. The reduced-order model (ROM), based on POD, can not only describe the problems, but also accelerate the calculations. Thus, this technique is studied extensively for heat transfer and is widely used in engineering.

Regarding the field of heat transfer, Banerjee et al. [12] established a POD-Galerkin ROM for heat transfer based on a finite element method, and Raghupathy et al. [13] established a boundary condition-independent ROM by combining a POD-Galerkin method with a finite volume method. The research of POD-Galerkin ROM are becoming increasingly mature, thus, it is widely applied to engineering [14–18].

The POD reduced-order models above, however, cannot be applied to the thermal simulation of the Secondary Dong-Lin pipeline. To the author's knowledge, the POD-based ROMs in the relevant literature are established for a fixed physical domain, although the boundary conditions and initial fields might vary, while, for the Secondary Dong-Lin crude oil pipeline, the physical domains vary along the pipeline since its diameter and buried depth is different from place to place. To solve this problem, the current research group first proposes a "body-fitted coordinate-based proper orthogonal decomposition reduced-order model" (BFC-POD-ROM) for the heat transfer problem [19,20], in which physical domains with different shapes or sizes can be mapped to the same computational domain.

Therefore, in this paper, BFC-POD-ROM is adopted to obtain the fast thermal simulation of China's Secondary Dong-Lin crude pipeline. Then, the detailed thermal scheme is determined based on the simulations. Finally, the predicted oil temperature distributions are verified through the field data of the Secondary Dong-Lin crude pipeline.

2. Oil Transportation Scheme and Thermal State of Secondary Dong-Lin Pipeline

2.1. Basic Situation

The Secondary Dong-Lin pipeline (constructed in 1999) was originally designed as a supplementary crude pipeline for the Old Dong-Lin pipeline (constructed in 1979), since the old one could not accomplish the crude transportation task by itself.

Prior to October 2015, the main task of the Secondary Dong-Lin pipeline was to transport the SL oil from the Sheng-Li oilfield in Dongying to the consumers in Linyi. Additionally, it transported the SL oil produced in the Bin-Nan oil production factory (also called Bin-Nan oil) through injection (See Figure 1), while the old one was arranged to transport the imported OM oil.

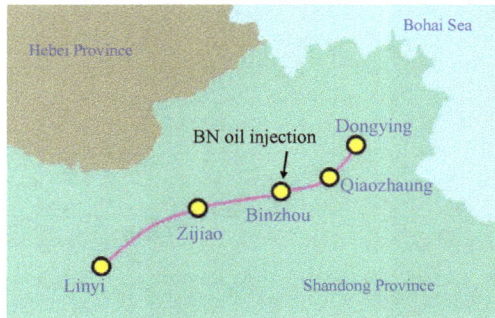

Figure 1. Route of the Secondary Dong-Lin crude pipeline.

The whole secondary pipeline is located in the Shandong Province of China. The route of the Secondary Dong-Lin pipeline is drawn in the map (See Figure 1) and its sketch map with details is found in Figure 2.

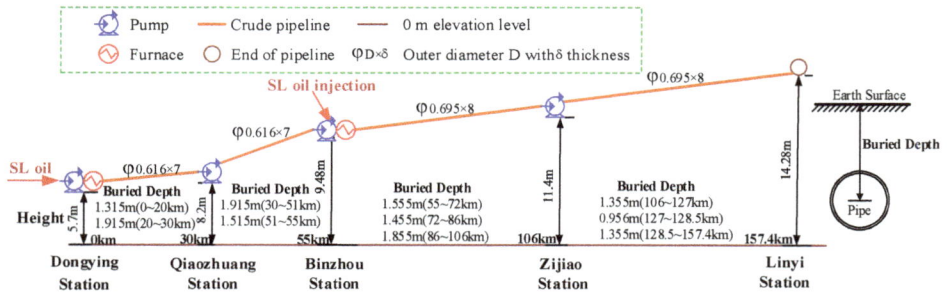

Figure 2. Sketch map of the Secondary Dong-Lin crude pipeline.

Figure 2 shows the total length of the pipeline is 157.4 km with a slight elevation change. Four pump stations (Dongying, Qiaozhuang, Binzhou, and Zijiao stations) are separately located in the positions of 0 km, 30 km, 55 km, and 106 km along the pipeline. Among them, the Dongying Station and Binzhou Station are equipped with furnaces, which means the crude oil can be heated in these two stations. The outer diameter of the pipeline is 0.616 m with a wall thickness of 7 mm before Binzhou station and becoming 0.695 m and 8 mm after Binzhou Station. The buried depth of the pipeline varies along the pipeline, with the minimum depth at 0.956 m and the maximum at 1.915 m.

Since the old Dong-Lin pipeline is too old and should be decommissioned as per regulations, the Secondary Dong-Lin pipeline was planned, by Sinopec Company, to take over the transportation tasks of the Old Dong-Lin pipeline after November 2015. Thus, the SL and OM oils would be transported in the same pipeline.

As demanded by the downstream consumer, the imported OM oil is of much better chemical quality and should not be mixed with SL oil. Therefore, the batching transportation is the only choice for the Sinopec Company (See Figure 3). Additionally, to decrease the viscosity of SL oil, the company decided to blend some OM oil into the SL oil at the beginning of the pipeline. The blended oil is named SL_{OM} oil in this paper. The optimal amount of SL_{OM} compositions also needs to be determined by thermal simulation. The rough scheme of batch transportation is shown in Figure 3.

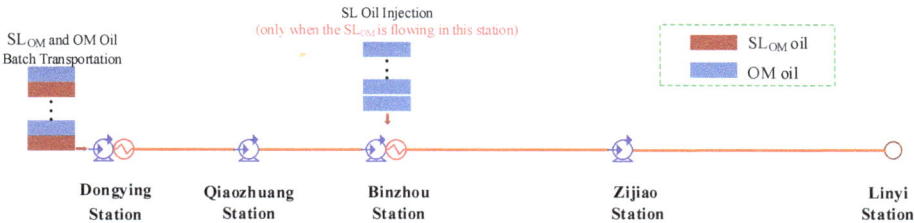

Figure 3. Sketch map of the batch transportation scheme.

Figure 3 shows the SL_{OM} oil and the OM oil are pumped alternately into the pipeline from the Dongying station, which is the first station and the beginning of the pipeline. To avoid degrading the quality of OM oil, the SL oil can only be injected when the oil flowing in Binzhou station is SL_{OM} oil.

Even though the rough scheme is easy to be pictured, to determine the safe and economic detailed transport scheme, many thermal analyses should be performed. The thermal state before and with batch transportation are introduced as follows.

2.2. Thermal State of the Dong-Lin oil Pipeline

2.2.1. Thermal State before the Batch Transportation

Prior to the batch transportation, the crude oils in the Secondary Dong-Lin oil pipeline were SL oil. Since the condensation point of SL is 11 °C, as shown in Table 1, it might be gelled in the pipeline in the winter when the environmental temperature is lower than the condensation point [21]. This could lead to scrapping the pipeline, which is unacceptable for the Sinopec Company.

Table 1. Basic physical properties of Shengli (SL) and Oman (OM) oil.

Oil	ρ_o (kg/m³)	$c_{p,o}$ (J/kg·°C)	θ_{cp} (°C)	μ (Pa·s)
SL oil	937	2000	11	Polynomial $P_n(T)$
OM oil	868	2100	0	See Figure 4
SL_{OM} oil	Can be predicted through properties of SL oil and OM oil by equations in Ref. [22]. Sinopec also did extensive testing on the SL_{OM} oil with different ratios of SL and OM oil.			

Thus, before October 2015, the heating process was adopted in the Secondary Dong-Lin pipeline to ensure the safety of the pipeline. Additionally, the heat process can also lower the viscosity and improve the fluidity of SL oil, which can reduce the power cost at pumps. Since the temperature of oil flowing out of Dongying Station and Binzhou station was kept at a certain value, the thermal state was approximately steady. Figure 5 gives the oil temperature distribution along the pipeline in October 2015.

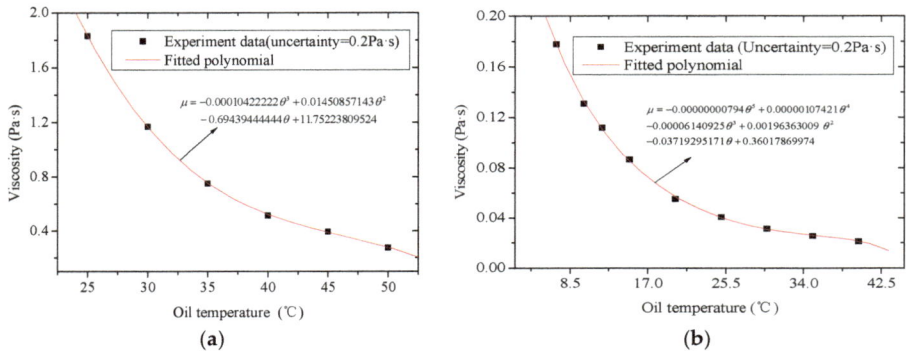

Figure 4. Viscosity-temperature curves and fitted equations of: (**a**) Shengli (SL) oil; and (**b**) Oman (OM) oil.

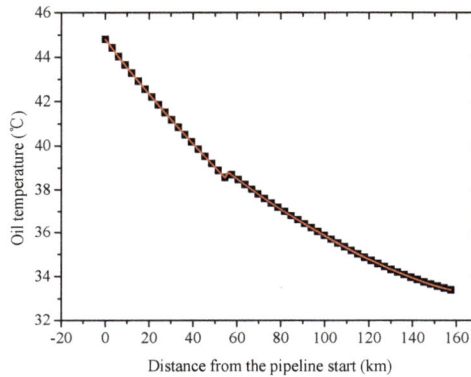

Figure 5. Oil temperature distribution along the pipeline in October 2015.

2.2.2. Thermal State of the Batch Transportation

Table 1 and Figure 4 show the fluidity differences of the SL_{OM} and OM oils are very significant at the same temperature. To save the energy consumed by furnaces, the SL_{OM} and OM oil are transported at different temperatures (See Figure 6). The SL_{OM} oil with high temperature is also called "hot oil" while the OM oil is called "cool oil". Figure 6 gives the typical oil temperature distribution flowing out of the Dongying station during a certain time frame.

Since the oil temperature varies, the thermal state of the whole pipeline is unsteady and changes dramatically when the SL_{OM} oil and OM oil alternate. Additionally, the thermal behaviors are so complicated that the thermal analysis to determine the detailed transportation scheme must be done by numerical simulations. Considering the changing environmental temperature from month to month, to clarify the thermal behavior of a certain case, the current authors must simulate the pipeline's thermal behavior from the beginning to the coldest month. This can be very time consuming.

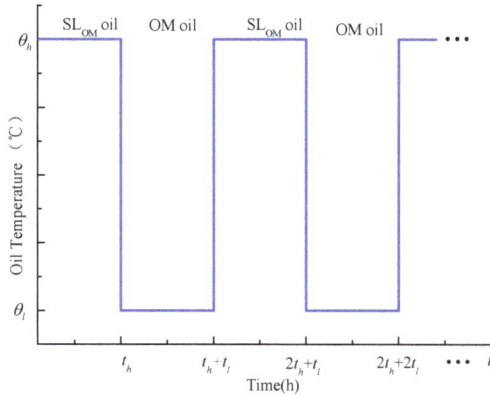

Figure 6. Oil temperature flowing out of Dongying station versus time in a certain time frame.

2.3. Detailed Batch Transportation Scheme to be Determined

The thermal-related operating parameters in batch transportation are as follows: The compositions of SL_{OM} (critical to its fluidity), flow fluxes of SL_{OM} and OM oils, transportation time of each patch (value of t_h and t_l in Figure 6), temperatures of SL_{OM} and OM oil flowing out of the Dongying station (value of θ_h and θ_l in Figure 6), how much and when the oil should be heated by the furnaces in Binzhou station, and the temperature and flux of SL oil injected in Binzhou station.

Thus, thousands of cases should be simulated to find a safe, economical, and relatively optimal transportation scheme. The time consumption of simulations by the frequently-used FVM can be hundreds of days on a personal computer, which is unacceptable for engineering practices. To quickly determine the thermal scheme for this paper, the current authors applied the body-fitted coordinate-based POD reduced-order model developed by their research group to the pipeline thermal simulation. The method is elaborately introduced in Section 4.

3. Physical and Mathematical Model for Secondary Dong-Lin Pipeline

Regarding the batch transportation of the Secondary Dong-Lin pipeline, as shown in Figure 3, sometimes the oil temperature can be higher than that of the surrounding pipe wall and soil, while the opposite is true at other times. This leads to the complicated, unsteady thermal state of the pipeline. Since the pipeline is 154.7 km long, the full 3-D simulation is impractical for the unsteady heat transfer in the pipeline. Thus, the problem is simplified by splitting the pipeline into a series of thermal elements, as shown in Figure 7.

Figure 7 shows the thermal element consists of a pipeline cross-section and a segment of the axial pipeline. The physical and mathematical model of the pipeline cross-section and the axial pipeline (namely the oil stream) are introduced in Sections 3.1 and 3.2, respectively.

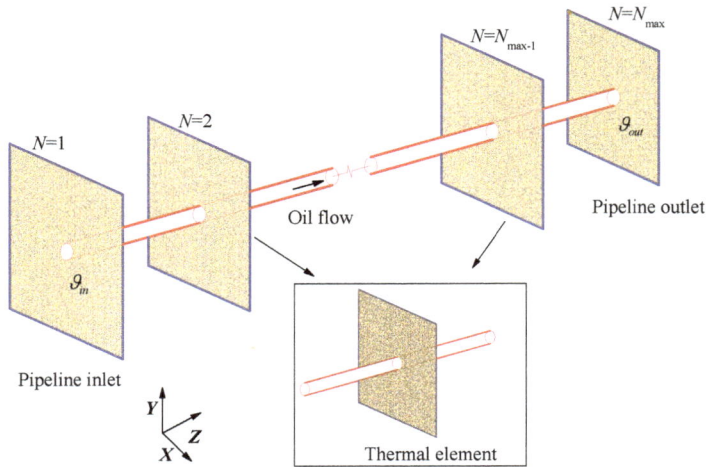

Figure 7. The simplification of the whole 3-D pipeline.

3.1. Physical and Mathematical Model of the Pipeline Cross-section

The sketch map of the pipeline cross-section is shown in Figure 8a. Since an oil pipe is buried in a certain place, it will influence the local temperature field in the soil. The further the soil from the pipe the smaller the influence can be, so it can be neglected for soil very far from the pipe. Thus, to simplify the simulation, it is believed there is a "thermal influence region" (See Figure 8a) of the oil pipe. It is assumed the oil pipe has no impact on the soil temperature field outside the influence region. According to the literature [6] and engineering experience, the thermal influence region of the hot crude oil pipeline is within 10 m, which means $L = 10$ m and $H = 10$ m, as shown in Figure 8a.

Considering the symmetry of the pipeline section (See Figure 8a), the physical model is obtained and is shown in Figure 8b. The physical domain is governed by heat conduction and the whole boundary can be divided into six parts, which are Lines O-A, A-B, B-C, C-D'-D, F-F'-O and semicircle D-E-F (See Figure 8b).

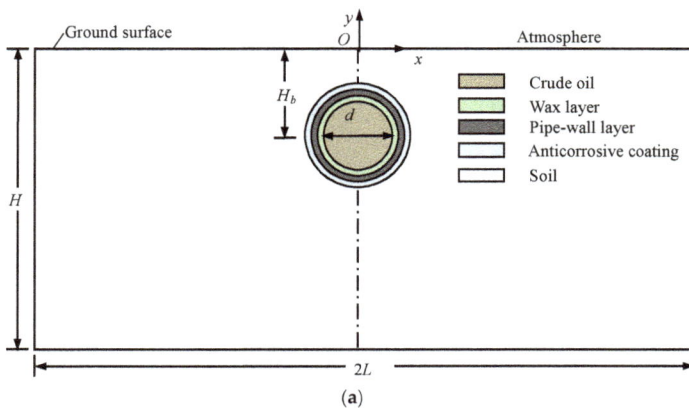

(**a**)

Figure 8. *Cont.*

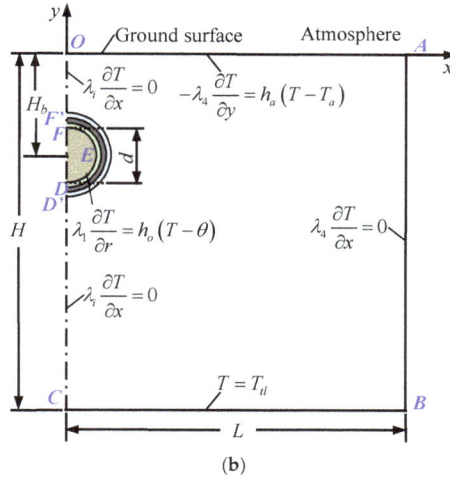

Figure 8. Thermal influence region and physical model of the pipeline cross-section on Cartesian coordinates: (**a**) Thermal influence region; (**b**) physical model.

Considering line O-A and semicircle D-E-F, they are convective heat transfer boundaries that describe the heat convection between the thermal region and the outside (air and oil, respectively, for Line O-A and semicircle D-E-M). Regarding Line A-B, it is the thermal adiabatic boundary, which means the oil pipe has no influence on the soil outside the thermal region. Looking at Line B-C, the temperature has a fixed value, which is the temperature of the thermostat soil layer. Considering Line C-D'-D and Line F-F'-O, they are symmetric boundary conditions.

Using Cartesian coordinates, the governing equation is shown in Equation (1):

$$\frac{\partial(\rho_i c_{p,i} T)}{\partial t} = \frac{\partial}{\partial x}\left(\lambda_i \frac{\partial T}{\partial x}\right) + \frac{\partial}{\partial y}\left(\lambda_i \frac{\partial T}{\partial y}\right) \tag{1}$$

where T stands for the temperature field in the cross-section. The subscript, $i = 1, 2, 3, 4$, stands for the wax layer [23], pipe-wall layer, anticorrosive layer (also called "the three layers"), and soil region around the pipe, respectively.

To take advantage of the BFC-based POD reduced-order model stated in the "Introduction", all the simulations in this paper are on body-fitted coordinates. To obtain the mathematical model on body-fitted coordinates, the physical domain is mapped to the calculation domain (See Figure 9) on body-fitted coordinates.

Correspondingly, Equation (1) is mapped to body-fitted coordinates and the governing equation for BFC is obtained as Equation (2) [24]:

$$J\frac{\partial(\rho_i c_{p,i} T)}{\partial t} = \frac{\partial}{\partial \xi}\left[\frac{\lambda_i}{J}\left(\alpha \frac{\partial T}{\partial \xi} - \beta \frac{\partial T}{\partial \eta}\right)\right] + \frac{\partial}{\partial \eta}\left[\frac{\lambda_i}{J}\left(\gamma \frac{\partial T}{\partial \eta} - \beta \frac{\partial T}{\partial \xi}\right)\right] \tag{2}$$

where, $\alpha = x_\eta^2 + y_\eta^2, \beta = x_\xi x_\eta + y_\xi y_\eta, \gamma = x_\xi^2 + y_\xi^2, J = x_\xi y_\eta - x_\eta y_\xi$.

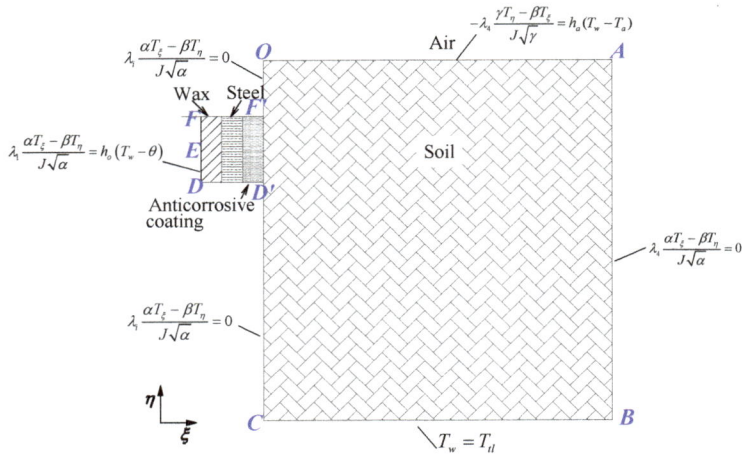

Figure 9. Physical model of pipeline cross-section on body-fitted coordinate.

Similarly, the boundary conditions on BFC also are obtained by mapping the boundary conditions on Cartesian coordinates, as shown in Figure 8b. The gained boundary conditions are as follows (See Figure 9):

$$\text{The boundary O-A}: \quad -\lambda_4 \frac{\gamma T_\eta - \beta T_\xi}{J\sqrt{\gamma}} = h_a(T_w - T_a) \tag{3}$$

$$\text{The boundary A-B}: \quad -\lambda_4 \frac{\alpha T_\xi - \beta T_\eta}{J\sqrt{\alpha}} = 0 \tag{4}$$

$$\text{The boundary B-C}: \quad T_w = T_{tl} \tag{5}$$

$$\text{The boundary C-D'-D and F-F'-O}: \quad \lambda_i \frac{\alpha T_\xi - \beta T_\eta}{J\sqrt{\alpha}} = 0 \, i = 1,2,3,4 \tag{6}$$

$$\text{The boundary D-E-F}: \quad \lambda_1 \frac{\alpha T_\xi - \beta T_\eta}{J\sqrt{\alpha}} = h_o(T_w - \theta) \tag{7}$$

Used in Equations (3)–(7), the subscript, w, stands for the wall in contact with the oil stream; subscripts, o and a, stand for oil and air, respectively; and the subscript, tl, stands for the thermostat layer.

Normally, the governing equation, Equation (2), is solved by FVM under the boundary conditions using Equations (3)–(7). Compared with the reduced order method introduced in Section 4, this currently presented method is called the "full order method".

3.2. Physical and mathematical model of the oil stream

The pipeline thermal simulation is the coupling of the cross-section and oil stream simulations. Since the physical and mathematical models for the cross-sections are already given above, the physical and mathematical model for the oil stream are shown in Equations (8)–(10):

Mass conservation equation:

$$\frac{\partial}{\partial t}(\rho_o A) + \frac{\partial}{\partial z}(\rho_o v A) = 0 \tag{8}$$

Energy conservation equation:

$$C_{p,o}\left(\frac{\partial \theta}{\partial t} + v\frac{\partial \theta}{\partial z}\right) - \frac{f v^3}{2d} = -\frac{4q}{\rho_o d} \tag{9}$$

Matching condition:

$$q = \lambda_1 \frac{\alpha T_\xi - \beta T_\eta}{J\sqrt{\alpha}} = h_o(T_w - \theta) \tag{10}$$

Among Equations (8), (9), and (10), θ denotes the hot oil temperature, v denotes the flow velocity of the oil stream, q denotes the heat flux between the oil stream and wax layer around the oil stream, and T_w stands for the temperature of the interface between the wax layer and oil flow, and its values are calculated by solving Equation (2). h_o represents the forced convection heat transfer coefficient of the oil flow and wax layer, which is a function of the oil temperature and must be determined by experimental data [22].

To solve Equations (7) and (8) under the matching conditions of Equation (9), the characteristics method is applied under the grid shown in Figure 10. Normally, the grid size along the pipe is between 0.5 km and 2 km.

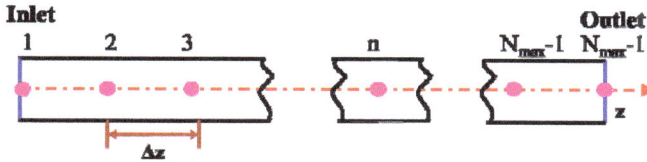

Figure 10. The grid along the pipe.

The final discretization of Equations (8) and (9) is shown in Equation (11):

$$\theta_i^n = \frac{\frac{f^{n-1}(v_i^{n-1})^3}{2d_i} - \frac{4q_{i-1}^{n-1}}{\rho d_i} - C_{p,o}\frac{\theta_i^{n-1} - \theta_{i-1}^{n-1}}{\Delta z}v_i^{n-1} + \frac{C_{p,o}\theta_i^{n-1}}{\Delta t}}{C_{p,o}/\Delta t} \tag{11}$$

Equation (11) shows the superscript, n, means the present value and $n-1$ stands for the value in the previous time step. q_{i-1}^{n-1} in Equation (11) is the key variable, which binds the pipeline cross-sections' thermal simulation with the oil temperature calculation along the pipe. Using Equation (10), the value of q_{i-1}^{n-1} can be calculated after the temperature field in the cross-section is obtained by solving Equation (2).

4. Model Reduction for Pipeline's Cross-section Thermal Simulation

To realize the thermal simulation of the crude oil pipeline, Equation (2) and Equations (8)–(9) must be solved. During the thermal simulation of the crude pipeline, the most time-consuming part is its cross-sections' temperature field calculation (solving of Equation (2)), which can occupy more than 99.99% of the whole process. The authors of this paper try to adopt the POD reduced-order model to significantly improve the calculation speed.

Using the POD reduced-order model [19], the temperature can be written as Equation (12):

$$T(\xi, \eta, t) = \sum_{k=1}^{M} a_k(t)\phi_k(\xi, \eta) \tag{12}$$

where, $\phi_k(\xi, \eta)$ $(k = 1, 2, \ldots M)$ are the POD basis functions which are dependent on space (ξ, η) and independent of time, t. The basis functions can be obtained by analyzing the sampling data by POD. $a_k(t)$ $(k = 1, 2, \ldots M)$ are the amplitudes, which are dependent on time and independent of space. $a_k(t)$ $(k = 1, 2, \ldots M)$ are the unknowns and their equations are called the "reduced-order model". M is the order of basis functions to describe the temperature field and the dimensions of the unknowns, $a_k(t)$.

Equation (12) shows there are two key points in the POD reduced-order model, the POD basis functions and the reduced-order model. The POD basis function and reduced-order models are introduced in Sections 4.1 and 4.2, respectively. The standard POD reduced-order model implementation procedure is introduced briefly in Section 4.3.

4.1. POD Basis Function

POD is a powerful mathematical method. Using the analysis on a set of simulation data (sampling matrix) obtained by full-order simulations, POD can extract a series of basis functions, which capture the dominant information of the physical problems. Regarding the unsteady-state heat transfer problem in this paper, the main process is as follows:

Suppose a two-dimensional unsteady-state heat transfer problem (such as the problem in this paper) of some specific conditions, which has N_ξ control points in the ξ direction and N_η control points in the η direction. The sampling matrix of this physical problem would be constructed.

The POD reduced-order model is applied to solve the unsteady problem with different conditions (boundary conditions and geometric shapes of the simulated domain) in this paper. The temperature fields at representative instances under various sampling conditions should be put into the sampling matrix. Suppose L kinds of sampling conditions are sampled and, for every condition, the temperature field at each time instance is obtained with a total number, K. Considering the ith condition, if the temperature fields in each time instance are sampled, the sampling matrix for the ith condition can be constructed as Equation (13). Similarly, for each condition, a sampling matrix for it can be obtained.

$$
\mathbf{T}_i = \begin{bmatrix}
T(\xi_1, \eta_1, t_1) & T(\xi_1, \eta_1, t_2) & \cdots & T(\xi_1, \eta_1, t_{K-1}) & T(\xi_1, \eta_1, t_K) \\
\cdot & \cdot & \cdots & \cdot & \cdot \\
T\left(\xi_{N_\xi}, \eta_1, t_1\right) & T\left(\xi_{N_\xi}, \eta_1, t_2\right) & \cdots & T\left(\xi_{N_\xi}, \eta_1, t_{K-1}\right) & T\left(\xi_{N_\xi}, \eta_1, t_K\right) \\
T(\xi_1, \eta_2, t_1) & T(\xi_1, \eta_2, t_2) & \cdots & T(\xi_1, \eta_2, t_{K-1}) & T(\xi_1, \eta_2, t_K) \\
\cdot & \cdot & \cdots & \cdot & \cdot \\
T\left(\xi_{N_\xi}, \eta_2, t_1\right) & T\left(\xi_{N_\xi}, \eta_2, t_2\right) & \cdots & T\left(\xi_{N_\xi}, \eta_2, t_{K-1}\right) & T\left(\xi_{N_\xi}, \eta_2, t_K\right) \\
\cdot & \cdot & \cdots & \cdot & \cdot \\
T\left(\xi_1, \eta_{N_\eta}, t_1\right) & T\left(\xi_1, \eta_{N_\eta}, t_2\right) & \cdots & T\left(\xi_1, \eta_{N_\eta}, t_{K-1}\right) & T(\xi_1, \eta_1, t_K) \\
\cdot & \cdot & \cdots & \cdot & \cdot \\
T\left(\xi_{N_\xi}, \eta_{N_\eta}, t_1\right) & T\left(\xi_{N_\xi}, \eta_{N_\eta}, t_2\right) & \cdots & T\left(\xi_{N_\xi}, \eta_{N_\eta}, t_{K-1}\right) & T\left(\xi_{N_\xi}, \eta_1, t_K\right)
\end{bmatrix} \tag{13}
$$

What needs to be noted is the temperature of all the moments is not necessarily put into the sampling matrix, if the dominant information has been contained in the sampling matrix. Subsequent to a sampling matrix for each condition being obtained, all the matrices are combined to produce a larger sampling matrix, $\mathbf{S} \in \mathbb{R}^{m \times n}$, shown in Equation (14), where $m = N_\xi \times N_\eta$ and $n = \sum_{i=1}^{L} K_i$ (K_i is column number of $\mathbf{T}_i$). Thus, the matrix, $\mathbf{S}$, contains the information of the temperature evolution at time-bearing different conditions.

$$
\mathbf{S} = \begin{bmatrix} \mathbf{T}_1 & \mathbf{T}_2 & \cdots & \mathbf{T}_L \end{bmatrix} \tag{14}
$$

Using the "snapshot method" or "singular value decomposition (SVD) method" [9], the basis functions matrix can be obtained. Usually, to save the time consumption, the SVD method is adopted as $m < n$ and the "snapshot method" is used as $m > n$. Both methods are proper for $m = n$. In this paper, the SVD method is applied in Section 5.2. Thus, the SVD method is introduced as follows:

Consider the sampling matrix, $\mathbf{S} \in \mathbb{R}^{m \times n}$, with rank, $d \leq min(m, n)$. Normally, in most engineering problems, there are no two same samplings put into a matrix, $\mathbf{S}$, $d = min(m, n)$. Since the SVD method is only used as $m \leq n$, d is equal to m. The SVD method guarantees real numbers, $\sigma_1 \geq \sigma_2 \geq \ldots \geq \sigma_d > 0$, orthogonal matrices, $\mathbf{U} \in \mathbb{R}^{m \times m}$ and $\mathbf{V} \in \mathbb{R}^{n \times n}$, which satisfy Equation (15):

$$\mathbf{U}^T \mathbf{S} \mathbf{V} = \begin{pmatrix} \mathbf{D} & 0 \\ 0 & 0 \end{pmatrix} \tag{15}$$

where, $\mathbf{D} = diag(\sigma_1, \ldots, \sigma_d) \in \mathbb{R}^{d \times d}$, U, and V are eigenvectors of $\mathbf{S}\mathbf{S}^T$ and $\mathbf{S}^T\mathbf{S}$, respectively. **U** and **V** are also called left singular vectors and right singular vectors. The first d columns of **U** and **V** are eigenvectors with eigenvalues, $\lambda_i = \sigma_i^2$, and the other columns are eigenvectors with eigenvalues, $\lambda_i = 0$. The first d columns in **U** are the POD basis functions required in the POD reduced-order model. Thus, the basis functions matrix, ψ, can be written as Equation (16):

$$\psi = \begin{bmatrix} \boldsymbol{\phi}_1 & \boldsymbol{\phi}_2 & \cdots & \boldsymbol{\phi}_d \end{bmatrix} \tag{16}$$

where, $\boldsymbol{\phi}_k$ is the kth columns of **U**, and $\boldsymbol{\phi}_k$ can be expressed as Equation (17):

$$\boldsymbol{\phi}_k = \left[\phi_k(\xi_1, \eta_1), \ldots, \phi_k(\xi_{N_\xi}, \eta_1), \phi_k(\xi_1, \eta_2), \ldots, \phi_k(\xi_{N_\xi}, \eta_2), \ldots, \ldots, \phi_k(\xi_{N_\xi}, \eta_{N_\eta}) \right]^T \tag{17}$$

4.2. Reduced-order model (equations of $a_k(t)$)

The reduced order model is established by projecting the governing equation, Equation (2), onto the space spanned by the first M basis functions. Following a series of deductions, the current research group established the BFC-based POD-Galerkin reduced-order model for the heat conduction problem (find the details in Ref. [20]), shown in Equation (18):

$$\sum_{k=1}^{M} \frac{da_k}{dt} G_{ik} = -\left(\oint \sqrt{\alpha} q^{(\xi)} \phi_i d\eta - \sqrt{\gamma} q^{(\eta)} \phi_i d\xi \right) - \sum_{k=1}^{M} a_k H_{ik} \quad i = 1, 2 \ldots M \tag{18}$$

where,

$G_{ik} = \int_\Omega J \rho c_p \phi_k \phi_i d\Omega$.

$H_{ik} = \int_\Omega \left[\frac{\lambda}{J} \left(\alpha \frac{\partial \phi_k}{\partial \xi} - \beta \frac{\partial \phi_k}{\partial \eta} \right) \frac{\partial \phi_i}{\partial \xi} + \frac{\lambda}{J} \left(\gamma \frac{\partial \phi_k}{\partial \eta} - \beta \frac{\partial \phi_k}{\partial \xi} \right) \frac{\partial \phi_i}{\partial \eta} \right] d\Omega$

Where the domain, Ω, here stands for the calculation domain on BFC shown in Figure 9. Equation (17) is a system of linear equations with $a_k (k = 1, 2, \ldots M)$ as unknowns, which can be solved by LU decomposition. Usually, for heat conduction problems, the value of M is below 30, which means the number of unknowns is below 30, while, in a full-order model, the unknowns can be thousands (3761 in the problem of this paper) or more, depending on the number of grids. Thus, the reduced-order model can reduce the order from thousands to less than 30, which makes the simulation speed increase significantly.

The boundary conditions are the same with the full-order model given above. The discretization of them in the POD reduced-order model can be found in Ref. [19].

4.3. The standard POD reduced-order model implementation procedure

Figure 11 gives the standard implementing procedure of the POD reduced-order model, which includes sampling, basis function extraction, reduced-order model solving, and physical field reconstructing. The details can be found in Reference [19].

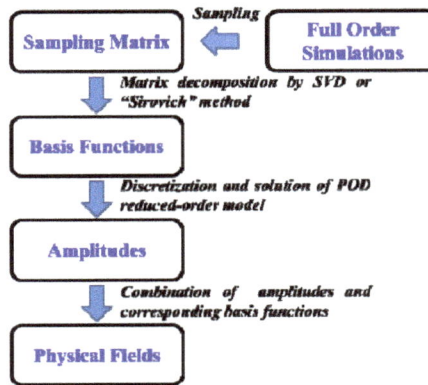

Figure 11. Procedure of proper orthogonal decomposition (POD) reduced-order model implementation.

5. Application and Discussions

The process of BFC-POD-ROM-aided fast thermal scheme determination for the Secondary Dong-Lin crude pipeline is elaborately introduced. First, the property and boundary variables of the pipeline are given. Subsequently, the implementation process and performance of the POD reduced-order model are illustrated. Finally, the thermal scheme for the Secondary Dong-Lin crude pipeline is determined and the results of the POD reduced-order model are compared with the field data.

5.1. Property and Boundary Variables for Secondary Dong-Lin crude Pipeline

To simulate the thermal behavior of the Secondary Dong-Lin pipeline, aside from the parameters given in Section 2, the property and boundary parameters should be given as well.

(1)   Property parameters

Figure 8 shows there are several different-property domains in the physical mode of the pipeline, which are the oil stream, three layers, and soil region. Since the properties of the oils have been given in Section 2, only the property parameters of the soil and the three layers are offered here (see Table 2.).

Table 2. Properties of the soil and the three layers.

Soil and the three layers	ρ_i (kg/m^3)	$C_{p,i}$ (J/kg·°C)	λ_i (W/m·°C)
Soil (0 km–56 km)	2235	1.67	943
Soil (56 km–157.4 km)	2235	1.35	943
Anticorrosive Layer	1000	0.4	1670
Pipe-wall Layer	7850	50	460
Wax Layer	1000	0.15	2000

As shown in Table 2, the soil thermal conductivity from 0 km to 55 km of the pipeline is 1.67 W/m·°C, while the soil thermal conductivity becomes 1.35 W/m·°C after 56km of the pipeline. The heat conductivities are obtained by inverse calculation through the operational data of the Sinopec Company, which is a frequently-used method in oil pipeline thermal simulations [7].

(2)   Boundary parameters

As shown in Figure 8, the boundary parameters of the pipeline cross-section include the temperature of the oil stream, θ, and the corresponding convective heat transfer coefficient, h_o, the temperature of air, T_a, the corresponding convective heat transfer coefficient, h_a, and the temperature of the thermostat layer, T_{tl}.

θ can be determined through the match condition between the pipeline cross-section and oil stream, as shown in Equation (10), and h_o can be found by the empirical formula in Ref. [22] using the above boundary parameters. T_a is not the real temperature of air, rather a pseudo air temperature interpolated by the temperature of the thermostat layer (T_{tl}) and the measured value of soil temperature, T_b, in the buried depth of the crude oil pipeline, which has been a typical method in engineering for easy simulation [22]. The interpolation equation is shown in Equation (19):

$$T_a = T_b + \frac{T_b - T_{tl}}{H} H_b \tag{19}$$

where, H is the depth of the thermostat layer, and H_b is the buried depth of the oil pipe (See Figure 8).

Thus, the thermostat layer temperature, T_{tl}, soil temperature in a buried depth, T_b, and convective heat transfer coefficient, h_a, are the boundary parameters needed (see Table 3). Using field measurements for this study, T_{tl} is 12.3 °C and h_a is 20 (W/m²·°C). T_b varies from day to day and mile to mile. Table 3 gives a series of field data in five typical spots measured by the Sinopec Company. T_b for other days and other spots are interpolated by the data from Table 3.

Table 3. Boundary parameters.

	Soil Temperature in Buried Depth T_b (°C)				
Date	0 km	30 km	55 km	106 km	157.4 km
Oct. 31st 2015	22.67	22.44	20.28	20.75	24.76
Nov. 30th 2015	15.90	19.39	14.89	17.31	21.09
Dec. 31st 2015	7.39	13.39	9.94	12.71	15.68
Jan. 31st 2016	5.86	11.27	8.03	9.33	13.02
Feb. 29th 2016	5.23	10.14	7.24	8.32	11.31

5.2. POD-ROM-based fast thermal simulation for Secondary Dong-Lin crude pipeline

Figure 11 shows that to obtain the POD-ROM-based fast thermal simulation, there are two main steps. The first is sampling and basis function extracting. The second is reduced-order model solving and physical field reconstructing.

5.2.1. Sampling and Basis function

The quality of basis functions has a significant influence on the accuracy of POD reduced-order model-based fast thermal simulation. The acceptable basis functions should contain the main characters of the temperature field evolution of the Secondary Dong-Lin crude pipeline. This mainly depends on the sampling process, in which the samplings should be representative and, for sake of time consumption, as few as possible.

Figure 7 shows the whole pipeline is separated into a series of slices and the main differences among the slices are geometrics (see Figure 2) and boundary conditions (see Table 3). Thus, for this particular problem, the obtained temperature basis function must be capable of depicting the temperature field under different combinations of geometrics and boundary conditions. The main process for sampling and basis function extraction is as follows:

First, the sampling conditions are given in Tables 4–6. The sampling conditions are a thermal analogy of the real conditions. Take "Sampling 1" as an example: Sampling 1 is set as an analog of the thermal situation of the 0–20 km part in the Secondary Dong-Lin crude pipeline. To obtain such an analog, the geometry and heat conductivity of Sampling 1 are set the same within the 0–20km part in the Secondary Dong-Lin crude pipeline. To save time in the samplings' calculation, the samplings' simulation times are set at 50 days, which is much shorter than the real conditions (five months). The alternate frequency (1.5 d/3.5 d one time) is also much higher than the real conditions. The real

T_b (soil temperature in buried depth) changes from month to month so are analogized by changing it every 10 days (see Table 6).

Table 4. The samplings.

Sampling No.	Geometry (See Table 5)	T_b (°C)	λ_4 (W/m·°C)	θ_h (°C)	θ_l (°C)	t_h/t_l (d/d)	$T_{initial}$ (°C)	t_s (d)
1	Geo1	See Table 6	1.67	23	38	1.5/3.5	0	50
2	Geo2			20	38		45	
3	Geo3			15	35		75	
4	Geo4		1.35	14	36	0.5/1.5	105	
5	Geo5			13	35		157	

Note: $T_{initial}$ (0) stands for the temperature field in the 0 km zone of the crude pipeline before the commission of the batching transportation scheme. $T_{initial}$ (45) stands for the temperature field at 45 km. The others follow in kind.

Table 5. Geometry parameters of Geo1–Geo5.

Geo No.	d (m)	H_b (m)	δ_w (m)	δ_{ac} (m)
Geo 1	φ616 × 7	1.315	0.003	0.007
Geo2	φ616 × 7	1.915	0.003	0.007
Geo3	φ695 × 8	1.3555	0.003	0.007
Geo4	φ695 × 8	1.5555	0.003	0.007
Geo5	φ695 × 8	1.8555	0.003	0.007

Table 6. T_b at different times.

Time	0 d–10 d	10 d–20 d	20 d–30 d	30 d–40 d	40 d–50 d
T_b (°C)	22.67	15.9	7.39	5.86	5.23

Subsequently, using FVM, the temperature field in each slice is calculated with time steps at 600s on body-fitted grids (grid number is 3761), as shown in Figure 12.

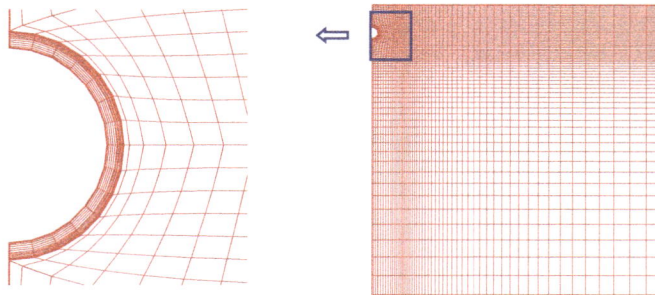

Figure 12. Body-fitted coordinate-based (BFC) grid of the pipeline cross-section.

Based on the temperature fields obtained by the FVM, the sampling matrix shown in Equation (14) can be constructed. To reduce the dimension of the sampling matrix (which can cause problems for the matrix decomposition) and avoid unnecessary information noise, sampling is dense when the temperature field changes quickly. Quite the reverse, it is sparse when the temperature field is changing slowly.

Thus, for this particular problem, the temperature field in every time step is put into the sampling matrix in time intervals [0 h, 5 h] after the alternates of θ_h and θ_l. One temperature field is adopted every five and 10 time steps in the time interval [5 h, 10 h] and [10 h, t_h or t_l], respectively. To summarize, the sampling matrix, **S**, shown in Equation (14) consists of 8330 temperature fields, which makes **S** a matrix with 3761 rows and 8330 columns.

Finally, the basis functions are extracted from the sampling matrix, **S**, by SVD. The energy distribution of basis functions is shown in Figure 13. The first 23 basis functions, $M = 23$ in Equation (12), are applied into the POD reduced-order model for fast simulation. Figure 14 gives contours of four typical basis functions based on Geo 1.

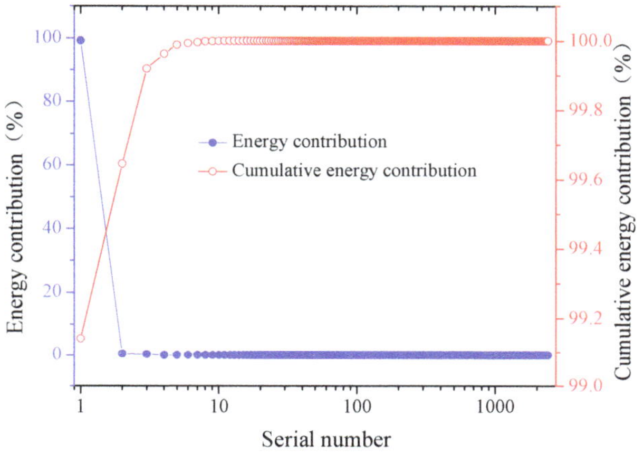

Figure 13. Energy distribution of basis functions.

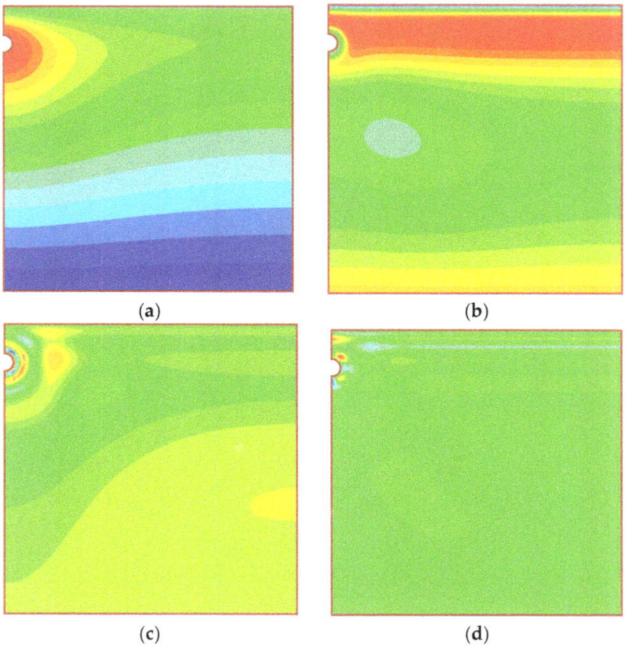

Figure 14. Contours of four typical basis functions based on Geo 1: (**a**) 1st; (**b**) 5th; (**c**) 15th; (**d**) 23rd.

5.2.2. POD Reduced-order model validation and thermal characteristics of batch transportation

To test the accuracy of the POD reduced-order model-based fast simulation, the first three potential schemes (see Table 7) given by Sinopec are simulated by a POD reduced-order model as well as the FVM.

Table 7. Parameters for Schemes 1–3.

Scheme No.	Q_h^D/Q_l^D (m^3h^{-1}/m^3h^{-1})	Q_h^{BI} (m^3h^{-1})	θ_h/θ_l $(^{\circ}C/^{\circ}C)$	t_h/t_l (d/d)	r_{SL}/r_{OM}
1	2161/2470			7.56/2.44	90:36
2	2322/2321	172	See Table 8	8.11/2.83	90:26
3	1872/2470			7.56/2.44	90:20

Table 8. Values of θ_h and θ_l in different months.

Time	θ_l (°C)	θ_h (°C)			Furnaces
		Scheme 1	Scheme 2	Scheme 3	
Oct. 2015	23.20	37.03	38.22	39.04	Both furnaces in Dongying and Binzhou are closed
Nov. 2015	20.00	36.37	37.78	38.75	
Dec. 2015	14.88	33.77	35.39	36.51	
Jan. 2016	13.60	33.89	35.64	36.84	
Feb. 2016	12.90	33.69	35.48	36.71	

Regarding the Secondary Dong-Lin crude pipeline, the coldest month each year is February. The environmental temperature decreased progressively from October 2015 to February 2016, as shown in Table 3, which made February 2016 the riskiest month. The oil temperature distribution has great significance for engineering. Considering Scheme 1, the oil temperature distributions during the last cycle (the last 2.44 d/7.56 d) in February 2016 are shown in Figure 15.

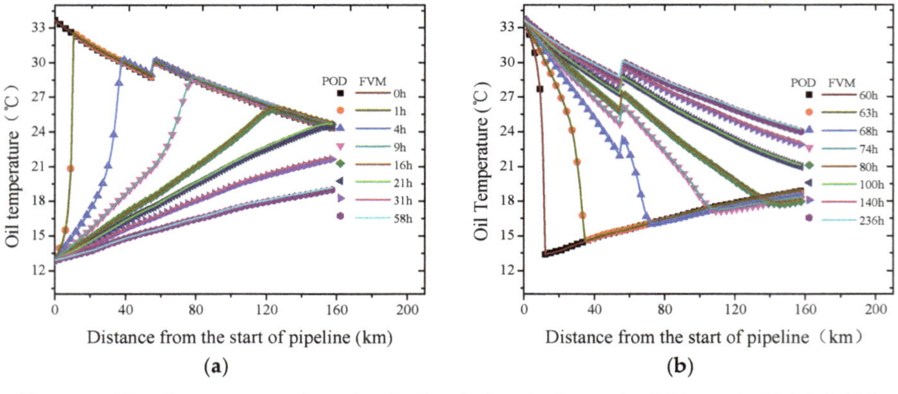

Figure 15. The oil temperature along the pipeline during the last cycle of Scheme 1: (a) 0 d–2.44 d; (b) 2.44 d–10 d.

The curves in Figure 15 are explained as follows: Beginning with the last cycle, $t = 0$ h in Figure 15a, the whole pipeline is filled with SL_{OM} oil. Table 8 shows the temperature of the oil in $z = 0$ km is 33.69 °C and, due to the heat loss to the environment, the temperature decreases along the pipeline from 0 km to 55 km. Since the injected SL_{OM} oil in $z = 55$ km is at 50 °C (higher than the upstream SL oil's temperature of 28.73 °C), the mixed oil becomes 30.18 °C, which shows a temperature jump at $z = 55$ km (See Figure 15a). The temperature after $z = 55$ km decreases for the earlier stated reason before $z = 55$ km.

Then, the OM oil, at 12.9 °C, is pumped into the pipeline, which is at a lower temperature than the SL oil's temperature. Thus, the OM oil is warmed when it flows along the pipeline, which can be shown by the curves at $t = 1$ h, 4 h, and 9 h. What should be noted is that the oil pipeline is not occupied fully by the OM oil until $t = 16$ h. Thus, the curves of $t = 1$ h, 4 h, and 9h show an up-and-down trend. The "uptrend-curves" part of the pipeline is filled with OM oil, which absorbs energy from the environment. The "downtrend-curves" part of the pipeline is filled with SL_{OM} oil, which releases energy to the environment. When at $t = 16$ h, the SL oil is completely driven out of the pipeline by the OM oil behind.

While the OM oil keeps absorbing the energy from the environment, the temperature of the environment decreases, leading to the OM oil temperature decreasing with time, as shown in Figure 15a. The "reduction" lasts until the oils alternate at the beginning of the pipeline. Figure 15b shows the temperature curves after the alternation. It illustrates the opposite thermal characteristics of Figure 15a, when SL_{OM} oil with a higher temperature than OM oil is pumped into and fully occupies the pipeline.

Figure 15a,b have the symbols and lines representing the results of the POD reduced-order model and the FVM, respectively. Even though the thermal characteristics of the Dong-Lin crude pipeline are very complicated, as stated above, the results of the POD reduced-order model agree well with those of the FVM. The main thermal characteristics along the pipeline of Schemes 2 and 3 are similar to those of Scheme 1, given in Figure 15, and the differences are just the values. Aside from the oil temperature distribution along the pipeline, the engineers are also concerned about the oil temperature flowing out of the pipeline. Thus, for the three schemes, Figure 16 gives the outflow oil temperature (in the end of the pipeline) versus time curves.

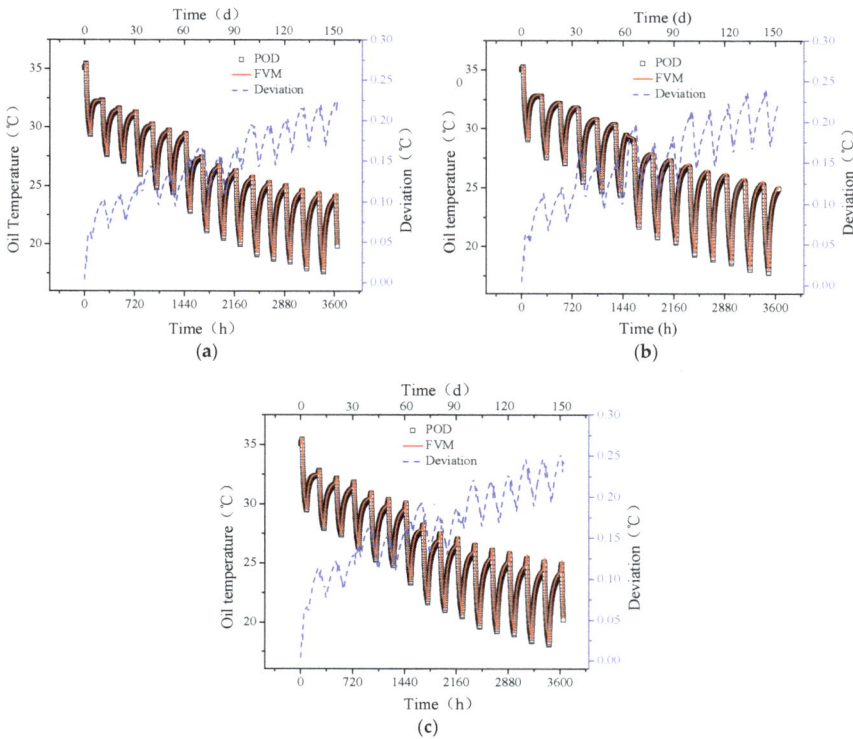

Figure 16. The oil temperature flowing out of the pipeline in Linyi Station: (**a**) Scheme 1; (**b**) Scheme 2; and (**c**) Scheme 3.

Figure 16 shows the oil temperature in the end of the pipeline periodically declines from October 2015 to February 2016. The periodic alternating of the oils pumped into the pipeline and the increasingly colder weather are responsible for the two trends. Looking at Figure 16, it can be found that the POD reduced-order model has good accuracy. The largest errors (compared with the FVM) for Schemes 1, 2, and 3 are 0.22 °C, 0.24 °C, and 0.25 °C, respectively. The mean errors (compared with FVM) for Schemes 1, 2, and 3 are 0.15 °C, 0.15 °C, and 0.16 °C, respectively.

To illustrate the accuracy of the POD reduced-order model more vividly, for Scheme 1, Figures 17 and 18, respectively, give the temperature fields at the beginning and end of the pipeline. Figures 17 and 18 have dashed lines representing the POD reduced-order model results and solid lines representing the FVM. The results agree well with each other.

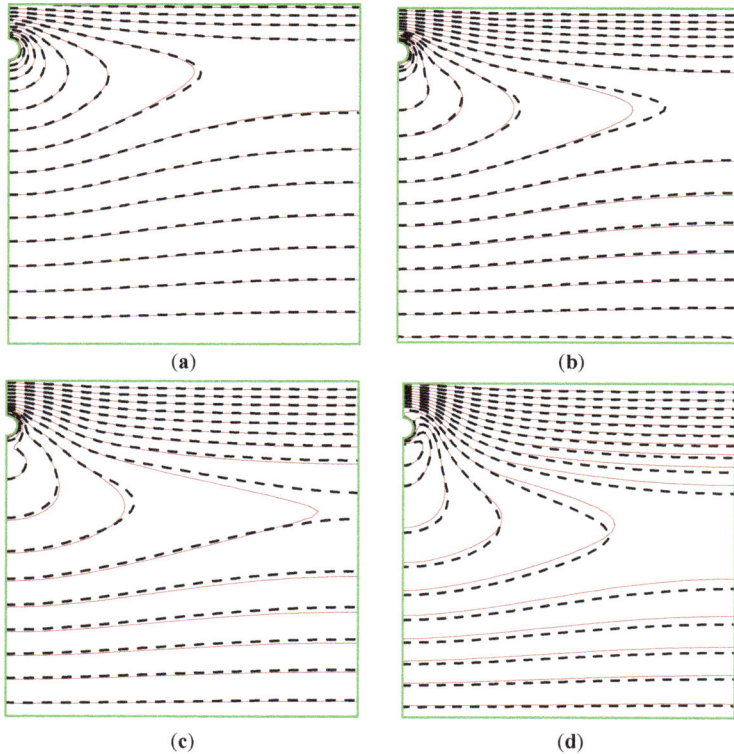

(a) (b)

(c) (d)

Figure 17. The cross-section's temperature field in the beginning of the pipeline (dashed lines: POD reduced-order model. Solid lines: Finite volume method (FVM)): (**a**) 38th day; (**b**) 76th day; (**c**) 114th day; (**d**) 152nd day.

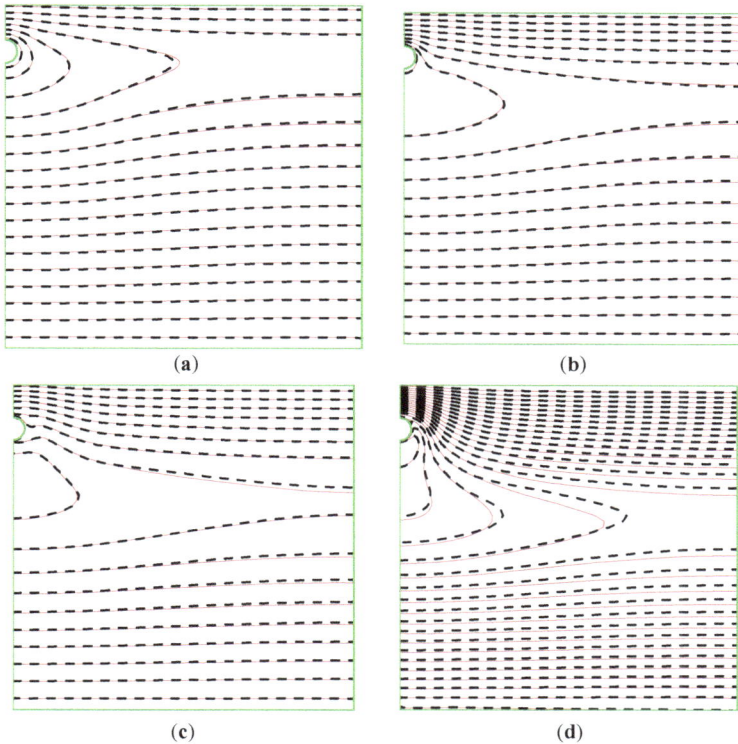

Figure 18. The cross-section's temperature field at the end of the pipeline (Dashed lines: POD reduced-order model. Solid lines: FVM): (**a**) 38th day; (**b**) 76th day; (**c**) 114th day; (**d**) 152nd day.

To illustrate the speed advantage of the POD reduced-order model, Table 9 shows the time consumption of the POD reduced-order model and the FVM. The simulation speed is more than 100 times faster than the FVM, which means much time is saved in the engineering.

Table 9. Time consumptions of the FVM and POD reduced-order models.

Scheme No.	FVM (h)	POD (h)	Acceleration Factor
Scheme 1	62.7	0.49	128
Scheme 2	64.5	0.52	124
Scheme 3	64.0	0.56	114

Considering the above analysis, it can be found that, for the thermal simulation of the Dong-Lin crude pipeline with oil batching transportation, the POD reduced-order model has good accuracy and significant efficiency. Thus, rather than the commonly-used FVM, the POD reduced-order model is adopted to determine the operational scheme in this paper.

5.3. Detail Batch Transportation Scheme Determination and Field Verification

Through the first three potential schemes given by Sinopec, the performance of the POD reduced-order model was verified. Following that, considering the capacity of oil stocks, blend ratios of SL_{OM} oils, behaviors of furnaces, and other factors, the Sinopec Company drew 1024 potential schemes in sum. All the schemes were simulated by the POD reduced-order model to find their thermal performance critical to the energy cost of the oil heated and oil pumped because oil fluidity is related to oil temperature (see Figure 4).

Considering the thermal and hydraulic consumption of each scheme given by the simulations, the Sinopec Company chose Scheme No. 124 (shown in Table 10) as the final operating scheme based on the considerations of power consumption and transportation risk. The risk means the restart-ability of the oil pipeline after being shut-down for an accident. The lower the temperature and the longer the shut-down time, the higher the risk is.

Table 10. Parameters for the determined scheme.

Scheme No.	Q_h^D/Q_l^D (m^{3}h^{-1}/m^{3}h^{-1})	Q_h^{BI} (m^{3}h^{-1})	θ_h/θ_l (°C/°C)	t_h/t_l (d/d)	r_{SL}/r_{OM}
124	1370/2064	172	See Table 11	4.58/1.17	9:2

Table 11. The determined values of θ_h and θ_l in different months.

Time	θ_l (°C)	θ_h (°C)	Furnace in Dongying	Furnace in Linyi
Oct. 2015	23.5	41.5		Open for both OM and SL$_{OM}$ oil Raise OM oil 8 °C Raise SL$_{OM}$ oil 5 °C
Nov. 2015	23	41.5	Close for OM oil.	
Dec. 2015	21	41.5	Open for SL$_{OM}$ oil and keep	
Jan. 2016	19	41.5	heating it to 41.5 °C	
Feb. 2016	17	41.5		

To ensure the success of the new scheme commissioning, the Sinopec Company did much more preparation than was expected. Thus, the company began to execute the chosen scheme on November 4, 2015. Figure 19 gives the comparisons between the simulated results and the operational field data.

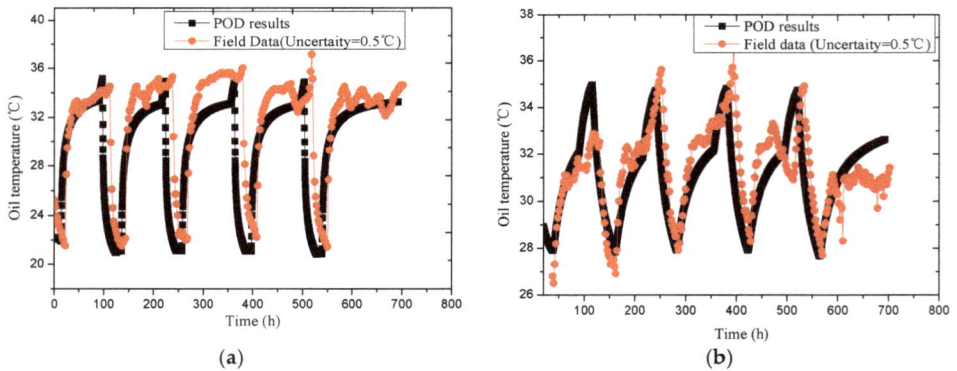

(a)

(b)

Figure 19. The comparisons between field data and the results predicted by the POD reduced-order model: (**a**) The oil temperature flowing into Binzhou station versus time on February 2016; (**b**) the oil temperature flowing into Linyi station versus time on February 2016.

Figure 19 demonstrates the thermal behavior under the chosen scheme can be predicted with an error acceptable to crude oil engineering. There are some spikes in the field data shown in Figure 19. The flow flux and the temperature are not controlled precisely during real-time engineering, which means the pipeline is operated around the chosen scheme, and not exactly on the scheme. To operate a crude oil pipeline exactly according to the planning scheme is impossible due to unpredictable factors (such as the flow shock of the upstream, and the temporary change of oil transportation task) existing in practical engineering.

The oil temperature flowing into other stations shows the same behavior. To properly describe the deviations between the field data and simulation results, the mean oil temperatures flowing into Binzhou and Linyi station are calculated. It should be noted that the mean temperature of OM oil and SL_{OM} oil are calculated independently (see Tables 12 and 13). The deviations are also shown in Tables 12 and 13.

Table 12. The mean temperature of OM oil in different months.

Oil	Data	Nov. 2015		Dec. 2015		Jan. 2016		Feb. 2016	
θ_{OM} in BZ	Field Data	28.89	(0.24, 0.8%)	26.78	(0.48, 1.8%)	25.66	(1.21, 4.7%)	23.54	(0.42, 1.8%)
	POD Results	29.13		26.30		24.45		23.12	
θ_{OM} in LY	Field Data	32.18	(1.27, 3.9%)	32.93	(0.15, 0.4%)	29.62	(0.48, 1.6%)	29.98	(0.32, 1.1%)
	POD Results	33.45		33.08		30.10		30.30	

Note: BZ is for Binzhou station and LY is for Linyi station. The values in the brackets are absolute and relative deviations, respectively.

Table 13. The mean temperature of SL_{OM} oil in different months.

Oil	Data	Nov. 2015		Dec. 2015		Jan. 2016		Feb. 2016	
$\theta_{SL_{OM}}$ in BZ	Field Data	36.02	(0.12, 0.3%)	35.66	(0.48, 1.3%)	33.89	(1.21, 3.6%)	33.64	(0.06, 0.2%)
	POD Results	36.14		36.35		34.63		33.58	
$\theta_{SL_{OM}}$ in LY	Field Data	33.99	(0.99, 2.9%)	34.48	(1.37, 4.0%)	31.29	(0.79, 2.5%)	31.72	(0.63, 2.0%)
	POD Results	34.98		35.85		32.08		32.35	

Note: BZ is for Binzhou station and LY is for Linyi station. The values in the brackets are absolute and relative deviations, respectively.

Tables 12 and 13 show the mean oil temperature errors are less than 1.27 °C and 1.37 °C for OM and SL_{OM} oil, respectively, which is acceptable for oil transportation engineering. There are three reasons to generate such a deviation. First, the physical model used in the current study is an approximate description and the POD reduced-order model itself is also an approximate mathematical model. Second, there are some inevitable errors in parameter values in the physical model, such as the heat capacity, density and viscosity of oil, the heat conductivity in the regions of the wax layer, pipe-wall layer, anticorrosive layer, and soil, and the forced convection heat transfer coefficient of the oil flow and the wax layer. Third, the pipeline is operated around the chosen scheme, not exactly on the scheme, as stated above, which is believed to be the biggest reason.

To summarize, the BFC-based POD-ROM is adopted to determine the detailed scheme to improve the efficiency more than a hundred times. Compared with the FVM, the POD reduced-order model reduces the simulation time from 264 days, using 10 computers' parallel computing, to 2.2 days. The Dong-Lin crude oil pipeline has been safely operating for more than two years using the determined scheme.

6. Conclusions

The determination of the crude pipeline's detailed batch transportation scheme could cost tremendous thermal simulation time, which is an unsolved problem in oil transportation engineering. To solve this problem for China's Secondary Dong-Lin crude pipeline, a fast scheme determination strategy was developed for the first time.

The main idea of the strategy was that, rather than the traditional FVM, the BFC-based POD reduced-order model was adopted to increase the speed of the thermal simulation. The whole strategy included three main steps, which are summarized as follows:

(1) Sampling matrix construction and basis function obtainment.

The quality of basis functions has significant influence on the accuracy of the POD reduced-order model-based fast thermal simulation. Thus, the corresponding samplings should be representative and, for the sake of time consumption, as few as possible. Regarding crude oil batch transportation, the following point is recommended: Design the sampling conditions by using the "analogy method", which means the geometry and heat conductivity should be the same within the pipeline and the boundary condition can be designed as an analogy of the real condition (see Tables 4–6). To reduce the dimensions of the sampling matrix, sampling can be dense as the temperature field changes quickly and sparse as the temperature field is slowly changing.

(2)   The validation of the POD reduced-order model.

The POD reduced order model is an approximate method. Its accuracy is dependent on the quality of obtained basis functions, the number of basis functions adopted in the POD reduced-order model, and the complexity of the problem. Thus, the POD reduced-order model should be validated by the full order model (FVM in this paper) before being applied to the engineering.

Regarding the specific problem in this paper, through the comparisons between the POD reduced-order model and FVM, it was found that the POD reduced-order model had good accuracy with a mean error of 0.16 °C, which is acceptable for engineering. Thus, it is believed the POD reduced-order model can be applied to thermal simulations of crude oil batch transportation. The obtained basis function is feasible for China's Secondary Dong-Lin crude pipeline and the number of adopted basis functions is appropriate.

(3)   The determination of the transportation scheme.

It was found that the POD reduced-order model can be more than one hundred times faster than the FVM. Therefore, to find a proper operational scheme, it is feasible to simulate hundreds, or thousands of schemes with reasonable time consumption. Moreover, this method can be combined with some optimization methods to find an optimized operating scheme.

Aided by the body-fitted coordinate-based POD reduced-order model, the details of the batch transportation scheme were determined and can be found in Tables 10 and 11. The Dong-Lin crude oil pipeline has been safely operating for more than two years using the determined scheme. Compared with the field data, the predicted results by the POD reduced-order model are of an acceptable accuracy for crude oil engineering. The mean oil temperature errors were less than 1.27 °C and 1.37 °C for OM and SL_{OM} oil, respectively.

Author Contributions: Conceptualization, B.Y.; Methodology, D.H.; Code, D.H.; Validation, Q.Y., D.C. and G.Z.; Writing—Original Draft Preparation, D.H.; Writing—Review & Editing, B.Y., D.H., and Q.Y.

Funding: The study is supported by National Science Foundation of China (No. 51706021), The Beijing Youth Talent Support Program (CIT&TCD201804037), the Jointly Projects of. Beijing Natural Science Foundation and Beijing Municipal Education Commission (KZ201810017023) and the Project of Construction of Innovative Teams and Teacher Career Development for Universities and Colleges Under Beijing Municipality (No. IDHT20170507).

Conflicts of Interest: The authors declare no conflict of interest.

Nomenclature

Roman Symbols

a_k	Amplitude of the kth POD basis function
A	Flowing area of the pipeline (m^2)
$c_{p,i}$	Specific heat capacity of region i (J/(kg·°C))
$c_{p,o}$	Specific heat capacity of oil (J/(kg·°C))
d	Diameter of the pipeline (mm)
f	The Darcy coefficient
h_o	Heat convection coefficient between the wax layer and oil stream (W/(m^2·°C))
h_a	Heat convection coefficient between the soil and air (W/(m^2·°C))
H	Thermal influence region on the vertical direction (m)
H_b	The buried depth of crude oil pipeline (m)
L	Thermal influence region on the horizontal direction (m)

q	The heat flux between the wax layer and oil stream (W/m^2)
Q_h^D, Q_l^D	Flow flux of hot and cool oil in Dongying station (m^3/h)
Q_h^{BI}	The flow flux injected in Binzhou station (m^3/h)
r_{SL}, r_{OM}	Ratio of SL oil and OM oil in SL$_{OM}$ oil
S	The sampling matrix
t_h	Transportation time of hot oil during one period (h)
t_l	Transportation time of cool oil during one period (h)
t_s	The total transportation time of hot and cool oils (h)
T	Temperature of the pipeline's cross-section
T_a	Temperature of the air (°C)
T_c	Temperature of the soil thermostat layer (°C)
v	Flow velocity of the oil stream (m/s)
x, y	Cartesian coordinate in the cross-section of pipeline (m)
z	Coordinate along the cross-section of pipeline (m)

Greek Symbols

δ_w	Thickness of the wax layer (m)
δ_{ac}	Thickness of the anticorrosive layer (m)
ϕ_k	The kth POD basis function
$\mathbf{\Phi}_k$	Vector of the kth POD basis function
λ_i	Heat conductivity coefficient of region i (W/(m·°C))
μ	Viscosity (Pa·s)
θ	Temperature of the oil (°C)
θ_{cp}	Condensation point of crude oil (°C)
θ_h	Temperature of the hot oil, namely SL$_{OM}$ oil (°C)
θ_l	Temperature of the cool oil, namely OM oil (°C)
$\theta_{SL_{OM}}$	Temperature of SL$_{OM}$ oil (°C)
θ_{OM}	Temperature of OM oil (°C)
ρ_i	Density of region i (W/(m·°C))
ρ_o	Density of oil (kg/m^3)
ξ, η	Body-fitted coordinate in the cross-section of pipeline (m)

Subscripts

1,2,3,4	Regions of wax layer, pipe-wall layer, anticorrosive layer and soil respectively
ac	Anticorrosive layer
h	Hot oil
l	Cool oil, namely low temperature oil
o	oil
OM	Oman oil
SL	Shengli oil
SL$_{OM}$	Mixture of SL oil and OM oil
tl	Thermostat layer
w	Wax layer
ξ, η	Partial derivatives of the variable

References

1. Yu, Y.; Wu, C.; Xing, X.; Zuo, L. Energy saving for a Chinese crude oil pipeline. In Proceedings of the ASME 2014 Pressure Vessels and Piping Conference, Garden Grove, CA, USA, 20–24 July 2014. [CrossRef]
2. Zhang, H.R.; Liang, Y.T.; Xia, Q.; Wu, M.; Shao, Q. Supply-based optimal scheduling of oil product pipelines. *Petrol. Sci.* **2016**, *13*, 355–367. [CrossRef]
3. Shauers, D.; Sarkissian, H.; Decker, B. California line beats odds, begins moving viscous crude oil. *Oil Gas J.* **2000**, *98*, 54–66.
4. Mecham, T.; Wikerson, B.; Templeton, B. Full Integration of SCADA, field control systems and high speed hydraulic models-application Pacific Pipeline System. In Proceedings of the International Pipeline Conference, Calgary, AB, Canada, 1–5 October 2000. [CrossRef]

5. Cui, X.G.; Zhang, J.J. The research of heat transfer problem in process of batch transportation of cool and hot oil. *Oil Gas Stor. Trans.* **2013**, *23*, 15–19. [CrossRef]
6. Wang, K.; Zhang, J.J.; Yu, B.; Zhou, J.; Qian, J.H.; Qiu, D.P. Numerical simulation on the thermal and hydraulic behaviors of batch pipelining crude oils with different inlet temperatures. *Oil Gas Sci. Technol.* **2009**, *64*, 503–520. [CrossRef]
7. Yuan, Q.; Wu, C.C.; Yu, B.; Han, D.; Zhang, X.; Cai, L.; Sun, D. Study on the thermal characteristics of crude oil batch pipelining with differential outlet temperature and inconstant flow rate. *J. Petrol. Sci. Eng.* **2018**, *160*, 519–530. [CrossRef]
8. Lumley, J.L. *The Structure of Inhomogeneous Turbulent Flows in Atmospheric Turbulence and Radio Wave Propagation*; Nauka: Moscow, Russian, 1967.
9. Sirovich, L. Turbulence and dynamics of coherent structures, Part 1: Coherent structures. *Q. Appl. Math.* **1987**, *45*, 561–571. [CrossRef]
10. Huang, N.E.; Shen, Z.; Long, S.R.; Wu, M.L.; Shih, H.H.; Zheng, Q.; Yen, N.C.; Tung, C.C.; Liu, H.H. The empirical mode decomposition and Hilbert spectrum for nonlinear and nonstationary time series analysis. *Proc. Roy. Soc. London A* **1998**, *454*, 903–995. [CrossRef]
11. Schmid, P.J. Dynamic mode decomposition of numerical and experimental data. *J. Fluid Mech.* **2010**, *656*, 5–28. [CrossRef]
12. Banerjee, S.; Cole, J.V.; Jensen, K.F. Nonlinear model reduction strategies for rapid thermal processing systems. *IEEE Trans. Semiconduct. Manuf.* **1988**, *11*, 266–275. [CrossRef]
13. Raghupathy, A.P.; Ghia, U. Boundary-condition-independent reduced-order modeling of complex 2D objects by POD-Galerkin methodology. In Proceedings of the IEEE Semiconductor Thermal Measurement and Management Symposium, San Jose, CA, USA, 15–19 March 2009. [CrossRef]
14. Fogleman, M.; Lumley, J.; Rempfer, D.; Haworth, D. Application of the proper orthogonal decomposition to datasets of internal combustion engine flows. *J. Turbul.* **2004**, *5*, 1–18. [CrossRef]
15. Ding, P.; Tao, W.Q. Reduced order model based algorithm for inverse convection heat transfer problem. *J. Xi'an Jiaotong Univ.* **2009**, *43*, 14–16. [CrossRef]
16. Thomas, A.B.; Raymond, L.F.; Pau, G.A.; Thomas, J.; Breault, R.W. A reduced-order model for heat transfer in multiphase flow and practical aspects of the proper orthogonal decomposition. *Theor. Comp. Fluid Dyn.* **2012**, *43*, 68–80. [CrossRef]
17. Gaonkar, A.K.; Kulkarni, S.S. Application of multilevel scheme and two level discretization for POD based model order reduction of nonlinear transient heat transfer problems. *Comput. Mech.* **2015**, *55*, 179–191. [CrossRef]
18. Selimefendigil, F.; Öztop, F. Numerical study of natural convection in a ferrofluid-filled corrugated cavity with internal heat generation. *Numer. Heat Tr. A-Appl.* **2015**, *67*, 1136–1161. [CrossRef]
19. Han, D.; Yu, B.; Yu, G.; Zhao, Y.; Zhang, W. Study on a BFC-based POD-Galerkin ROM for the steady-state heat transfer problem. *Int. J. Heat Mass Tran.* **2014**, *69*, 1–5. [CrossRef]
20. Han, D.; Yu, B.; Zhang, X. Study on a BFC-Based POD-Galerkin Reduced-Order Model for the unsteady-state variable-property heat transfer problem. *Numer. Heat Tr. B-Fund.* **2014**, *65*, 256–281. [CrossRef]
21. Yu, G.; Yang, Q.; Dai, B.; Fu, Z.; Lin, D. Numerical study on the characteristic of temperature drop of crude oil in a model oil tanker subjected to oscillating motion. *Energies* **2018**, *11*, 1229. [CrossRef]
22. Yang, Y.H. *Design and Management of Oil Pipeline*, 1st ed.; Press of China Petroleum University: Qing Dao, China, 2006.
23. Cheng, Q.; Gan, Y.; Su, W.; Liu, Y.; Sun, W.; Xu, Y. Research on exergy flow composition and exergy loss mechanisms for waxy crude oil pipeline transport processes. *Energies* **2017**, *10*, 1956. [CrossRef]
24. Tao, W.Q. *Numerical Heat Transfer*; Xi'an Jiaotong University Press: Xi'an, China, 2001.

MDPI

St. Alban-Anlage 66

4052 Basel

Switzerland

Tel. +41 61 683 77 34

Fax +41 61 302 89 18

www.mdpi.com

Energies Editorial Office

E-mail: energies@mdpi.com

www.mdpi.com/journal/energies

www.ingramcontent.com/pod-product-compliance
Lightning Source LLC
Chambersburg PA
CBHW051838210326
41597CB00033B/5693